Sanctorius Sanctorius and the Origins of Health Measurement

Teresa Hollerbach

Sanctorius Sanctorius and the Origins of Health Measurement

 Springer

Teresa Hollerbach
Max Planck Institute for the History of Science
Berlin, Germany

ISBN 978-3-031-30120-9   ISBN 978-3-031-30118-6   (eBook)
https://doi.org/10.1007/978-3-031-30118-6

This work was supported by Max Planck Institute for the History of Science (MPIWG)

This Springer imprint is published by the registered company Springer Nature Switzerland AG
The registered company address is: Gewerbestrasse 11, 6330 Cham, Switzerland

Foreword

Virtually all historians of science have dealt with the Renaissance era: historians of antiquity, because of the reception of ancient science; modern historians, because of the study of the roots of modernity; those in global history who make comparisons or explain why one should avoid them; postcolonialists explaining why they are "post"; and so on. It is practically impossible to avoid the Renaissance, even if only in a single article or in the frame of a lecture.

While avoiding talking about it emphatically, and especially without resorting to the now completely outdated concept of scientific revolution, the Renaissance remains on the desk for historians of science because, although it has long been the subject of study and research, it continues to present itself simultaneously as an era of destruction and recomposition. While the individual trajectories of these processes seem to be clearly understood, they nevertheless appear elusive when observed as a whole, because an understanding of their mutual influences requires broad contextualization, which in turn can be achieved only by studying the history of institutions, society and its culture, economics, technology, religion, and the laws and orders of politics.

Remaining in the purely scientific sphere, in the eyes of many historical actors and especially in the early stages of this period, up to the late sixteenth century, the Renaissance actually appeared as a natural continuation of the conceptual organization of science as arranged in the late Middle Ages, between the twelfth and thirteenth centuries. Institutionally crucial in ensuring this sense of continuity were the universities. First founded at the beginning of the thirteenth century, these institutions handed down the idea that harmony between science, nature, and the divine was the main purpose of intellectual endeavors. Scientific knowledge was organized according to the precepts of the quadrivium, although no longer exactly in the canonical forms established during Late Antiquity. This continuity was continuously supported and confirmed by another process that dominated and regulated the dialogue between scientific development, tradition, and the dissemination of scientific culture: the process of homogenizing knowledge. As the network of universities in Europe grew in number and relevance, the knowledge circulating among them, and disseminated by them to ever-broader sectors of society, was slowly being homogenized in terms of content and curricular standards. First, through the Paris model and then, following the model of studies conceived and implemented by Philipp Melanchthon at Wittenberg, European scientific culture, under the impetus of the universities, became an identity-shaping factor.

At the same time, however, a contrary thrust was taking place, namely, toward the dissolution of the quadrivium as an organizational pattern of knowledge. The reasons for this contrary process are better sought in the field of technology or, rather, in major social phenomena that involved a strong technological component. Chief, among these, are such phenomena as urbanization, voyages of exploration, and the reshaping of the art of war. In all cases, architecture—civil, military, and naval—as well as practical mechanics were the seedbeds of technological development, while the latter became increasingly prominent in the economic and political agenda of territorial entities, as their politics gradually moved toward an increasingly absolutist model.

Under this thrust, new disciplines emerged, such as nautical astronomy. While this enriched the quadrivium scheme by naturally associating with the classical discipline of astronomy, it did not actually contribute to upholding its ideal of showing and confirming the harmony of the cosmos. From this perspective, the new disciplines nullified the traditional scientific and ethical mission: in other words, they divested the quadrivium of its *raison d'être*.

The temporal perspective and vision of the future of Renaissance society likewise found itself caught between these opposing thrusts. On the one hand, the idea was perpetrated of the universe always being the same, with an ever-changing sublunar world—all destined to end, however. The "end of the world" or the "end of humanity" as conceived by medieval Christian eschatology remained, within the framework of cultural expectations, the fundamental parameter to which the behavior of most conformed. On the other hand, a different idea was creeping in more and more insistently, which in the centuries to come would be defined as the idea of progress. Accompanied by the ubiquitous rhetoric of *nova scientia*, the perception of acceleration, the fundamental component of progress, was spreading. The increased speed of military campaigns and the effectiveness of machines combined with the sense of a shrinking world fueled by each ship returning to port and each new contour marked on the map by the cosmographer that transformed portions of the world from unexplorable to eventually explored. Still dressed as a medieval scholar, Renaissance man gradually forgot the harmony of the cosmos and instead focused on exploring space and all the phenomena that can be observed in it.

Renaissance people became the mirror of this dynamic. They became universal figures capable of dealing with problems from metaphysics to principles of theoretical mechanics, from shipbuilding to the history of antiquity, from natural philosophy to telescopic astronomy. The universal Renaissance figure was a philosopher, engineer, architect, physician, artist, and, above all, increasingly a mathematician, measurer, and quantifier, equipped with a wealth of mathematical, mechanical, and optical tools and instruments.

The Renaissance figure was the agent caught between these opposing drives: those who rejected one of them, those who rejected both, those who sought reconciliation. It would be through the mathematization of practical mechanics and, therefore, the emergence of theoretical mechanics that the first great new systematization of knowledge would take place: the fusion of natural philosophy and theoretical mechanics which gave birth to classical physics and concluded the Renaissance era.

But before this final culmination, there was an unprecedented reshuffling of knowledge and scientific fields. Practical knowledge was codified and thus incorporated into the great structures of knowledge circulation and sharing. For a long time, the research agenda did not change but crucial problems were gradually addressed from different approaches. Violent and natural motions, the existence of the vacuum, the functioning of the human body, the nature of flight, and a thousand other topics were explored without regard for the connection dictated by the harmonious worldview, and this paved the way for the expression of ideas that were truly new, even if not yet elements of a systematic knowledge system.

Such ideas were not innovations, because they were not part of a new research framework. Only in hindsight, given the end of medieval science and the stabilization of modern science accompanied by the process of industrialization, do these new ideas appear to us historians as innovations, because we are now able to pin them down within a new system of knowledge. But the Renaissance figure hardly ever saw themselves as an innovator. This characterizes the background of Sanctorius Sanctorius, too, who is said to have revolutionized medicine through quantification. To us, he surely does appear as one of the founding fathers of modern medicine; but if Sanctorius were with us now, and aware of the developments in medicine over the centuries since his death, he would probably rebel against being called a revolutionary and instead declare himself a proud member of the Galenic medical tradition. The reasons he would do so can be learned in this book.

Research Group Leader   Matteo Valleriani
Department I Max Planck Institute for the
History of Science
Berlin, Germany

Hon. Prof. at Technische Universität Berlin
Germany

Prof. by Special Appointment at Tel Aviv University
Israel

Acknowledgments

This book is based on a dissertation written in Department 1 of the Max Planck Institute for the History of Science in Berlin (MPIWG), whose director, Jürgen Renn, has my sincere thanks for his support of my project. My greatest debt is to my supervisors, Matteo Valleriani and Friedrich Steinle, who kindly devoted their time to guiding and encouraging me throughout the entire project. They gave me the support and the freedom to complete the work to the best of my abilities. Not only did they accompany my dissertation from the start, but also my development as a researcher. I first listened to a lecture by Friedrich Steinle in 2009, when I had just begun my Bachelor course at the Technical University Berlin (TU) and he was embarking on his first semester there, as Professor of the History of Science. One year later, a fellow student and friend, Julia Jägle, recommended that I take a course taught by Matteo Valleriani; I did so and have never regretted it. With their university courses, Friedrich Steinle and Matteo Valleriani aroused my interest in the History of Science and certainly it is no exaggeration to say that, without them, I would not be where I am today.

The project to reconstruct the Sanctorian chair has been integral to my research on Sanctorius and his work. I thank Friedrich Steinle, Hans-Liudger Dienel, and Jürgen Renn, who made possible the collaboration between the MPIWG and the Institute for Vocational Education and Work Studies (TU Berlin). I would like to thank Katharina Wegener and Volker Klohe, without whose help I would have found it impossible to realize the replica. I am also immensely grateful to Jochen Büttner, who provided insight and expertise that greatly assisted my research into the Sanctorian chair and its replication. My thanks go also to Agnes Bauer, Nana Citron, Andrea Grimm, Alev Güzel, Leon Kokkoliadis, and Paul Weisflog, both for their hands-on participation and the many fruitful seminar discussions revolving around aspects of the reconstruction project. I would like to emphasize here that Matteo Valleriani played a crucial role throughout the research regarding the reconstruction of the Sanctorian chair—from the first idea to the subsequent experiments. I would also like to express my gratitude to Hansjakob Ziemer and Stephanie Hood, for supporting me in realizing the exhibition of the replica at two public events in Berlin:

the science fairs Long Night of Sciences, in 2018, and City of Knowledge, in 2021. In this way, they encouraged my interest in science communication and gave me an opportunity to hone my skills in this field. I am very grateful to them also for the invaluable workshops they organized at the Institute, and for giving me the chance to be part of the Max Planck Society's major science festival, the Max Planck Day, in 2018.

Special thanks are due also to Ohad Parnes, who kindly accompanied and supported my work in my capacity as external representative of the doctoral researchers at the MPIWG. He also gave me the benefit of his advice on Chapters 1 and 2, and I am thankful for the continual exchanges "in the corridors" of the MPIWG. My thanks go also to Elaine Leong for her precious advice on approaching the subjects of Renaissance Galenism and material culture studies in the History of Medicine. I am grateful to Anna Jerratsch for allowing me to read the manuscript of her dissertation "Der frühneuzeitliche Kometendiskurs im Spiegel deutschsprachiger Flugschriften" (The early modern discourse on comets, as reflected in pamphlets in German), before it went to press; and to my colleagues Juliane Schmidt, Carla Rodrigues Almeida, Elizabeth Merrill, and Bernadette Lessel for discussions, laughter, shared lunch breaks, and help when I needed it. In 2017, I had the privilege to be part of the Steering Group of the Max Planck PhDnet—a time that taught me a great deal about successful teamwork. Thanks to Leo, Jana, Rafa, Gabe, and Lisa for this great experience.

I would also like to thank Lindy Divarci for her editorial work on an article that is part of this book, for her untiring support regarding the book's publication, and for her yoga classes, which continue to help me keep my balance and stay well during stressful times. Through her ready grasp of the material and conscientious editing, Jill Denton made my English more readable and greatly improved the quality of the book. Simon Stolz patiently helped me format the manuscript, for which I am very grateful. I wish to acknowledge the assistance provided by Dirk Wintergrün, who developed and maintained the digital annotator that I used to study Sanctorius's books. In this regard, I would also like to thank Klaus Thoden for preparing the transcribed texts for embedding in the annotator; Ralf Hinrichsen for ordering the transcriptions; and Urte Brauckmann, Ellen Garske, and Sabine Bertram for acquiring the digitization of Sanctorius's books. The librarians at the MPIWG and the digitization group provided invaluable support in obtaining literature and many of the illustrations in this book. The MPIWG Library covered all the expenses this involved and has enabled me, through a publishing fund, to publish this work as an open access monograph. In this regard, I am very grateful to the Head of Library, Esther Chen. During my research stay in Padua and Venice, I had the opportunity to work in many libraries and archives. I would like to thank the staff of the following institutions: Archivio di Storia Universitaria Padovana, especially Francesco Piovan; the Biblioteca Civica Padova; the Biblioteca del Seminario Vescovile di Padova; the Biblioteca Pinali Antica, especially Giulia Rigoni Savioli; the Biblioteca Universitaria Padova; the Archivio di Stato di Venezia; the Biblioteca Marciana; and the Biblioteca del Museo Correr.

Finally, I would like to express special thanks to my family and friends for their wholehearted support throughout my entire PhD journey, including the many necessary distractions and incentives to complete this book. My parents, especially my mother, and my sister always had an open ear and have been there for me every step of the way. And Philip, I cannot thank you enough for keeping things going, cheering me up, believing in me, and always being there.

Contents

Abbreviations and Short Titles

Abbreviations

AAU   Archivio Antico dell'Università, Padua
ASVe   Archivio di Stato, Venice
BMCVe   Biblioteca del Museo Correr, Venice
BNMVe   Biblioteca Nazionale Marciana, Venice
BUP   Biblioteca Universitaria, Padua

Short Titles of Sanctorius's Works (listed in their first editions)

Methodi vitandorum errorum	*Methodi vitandorum errorum omnium, qui in arte medica contingunt, libri quindecim* (Venice, 1603)
Commentary on Galen	*Commentaria in Artem medicinalem Galeni* (Venice, 1612)
De statica medicina	*Ars Sanctorii Sanctorii Iustinopolitani de statica medicina, aphorismorum sectionibus septem comprehensa* (Venice, 1614)
Commentary on Avicenna	*Commentaria in primam Fen primi libri Canonis Avicennae* (Venice, 1625)
Commentary on Hippocrates	*Commentaria in primam sectionem Aphorismorum Hippocratis* (Venice, 1629)
De remediorum inventione	*De remediorum inventione* (Venice, 1629).

List of Figures

Chapter 1
Introduction

Abstract At the turn of the seventeenth century, the Venetian physician Sanctorius Sanctorius (1561–1636) developed instruments to measure and to quantify physiological change. As trivial as quantitative assessment with regard to health issues might seem to us today – in times of fitness trackers and smart watches – it was a highly innovative step at the time. With his instruments, Sanctorius introduced quantitative research into physiology and thus represents an early case of today's self-tracking, or self-quantifying, technology. Until now, no systematic research has been undertaken to investigate Sanctorius and his work from the broader perspective of processes of knowledge transformation in early modern medicine while including the entire range of his activities—intellectual and practical—rather than just a selection. This work aspires to fill that gap. As an introduction to the entire book, this chapter gives an overview of the aims, sources, methodologies and contents of the book.

How many steps have you taken today? How many calories did you burn? Is your smartwatch buzzing again, to remind you to leave your desk and get some exercise? Wearable technology in the form of smart watches or fitness trackers, for example, has become a familiar part of daily life for most of us. According to Meghann Chilcott, member of the Forbes Technology Council, the market value of fitness technology wearables is likely to grow to over \$23 billion by 2025.[1] The technology's rise illustrates the importance of quantitative assessment for society today, especially with regard to health issues; and it reveals how deeply integrated such asessment has become in our everyday lives. But of course, this has not always been the case. At the turn of the seventeenth century, when the Venetian physician Sanctorius Sanctorius (1561–1636) stepped into his famous steelyard to measure changes in weight, medicine had not yet been conceived of in quantitative terms. Not numbers, but the physician's *senses* were central to any diagnosis. By

---

[1] See: https://www.forbes.com/sites/forbestechcouncil/2020/03/09/wearing-it-well-the-next-steps-for-wearable-medical-technology/#76945c308d1a. Accessed 16 June 2020.

© The Author(s) 2023
T. Hollerbach, *Sanctorius Sanctorius and the Origins of Health Measurement*,
https://doi.org/10.1007/978-3-031-30118-6_1

developing several instruments to measure physiological change, Sanctorius introduced into the medical field a form of quantitative research that represents an early
iteration of today's self-tracking, or self-quantifying, technology.

Historical accounts of Sanctorius and his work tend to foreground the genius
who invented, almost out of the blue, a new medical science that profoundly influenced the modern age. This new science is known as iatrophysics, iatromechanics,
or sometimes iatromathematics (from the Greek "iatro," meaning "physician").
These terms by no means denote clear categories, but rather have been quite flexibly
applied, retrospectively, to developments in research on medicine and the philosophy of nature. The terms are comparable nevertheless: all of them reflect the importance of quantification in medical research, as well as the field's tendency to utilize
numerical values and mechanical factors.[2] Besides these heroic narratives, there are
a few critical voices who have emphasized instead Sanctorius's strong adherence to
the medical tradition of his day, namely Galenic medicine (Wear 1973, 1981; Farina
1975). Admittedly, these are merely the two ends or extremes of what amounts
overall to a more balanced spectrum of views of Sanctorius.[3] Yet, some commentators do conjure an image of an innovator who developed his novel approach *despite*
clinging to those traditional concepts frequently dismissed as old-fashioned
Galenism. In doing so, they overlook a decisive dimension of the complex process
through which Sanctorius generated new knowledge, as I will show in this book.

Until now, no systematic research has been undertaken to investigate Sanctorius
and his work from the broader perspective of processes of knowledge transformation in early modern medicine while including the entire range of his activities—
intellectual and practical—rather than just a selection. This work aspires to fill that
gap. By examining not only those parts of Sanctorius's works that are, or appear to
be, innovative, but also his work in its entirety, in the context of its day and in its
various facets, I try to shed light on the epistemic processes that led Sanctorius to
develop his quantitative approach to physiology. I hope thus to contribute to our
understanding of the ways in which knowledge was generated and transformed in a
period that was shaped by numerous historical developments of far-reaching significance in science and that is, indeed, often deemed a "scientific revolution." As will
be seen, in Sanctorius's undertakings, medicine and technology intersect. It is
essential, therefore, that any historical study of his work take into account knowledge and practices in both of these fields and their mutual impact. I do so here, by
examining scientific development through the twin lens of the histories of medicine
and technology. In doing so, I consider not only the intellectual but also and especially the practical dimensions of Sanctorius's activities. This is a marked departure

---

[2] Capello 1750, Vedrani 1920, Giordano and Castiglioni 1924, Castiglioni 1931, 1936, Baila 1936, Major 1938, Miessen 1940, Premuda 1947, 1950, Sanctorius and Lebàn 1950: 13–102, Ettari and Procopio 1968, Rothschuh 1968, Mattioli 1985: 253–62, Eknoyan 1999, Lemmer 2015.

[3] Del Gaizo 1889, Grmek 1952, 1967, 1975, 1990, Siraisi 1987, Dacome 2001, 2012, Sanctorius and Ongaro 2001: 5–47, Guidone and Zurlini 2002, Maclean 2002, Poma 2012, Bigotti and Taylor 2017, Bigotti 2018, Hollerbach 2018.

from other research to date, which has usually focused on Sanctorius's thinking, not on his making and doing.[4]

To put it in a nutshell, this book aims for a broad-ranging and yet integrative view of Sanctorius and his work that examines both innovation and tradition, as well as their complex interplay within the realms of theory and practice, and their social dimensions. It thus facilitates a reevaluation of Sanctorius's role in the wider process by which medical culture began to be transformed in the early modern period—a process that ultimately led to Galenic medicine being abandoned in favor of a new medical science based on the use of quantification in medical research.

Sources and Methodologies Around 2000 pages, often subdivided into columns, in six books: this is Sanctorius's written output in quantitative terms.[5] With the sole exception of his renowned *De statica medicina*, his work is available only in the Latin original. It is this, perhaps, which has prevented scholars from investigating all of Sanctorius's work. Moreover, three of his six books are lengthy commentaries on early medical works still authoritative in his day: Galen's *Ars medica*, Avicenna's *Canon,* and Hippocrates's *Aphorisms*. The *Commentary on Avicenna* has attracted attention, since it is the sole work in which Sanctorius published illustrations of his instruments. Contrary to the traditional historical approach to Sanctorius, which begins—and often also ends—with the *De statica medicina* and the *Commentary on Avicenna,* his major publications, I set out to find my way through the maze of words in the Venetian physician's lesser-known works—the *Commentary on Galen* and the *Commentary on Hippocrates*.

However, analysis of these medical commentaries involves other challenges besides the great masses of Latin text. As the historian Per-Gunnar Ottosson has pointed out in his study of late medieval commentaries on Galen's *Ars medica*, the topics here are discussed not in their own right, but always in relation to the original work commented upon. Thus, when interpreting the content of the commentaries, there is always the problem of determining whether a statement is merely a set phrase without any special significance, an effort to give objective expression to a medical authority, or an expression of the author's original personal convictions. According to Ottosson, the only way to solve this problem is to consider these texts in a broader historical context and compare them with earlier views; for only so can any significant changes in attitude be ascertained (Ottosson 1984: 65). This is the

---

[4] Only in recent times have the material dimensions of Sanctorius's undertakings been the subject of historical research. See: Bigotti and Taylor 2017, Hollerbach 2018.

[5] In Sanctorius's books the page numbering is either according to columns, or it is a foliated pagination (with *recto* and *verso* indicating front and back of each numbered folio leaf). For my total page count, I converted the pagination into regular, sequential pagination. With regard to the number of published books, I counted the *Commentary on Galen* that was published in two separate volumes as one book, whereas I counted the *Commentary on Hippocrates* and the *De remediorum inventione* as two separate books, even though they were published together in the same volume. For more information on Sanctorius's publications, see Chap. 2.

method I used when analyzing Sanctorius's two commentaries—contextualizing them in the framework of contemporary Galenic medicine.

But why all this effort? Medical historian Nancy Siraisi has convincingly shown that the study of medical commentaries is worthwhile, revealing their value as historical sources. Rather than being reactionary theoretical writings with little significance for Renaissance medicine, commentaries by academic physicians can offer important insight into the intellectual and scientific culture of the period. In fact, writing commentaries on authoritative texts fell within the mainstream of contemporary intellectual life. Accordingly, Sanctorius's commentaries illustrate his responses to contemporary intellectual currents and reveal how he adopted specific technical or practical innovations into a still largely traditional framework. Given that his commentaries originated in the lectures he gave as a professor of medicine at the University of Padua, they provide a window onto his university medical teaching; although of course they do not necessarily directly mirror his classroom practice. In addition to this, they reveal how much his lectures on authoritative texts reflected his own interests and, too, his encounters with the ideas, activities, and controversies of the intellectual environment in which he produced them (Siraisi 1987: 4–12). This is why I paid particular attention to those two commentaries by Sanctorius that had hitherto been largely overlooked. I was convinced that they were key to understanding Sanctorius's own intentions and to approaching Sanctorius in the light of his own era.

In order to navigate the masses of text, I worked with digitized versions of the first editions of Sanctorius's books, which were embedded in a digital annotator along with searchable transcripts of the original texts. While reading, I annotated text passages, highlighted the works, people, and locations cited, and defined certain keywords, such as "quantity" (*quantitas*), for example, as I show in Appendix I. This helped me get an overview of the contents and find my way through the many pages while writing up this research.

I complemented my analysis of Sanctorius's publications by research in the libraries and archives of Padua and Venice, the two cities where Sanctorius mainly lived and worked. This shed light on his biography as well as on his social and institutional setting: the milieu in which he moved.

Besides Sanctorius's writings, I focus in the book on the material aspects of his research. This accords with the greater attention placed by historians of science, in recent decades, on those practical and material dimensions of research endeavors that shape the processes of knowledge transformation.[6] In adopting this material culture approach, I gave particular consideration to the practical features of his projects, above all his instruments and their possible use. In order to further approximate Sanctorius's medical practice and thereby trace the mechanical and practical knowledge involved in his undertakings, I used the replication method.[7] Namely, as

---

[6] E.g., Cowan 1993, Pickering 1995, Heering 2008, Smith 2009, Breidbach et al. 2010, Anderson et al. 2013, Rabier 2013, Smith et al. 2014, Valleriani 2017, Leong 2018.

[7] For more details on how I applied the replication method to Sanctorius's weighing procedures, see Sect. 7.5.2.

part of the research undertaken for this book, I reconstructed his most famous instrument, the Sanctorian weighing chair, and sought to replicate his weighing procedures, so as to investigate the design, operation, use, and purpose of the instrument.

Plan of the Book The book is divided into eight chapters. After the introduction chapter, Chapter 2 opens with a biographical account of Sanctorius that situates him in his social, institutional, and professional context. It critically evaluates the existing biographies of the Venetian physician and complements them with my own research into the primary sources. Episodes of Sanctorius's life that have hitherto received little or no attention are discussed in more detail. This opens up a new perspective on the life and work of Sanctorius, setting the stage for the more comprehensive review of his work to be found in the following chapters.

Chapter 3, "Sanctorius's Galenism," deals with Sanctorius's intellectual background and places his book *De statica medicina* within the framework of contemporary Galenic medicine. Usually celebrated for its innovative, quantitative approach to medicine, the *De statica* is mostly read in isolation from the Galenic tradition. However, as I will show, an analysis of this context is crucial to understanding how Sanctorius developed his novel ideas and revised the then prevailing medical knowledge. Of particular importance in this regard are the dietetic doctrine of the "six non-natural things" and the concept of insensible perspiration, an invisible excretion of the human body. Potential links between Sanctorius's notions and the doctrine of the ancient medical school of the Methodists and corpuscular ideas are likewise scrutinized. The chapter concludes with an analysis of the *De statica medicina* itself, focusing on the conceptual backdrop against which Sanctorius developed the weighing procedures he presented in the book. References to Sanctorius's other publications help situate his ideas in the broader framework of his endeavors overall, and thus contribute to an understanding of the theoretical context from which the *De statica medicina* emerged.

Turning from the conceptual to the practical and material resources for Sanctorius's undertakings, Chap. 4, "Sanctorius's Work in its Practical Context," highlights the practical context of the *De statica medicina* and explores Sanctorius's use of instrumentation. Investigation of the form and style of the *De statica medicina* and its relation to the literary genre of *Regimina sanitatis*—a medieval tradition of rules of health—allows important conclusions to be drawn about how Sanctorius shared his practical experience, as well as about his intended audience, and more generally, the purpose of the publication. It offers insight into the way in which Sanctorius connected theory and practice. To complement established research on Sanctorius, the analysis here of his use of instrumentation focuses, not on the measuring instruments but rather on the various other, lesser-known devices that he developed, which range from surgical devices, to a special sickbed, to cupping glasses. The actual measuring instruments are treated in a later chapter. Here, I also examine the relation of these other devices both to Sanctorius's medical practice and his teaching activities at the University of Padua. Even though—or precisely because—they were not part of his quantitative approach to physiology, studying them helps complement the picture of Sanctorius as a practicing physician.

Moreover, it provides glimpses of the social context in which he developed and used his instruments and of how he used his head and hands in medicine. Finally, the findings of this chapter allow the *De statica medicina* to be situated anew within the broader practical context of Sanctorius's undertakings.

The central theme of Chap. 5, "Quantification in Galenic Medicine," is to identify and explore different forms of quantification in the medical tradition, on which Sanctorius may possibly have drawn for his quantitative approach to physiology. Firstly, I address theories and practices connected to dietetics and pharmacology, as well as the Galenic concept of a latitude of health that assumed certain graduations in a person's state of health. Secondly, I reconsider how the work of Sanctorius relates to that of two earlier authors who are commonly associated with him and his static medicine: the Alexandrian physician Erasistratus (third century BCE) and the German Catholic cardinal and scholar Nicolaus Cusanus (1401–1464). Both were proponents of early quantitative approaches to medical problems, which is why their undertakings have been often related to Sanctorius and his use of quantitative measurements. Thirdly, I outline instances of quantitative physiological reasoning in Galen's work, as well as in that of Renaissance scholars, and I analyze their possible connection to Sanctorius.

Before considering Sanctorius's measuring instruments in more detail, I examine more generally, in Chap. 6, "Quantification and Certainty," the context in which Sanctorius presented these devices in his works. Unlike previous studies of Sanctorius's measuring instruments, which often focused on the *Commentary on Avicenna*, this being the only work in which Sanctorius included illustrations of his instruments, I analyze the measuring instruments in the light of all of Sanctorius's publications. Furthermore, I scrutinize how the various instruments are related to one another and discuss Sanctorius's possible complementary use of them. Of particular interest in this context is the role of the *De statica medicina*, it having become exemplary of Sanctorius's quantitative approach to physiology. These considerations serve as an introduction to my in-depth study of Sanctorius's measuring instruments in Chap. 7; and they reveal the agenda behind his inventions and efforts at quantification—namely, to enhance the degree of certainty in medicine—particularly given that the conjectural character of medicine and thus of its certainty were much debated issues in the medical works of his day. While there is not a shadow of a doubt that Sanctorius departed from traditional views by introducing new quantitative procedures into medicine, investigation of the roles that he assigned, on the one hand, to logical reasoning and, on the other, to experience, empirical knowledge, and his new methods of quantification draws a more complex picture of the combination of theory and practice in all of his work.

As its title, "Measuring Instruments," suggests, Chap. 7 deals with Sanctorius's most famous devices—*pulsilogia*, thermoscopes, hygrometers, and balances—which he developed to measure physiological changes. Having attracted considerable scholarly attention over the centuries, they underpin the narrative that identifies Sanctorius as a great innovator and as the founder of a new medical science, whose integral components were mechanization, measurement, and numerical values. The findings of the foregoing chapters allow us now to move beyond these selective accounts of Sanctorius and his work and to take a closer look at, and reevaluate, his celebrated measuring instruments and their use. I explore their design and operation, the contexts in which they emerged, how Sanctorius possibly used them, and what exactly they measured for what purpose. Furthermore, I analyze the hitherto largely ignored two steelyards that Sanctorius devised to gauge climatic conditions, and thereby cover the entire range of his measuring procedures. Moreover, I present the results of the reconstruction of the Sanctorian weighing chair and of the replication of his experimental practice, showing how this approach opened up new perspectives on Sanctorius's work, his doctrine of static medicine, and the operation and purpose of his weighing chair.

The book concludes with a reflection on the epistemic processes that made the use of quantification and measurements in medicine at all *conceivable* to Sanctorius and which might also explain how these methods *made sense* to him in ways that they had not before. To this end, in "Sanctorius Revisited," Chap. 8, I bring into focus the relation between the categories of innovation and tradition in Sanctorius's work, as well as the interplay of the realms of theory and practice, so as to unify the main results of my research. Then, based on my analysis of the measuring instruments in Chap. 7, I reflect on what *quantifying health* meant to Sanctorius. Finally, I briefly outline how his measuring instruments were received. Building on the historical analyses of the previous chapters, I present a new and revised view of the Venetian physician, Sanctorius, which hopefully contributes to a better understanding not only of his own work but also, more generally, of how knowledge was transformed in the early modern period.

References

Anderson, Katharine, Mélanie Frappier, Elizabeth Neswald, and Henry Trim. 2013. Reading Instruments: Objects, Texts and Museums. *Science and Education* 22: 1167–1189.

Baila, Ernesto. 1936. Santorio Santorio, il precursore della medicina sperimentale. *Gazzetta Sanitaria* n.a: 13–14.

Bigotti, Fabrizio. 2018. The Weight of the Air: Santorio's Thermometers and the Early History of Medical Quantification Reconsidered. *Journal of Early Modern Studies* 7: 73–103.

Bigotti, Fabrizio, and David Taylor. 2017. The Pulsilogium of Santorio: New Light on Technology and Measurement in Early Modern Medicine. *Society and Politics* 11: 55–114.

Breidbach, Olaf, Peter Heering, Matthias Müller, and Heiko Weber. 2010. *Experimentelle Wissenschaftsgeschichte*. Paderborn: Fink Wilhelm.

Capello, Arcadio. 1750. *De Vita Cl. Viri Sanctorii Sanctorii … Accedit Oratio ab eodem Sanctorio habita in Gymnasio Patavino dum ipse primarium Theoricae Medicinae explicandae munus auspicaretur*. Venice: Apud Jacobum Thomasinum.

Castiglioni, Arturo. 1931. The Life and Work of Santorio Santorio (1561–1636). *Medical Life* 135: 725–786.

———. 1936. Santorio Santorio Capodistriano (1561–1636) nel terzo centenario della sua morte: commemorazione tenuta nell'aula della R. Università di Padova il 16 dicembre 1936. *Le forze sanitarie* 5: 1593–1604.

Cowan, Ruth Schwartz. 1993. Descartes's Legacy: A Theme Issue on Biomedical and Behavioral Technology. *Technology and Culture* 34: 721–728.

Dacome, Lucia. 2001. Living with the Chair: Private Excreta, Collective Health and Medical Authority in the Eighteenth Century. *History of Science* 39: 467–500.

———. 2012. Balancing Acts: Picturing Perspiration in the Long Eighteenth Century. *Studies in History and Philosophy of Biological and Biomedical Sciences* 43: 379–391.

Del Gaizo, Modestino. 1889. *Ricerche Storiche intorno a Santorio Santorio ed alla Medicina Statica. Memoria letta nella R. Accademia Medico-Chirurgica di Napoli il dì 14 Aprile 1889*. Naples: A. Tocco.

Eknoyan, Garabed. 1999. Santorio Sanctorius (1561–1636): Founding Father of Metabolic Balance Studies. *American Journal of Nephrology* 19: 226–233.

Ettari, Lieta Stella, and Mario Procopio. 1968. *Santorio Santorio: la vita e le opere*. Rome: Istituto nazionale della nutrizione.

Farina, Paolo. 1975. Sulla formazione scientifica di Henricus Regius: Santorio Santorio e il "De statica medicina". *Rivista critica di storia della filosofia* 30: 363–399.

Giordano, Davide, and Arturo Castiglioni. 1924. Centenarî e commemorazioni: Santorio Santorio. *Rivista di storia delle scienze mediche e naturali* 15: 227–237.

Grmek, Mirko D. 1952. *Santorio Santorio i njegovi aparati i instrumenti*. Zagreb: Jugoslav. akad. znanosti i umjetnosti.

———. 1967. Réflections sur des interprétations mécanistes de la vie dans la physiologie du XVIIe siècle. *Episteme: rivista critica di storia delle scienze mediche e biologiche* 1: 17–30.

———. 1975. Santorio, Santorio. In *Dictionary of Scientific Biography*, ed. Charles Coulston Gillispie, 101–104. New York: C. Scribner's Sons.

———. 1990. *La première révolution biologique: Réflexions sur la physiologie et la médecine du XVIIe siècle*. Paris: Payot.

Guidone, Mario, and Fabiola Zurlini. 2002. L'introduzione dell'esperienza quantitativa nelle scienze biologiche ed in medicina Santorio Santorio. In *Atti della XXXVI tornata dello Studio firmano per la storia dell'arte medica e della scienza, Fermo, 16–17–18 maggio 2002*, ed. Studio firmano per la storia dell'arte medica e della scienza, 117–137. Fermo: A. Livi.

Heering, Peter. 2008. The Enlightened Microscope: Re-enactment and Analysis of Projections with Eighteenth-century Solar Microscopes. *British Journal for the History of Science* 41: 345–367.

Hollerbach, Teresa. 2018. The Weighing Chair of Sanctorius Sanctorius: A Replica. *NTM Zeitschrift für Geschichte der Wissenschaften, Technik und Medizin* 26: 121–149.

Lemmer, Björn. 2015. Sanctorius Sanctorius–Chronophysiology in the Seventeenth Century. *Chronobiology International* 32: 728–730.

Leong, Elaine. 2018. *Recipes and Everyday Knowledge: Medicine, Science, and the Household in Early Modern England*. Chicago: University of Chicago Press.

Maclean, Ian. 2002. *Logic, Signs, and Nature in the Renaissance: The Case of Learned Medicine*. Cambridge/New York: Cambridge University Press.

Major, Ralph H. 1938. Santorio Santorio. *Annals of Medical History* 10: 369–381.

Mattioli, Mario. 1985. *Grandi indagatori delle scienze mediche*. Naples: Idelson.

Miessen, Hermann. 1940. Die Verdienste Sanctorii Sanctorii um die Einführung physikalischer Methoden in die Heilkunde. *Düsseldorfer Arbeiten zur Geschichte der Medizin* 20: 1–40.

Ottosson, Per-Gunnar. 1984. *Scholastic Medicine and Philosophy: A Study of Commentaries on Galen's Tegni (ca. 1300–1450)*. Naples: Bibliopolis.

Pickering, Andrew. 1995. *The Mangle of Practice: Time, Agency, and Science*. Chicago: University of Chicago Press.

Poma, Roberto. 2012. Santorio Santorio et l'infallibilité médicale. In *Errors and Mistakes. A Cultural History of Fallibility*, ed. Mariacarla Gadebusch Bondio and Agostino Paravicini Bagliani, 213–225. Florence: SISMEL-Edizioni del Galluzzo.

Premuda, Loris. 1947. *Intorno a Santorio Santorio ed alla medicina giuliana del passato: Orazione ufficiale del dottor Loris Premuda all'apertura del primo Convegno medico giuliano detta il 14 settembre 1946*. Trieste: F. Zigiotti.

———. 1950. Santorio Santorio. *Pagine istriane* 1: 117–124.

Rabier, Christelle. 2013. Introduction: The Crafting of Medicine in the Early Industrial Age. *Technology and Culture* 54: 437–459.

Rothschuh, Karl. 1968. Henricus Regius und Descartes. Neue Einblicke in die frühe Physiologie (1640–1641) des Regius. *Archives internationales d'histoire des sciences* 21: 39–66.

Sanctorius, Sanctorius, and Evaristo Lebàn. 1950. *De statica medicina: con un saggio introduttivo di Evaristo Lebàn*. Florence: Santoriana, A. Vallecchi.

Sanctorius, Sanctorius, and Giuseppe Ongaro. 2001. *La medicina statica*. Florence: Giunti.

Siraisi, Nancy. 1987. *Avicenna in Renaissance Italy: The Canon and Medical Teaching in Italian Universities After 1500*. Princeton: Princeton University Press.

Smith, Pamela H. 2009. Science on the Move: Recent Trends in the History of Early Modern Science. *Renaissance Quarterly* 62: 345–375.

Smith, Pamela H., Amy R.W. Meyers, and Harold J. Cook, eds. 2014. *Ways of Making and Knowing: The Material Culture of Empirical Knowledge*. Ann Arbor: The University of Michigan Press.

Valleriani, Matteo, ed. 2017. *The Structures of Practical Knowledge*. Cham: Springer.

Vedrani, Alberto. 1920. Santorio Santorio da Capo d'Istria: (1561–1636). *Illustrazione medica italiana* 2: 26–29.

Wear, Andrew. 1973. *Contingency and Logic in Renaissance Anatomy and Physiology*. Phd diss., Imperial College, London.

———. 1981. Galen in the Renaissance. In *Galen: Problems and Prospects*, ed. Vivian Nutton, 229–262. London: The Wellcome Institute for the History of Medicine.

Chapter 2
Sanctorius Sanctorius: Between Koper and Venice

Abstract By way of introduction, this chapter gives a biographical account of Sanctorius that situates him in his social, institutional, and professional context. The chapter critically evaluates the existing biographies of the Venetian physician and complements them with my own research on the primary sources. Episodes in Sanctorius's life that have hitherto received little or no attention are discussed in more detail. This opens up a new perspective on the life and work of Sanctorius, setting the stage for the more comprehensive reconsideration of his work to be found in the following chapters.

Keywords History of medicine · Sanctorius Sanctorius · University of Padua · Venetian republic

Many scholars have written biographical accounts of Sanctorius, often composed in the context of commemorations or in lexica.[1] They differ in terms of scope, detail, and precision, as well as in their choice of source material. Some include research on the primary sources, whereas others seem to be mere summaries of the existing secondary literature.[2] While some provide bibliographic information on the sources they use, others show little trace of this.[3] Apart from these mostly, brief biographies, there are also studies that comprehensively analyze the life of the famous physi-

---

[1] Mangeti 1731: 154 f., Renauldin 1825: 308 ff, Stancovich 1829: 235–59, Vedrani 1920, Giordano and Castiglioni 1924, Capparoni 1925–1928: 55–9, Baila 1936, Del Gaizo 1936, Major 1938, Premuda 1950, Sanctorius and Lebàn 1950: 23–38, Grmek 1975, Mattioli 1985: 253–62, Eknoyan 1999, Gedeon 2006: 18, 36 ff., 48 ff., 54 f. This is not a comprehensive list, but only a selection of the many biographical accounts of Sanctorius.

[2] Examples of biographical accounts that include research on the primary sources are Mangeti 1731: 154 f., Grmek 1975. Biographical accounts that merely summarize the existing secondary literature include, e.g., Stancovich 1829: 235–59, Vedrani 1920, Capparoni 1925–1928: 55–9, Baila 1936, Major 1938, Premuda 1950, Sanctorius and Lebàn 1950: 23–38, Mattioli 1985: 253–62, Eknoyan 1999, Gedeon 2006: 18, 36 ff., 48 ff., 54 f.

[3] The following accounts provide bibliographic data.g., Stancovich 1829: 235–59, Vedrani 1920, Major 1938, Premuda 1950, Grmek 1975, Mattioli 1985: 253–62, Eknoyan 1999. Examples of accounts that contain little bibliographic data are Renauldin 1825: 308 ff, Giordano and Castiglioni

© The Author(s) 2023

T. Hollerbach, *Sanctorius Sanctorius and the Origins of Health Measurement*,
https://doi.org/10.1007/978-3-031-30118-6_2

cian.[4] Moreover, biographical data on Sanctorius can be gleaned also from works on other topics, which are not always included in the literature on Sanctorius himself.[5] The following chapter critically evaluates this existing literature and complements it with my own research on the primary sources. Wrong or insufficiently documented claims are identified and, whenever possible, clarified. Episodes in Sanctorius's life that have hitherto received little or no attention are discussed in more detail. Most people's image of Sanctorius is of him sitting in a huge balance. They know him as an outstanding doctor with a splendid career, as a genius, who, almost out of the blue, invented a new medical science that profoundly influenced the modern age. But does this image match the biographical evidence? Is it still a valid view of Sanctorius? In the following account of his biography, I try to find the answers to these questions.

2.1   Childhood and Education

Sanctorius Sanctorius (Fig. 2.1) was born on March 29, 1561, in the town of Koper, in a region which at the time was in the Venetian Republic and is today a part of Slovenia.[6] His father, Antonio, a Friulian nobleman, had been called to Koper as a high official of the Venetian Republic.[7] While serving there, he met and married Elisabetta Cordonia, a local noble heiress. Sanctorius was the firstborn of their four children. In keeping with an Istrian fashion of the time, he received his family name as his given name. Together with his younger siblings, Isidoro, Diana, and Franceschina, Sanctorius spent his childhood in Koper, completing his early schooling there. But soon his father took him to Venice and had him enter into the highest circles of Venetian society. One friend of Sanctorius's father was Giacomo Morosini, a descendant of a long-established, noble Venetian family, who enabled Sanctorius to study under the private tutors of his sons, Paolo (1566–1637) and Andrea (1558–1618). Thus, Sanctorius received excellent training in classical languages, literature, philosophy, and mathematics (Castiglioni 1931: 733 f.; Grmek 1975: 101).

1924, Capparoni 1925–1928: 55–9, Baila 1936, Del Gaizo 1936, Sanctorius and Lebàn 1950: 23–38, Gedeon 2006: 18, 36 ff., 48 ff., 54 f.

[4] Capello 1750, Del Gaizo 1889, Castiglioni 1931, Grmek 1952, Ettari and Procopio 1968, Sanctorius and Ongaro 2001: 5–16.

[5] E.g., Rossetti 1984, Sarpi 1969, Anonym 1882, Ziliotto 1944.

[6] In the present work, I use the Latinized version of Sanctorius's name, which Sanctorius himself used in the first editions of his works. See: Sanctorius 1603; Sanctorius 1612a, b; Sanctorius 1614; Sanctorius 1625; Sanctorius 1629a. In general, however, personal names of Italian origin appear in the present work in their Italian form. In cases where the Latin forms are more familiar or the Italian forms are uncertain, Latin forms have been used. With regard to place names, I use Sanctorius's designations (whenever possible, in English translation) and try to match historical regions with today's regions.

[7] The area of Friuli was under the dominion of the Venetian Republic at this time.

Fig. 2.1 Portrait of Sanctorius Sanctorius (date and author unknown) (Biblioteca Civica Padova, RIP.II.309). (By kind permission of Comune di Padova—Assessorato alla Cultura)

Santorius Santorio Justinopolitanus —
Celeberrimus Statice Medicine Professor.
Nat 1561. Mort Venetiis 1636.

In 1575, Sanctorius enrolled at the University of Padua, where he followed the traditional curriculum of the arts faculty, which consisted of logic and philosophy, followed by medical studies. At only fourteen years of age, he was three or four years younger than the average freshman at an Italian university. The University of Padua was flourishing at the time and was a notable center of Aristotelian natural philosophy. Medical teaching there dated back to the thirteenth century and comprised three subjects: medical theory, medical practice, and surgery.[8] Among Sanctorius's teachers in the field of philosophy were Francesco Piccolomini (1520–1604) and Giacomo Zabarella (1533–1589) and, in the field of medicine, Bernardino Paterno (fl. second half of the sixteenth century), Girolamo Fabrici d'Acquapendente (1533–1619), and Girolamo Mercuriale (1530–1606) (Ettari & Procopio 1968: 41; Grmek 1975: 101; Schmitt 1985: 1, 4; Sanctorius & Ongaro 2001: 6; Grendler 2002: 4, 148).

---

[8] The distinction between medical theory (*theoria*) and medical practice (*practica*) in the context of the medical university curriculum is somewhat misleading for the modern reader. Both dealt with a combination of theoretical and practical issues and their differences lay more in context, in their direct relevance to treatment, and, probably, in the amount of concrete physical detail that they presented. Thus, textbooks used for the teaching of *practica* were methodologically not necessarily different from those used for the teaching of *theoria*. What set them apart was their focus on anatomical, pathological, or therapeutic factual detail. See: Siraisi 1987: 54, Bylebyl 1979: 338.

2.2 Sanctorius's Early Practice: Travels, Relations, and Much Uncertainty

Sanctorius graduated in 1582, after seven years of study, and began to devote himself to the practice of medicine. Little is known about his whereabouts and activities over the next twelve years, up to the turn of the seventeenth century. However, I follow the clues that I have. Sanctorius mentions that he launched his static experiments—a systematic study of changes in weight, which he used to quantify the insensible perspiration of the human body—in 1584 or 1590.[9] Thus, the weighing procedures and his special weighing chair, both of his own invention and the reason for his later fame as the founder of a new medical science, accompanied his medical practice quite early on. I will return to this later in more detail.

Writing in 1750, Arcadio Capello referred to a letter of October 20, 1587, in which the Paduan vicar Nicolò Galerio recommended Sanctorius, in the name of the university, to "a certain Polish prince," who had asked the "very renowned faculty" to send him a "very good" medical man (Castiglioni 1931: 735).[10] The original letter seems to have been lost, just like the copy Capello claimed to have seen. While there is no reason to doubt the authenticity of the letter, there is no evidence that Sanctorius actually left for Poland as most of his biographers assert.[11] The fact that Capello did not give the name of the intended recipient of the letter suggests that the copy did not bear a name. It may have been addressed to Sigismund III Vasa, but this is mere speculation (Grmek 1975: 101; Grmek 1952: 13; Bigotti 2016: 2). According to Arturo Castiglioni, nothing in the Polish archives suggests that Sanctorius ever stayed in Poland (Castiglioni 1931: 779 fn. 10). New archival research must be undertaken to clarify whether Castiglioni's findings of 1931 are still tenable (Castiglioni 1931: 779 fn. 10; Grmek 1952: 13; 1975: 101; Bigotti 2016: 2).

Two years later, in 1589, Sanctorius was recommended also to the governors of Koper, who were likewise in search of a good physician. Leandro Zarotti (1515–1596) and Zuanne Vittorio (life dates unknown) wrote from Venice that they had had the chance to meet Sanctorius only once or twice, because he was so often

---

[9] With regard to Sanctorius's weighing procedures, conducted in order to quantify insensible perspiration, I use the term "experiment" since he meant his static experiments (*staticis experimentis*) in the sense of repeated and controlled observations, see Sect. 6.2.5. In the preface to his work *De statica medicina*, Sanctorius stated that he had conducted the experiments over the course of thirty years, see: Sanctorius 1614: Ad lectorem. However, in a letter Sanctorius sent to Galileo Galilei with a copy of his *De statica medicina* in 1615, he mentioned that he had carried out the experiments over a span of twenty-five years, see: Sanctorius 1902.

[10] "… ad Principem quemdam Polonum …," see: Capello 1750: IX, fn. a. "… cum Poloniae Regulus quidam ex Patavino Archilyceo Virum Jatrices peritissimum exoptaret, Sapientissimi illius Collegii Patres Sanctorium illuc mittendum unanimi sententia decreverint." See: ibid.: IX.

[11] E.g., ibid.: IX, Del Gaizo 1889: 7, Giordano and Castiglioni 1924: 237, Capparoni 1925–1928: 55, Castiglioni 1931: 735, Premuda 1950: 119, Ettari and Procopio 1968: 24. Only Grmek doubts that Sanctorius spent some years in Poland (Grmek 1952: 13 f., Grmek 1975: 101). His allusion to the lack of primary sources is, however, important and leads to the conclusion that the aforementioned authors based their assumptions on conjecture, or on quotations of other secondary literature.

away, but were convinced of his skills, as others were, too.[12] Thus, Sanctorius was still based in Venice at the time, and if ever he did depart for Poland, then only later. However, the position in Koper seems to have gone to another physician, Pietro Antonio Giusti (life dates unknown), who was recommended to the governors in the same letter as Sanctorius.[13]

It is certain nevertheless that Sanctorius spent some time in his hometown. He was a member of the Accademia Palladia, which represented an important meeting place for the intellectual Istrian elite. Consisting of mainly young scholars (Ziliotto 1944: 144 fn.), the academy in the late sixteenth century was especially engaged in discussions of love. In the work *De cento dubbi amorosi* (On One Hundred Amorous Doubts), Girolamo Vida (1563–91) compiled public talks held at the Accademia Palladia, including a lecture of Sanctorius's on the meaning of colors (Vida 1621: 76r–86v).[14] According to Baccio Ziliotto, author of a work on the academies and academics of Koper, Sanctorius presided over the academy for several years during the 1580s (Ziliotto 1944: 144); and in any case he must have held his lecture before Girolamo Vida died in 1591. Presumably this was also the time when Sanctorius met the physician Marc'Antonio Valdera, another *Palladiano*. They seem to have been close friends, as Sanctorius posthumously published Valdera's work *L'Epistole d'Ovidio* (The Epistles of Ovid), in which he referred to him as "my such dear friend … [who] from early youth onwards pursued the sciences with all diligence, so that he won great admiration as a most excellent philosopher, and physician…" (Valdera 1604: 7).[15] Thus, besides his medical practice, Sanctorius fostered acquaintance with young intellectuals in his hometown and dedicated himself, with them, to poetry and literature.

Moreover, there is evidence of Sanctorius spending time in Croatia and Hungary: he referred in some of his works to experiences he had had in those countries. In Hungary, Sanctorius wrote, he had to accustom himself to the unleavened bread served there, and to the wine that seemed less mellow to him than the Italian variety.[16] He practiced medicine for five years in Pannonia, a region named for a

---

[12] A transcription of the letter is printed in: Anonym 1882: 90 f. Castiglioni 1931: 735, Grmek 1952: 9, 14 and Ettari and Procopio 1968: 24 misdated the letter to 1599.

[13] Pietro Antonio Giusti is listed as a physician in Koper for the year 1589. See: Pusterla 1891: 64.

[14] The work was published posthumously in 1621 by Agostino Vida, a relative of Girolamo Vida. See: Vida 1621: dedication. Sanctorius's discourse exemplifies the influence of Renaissance Humanism on the members of the Accademia Palladia. Medical and natural philosophical authors are mostly replaced by poets like Vergil, Ovid, Horaz, or Boccaccio. A discussion of their opinions on colors and the metaphorical meaning of the latter are the main part of the discourse. See: ibid.: 76r–86v.

[15] "… mio cosi caro amico, …; egli dalla prima giovenezza attese con ogni sollecitudine alle scientie, onde con grand' ammiratione riuscì Filosofo, & Medico Eccellentissimo: …" See: Valdera 1604: 7. The English translations of quotations are mine unless otherwise indicated.

[16] "… tale quippiam mihi contigit dum in Hungariam fecessi; quia primis mensibus panem illum azimum Hungaris assuetum abhorrui, attamen paullo post, dum assuescerem dulcior mihi est visus; Similiter vinum, quod Italico erat aliquantulum dissimile mihi videbatur minus suave, itidem de omni ferculo, demum tamen acquisita illorum consuetudine." See: Sanctorius 1603: 86v.

province of the former Roman Empire and which extended over the territory of present-day western Hungary, parts of eastern Austria, and parts of several Balkan states, primarily Slovenia, Croatia, and Serbia (Encyclopaedia Britannica 2018b).[17] In Croatia, Sanctorius tells, he designed and used two kinds of steelyard (*statera*), a pair of scales with unequal arm lengths. One was an anemometer, to measure the impetus of the wind. The other was an early type of hydrodynamometer, to measure the force of water currents.[18] The earliest biographer of Sanctorius, Giacomo Grandi, wrote that Sanctorius practiced medicine for several years in Karlovac, in Croatia, and traveled also to the German territories (Grandi 1671: 10 f.).[19] Indeed, Sanctorius himself mentioned the city of Karlovac, where he made observations regarding venomous diseases (Sanctorius 1603: 163r–163v).

The lack of references to Poland on Sanctorius's part has led Mirko Grmek to suggest that he was in the service not of a Polish prince, but rather of a Croatian or Hungarian nobleman, and therefore resided in Croatia and Hungary. According to Grmek, Sanctorius left Croatia when a lethal plague was raging there (Grmek 1952: 14 f.; 1975: 101). While it cannot be clarified whom Sanctorius served, whether or not he was ever in Poland, or when and why he returned to Padua or Venice, it can be assumed that he was by then already a well-known and highly appreciated physician. The fact that Nicolò Galerio recommended him in the name of the University of Padua as early as 1587 shows—in combination with the travels to Pannonia, Croatia, and Hungary, to which he himself bore witness—that he was very probably consulted by noblemen all over the Venetian Republic and the Balkans.

There is proof that Sanctorius was in Venice on October 5, 1607, being one of the first to have aided Fra Paolo Sarpi (1552–1623), who was injured in the famous assassination attempt (Castiglioni 1931: 735; Sanctorius & Ongaro 2001: 8). In 1603, in Venice, Sanctorius published his first book, *Methodi vitandorum errorum omnium qui in arte medica contingunt* (Methods to avoid all errors occurring in the medical arts).[20] It was evidently well received, since further editions appeared in

---

In another passage of the same work, Sanctorius stated: "… audias pro huius rei confirmatione, quid mihi contigit, dum in Hungaria Medicum agerem; …" See: ibid.: 92r. Further references to Sanctorius's stay in Hungary can be found in ibid.: 125r, 135v, 136r, 159v, 163v, 211v, 222v, 225v.

[17] "… quod certè mihi contigit, dum cursu quinque annorum medicinam facerem in Panonia, …." See: Sanctorius 1612b: 131.

[18] "… sed libet referre quod in Croatia observavimus: erat locus ventorum strepitu, & magno fluminum impetu insignitus: incolę vero aliquando illo strepitu à somno avocabantur, aliquando vero ad somnum proclives reddebantur: proposui, ut subtiliter causam inveniremus, lance ponderari posse utrumque impetū, quod ab amicis coactus, ut id ostenderem pręstiti duobus stateris, per primam ventorum, per secundam vero aquę impetum, utriq; …" See: Sanctorius 1625: 246.

[19] "Porrò qua laude Medicinam exercuerit, dicant Germaniae loca, quae peregrinationis utilitate captus lustravit; dicant Carlostati Cives, qui operam eius verè opiferam aliquot annos admirati sunt; …" See: Grandi 1671: 10 f.

[20] I refer to this work henceforth as *Methodi vitandorum errorum*.

1630 and 1631.[21] This work, probably conceived during Sanctorius's stay abroad, was dedicated to Ferdinand of Austria (1578–1637), the later Holy Roman Emperor, Ferdinand II, which leads to yet another suggestion: that Sanctorius was in fact in his service (Castiglioni 1931: 736; Sanctorius & Ongaro 2001: 8). Still, the available source material permits nothing but speculation.

2.3   Professorship at the University of Padua

The next period of Sanctorius's life is better documented, so that the hazy realm of ambiguity can be left behind. Owing to the success of the *Methodi vitandorum errorum* as well as to the fame he had gained as a practicing physician, Sanctorius was appointed first ordinary professor of *theoria* at the University of Padua—by a ducal degree of October 6, 1611. The position had been vacant for eight years, since the death of Orazio Augenio (1527–1603). Sanctorius was granted a six-year tenure and an annual stipend of 800 florins (ASVe-b: f. 319v–320r; ASVe-c).[22] This generous salary was not unusual for the leading ordinary professor of medical theory, who generally ranked among the highest paid members of the arts and medicine faculty.[23] What was unusual, was that Sanctorius accepted a professorship after nearly thirty years of medical practice (Grendler 2002: 160, 319).

Given the high esteem Sanctorius had long enjoyed as a practicing physician, a university position with strict duties and harsh competition seems an unlikely choice for him. Apart from regular public lectures, professors at Padua usually also gave private lessons. Even during the vacation periods, they had to ask for permission to leave the city. Moreover, they had to attract a minimum number of students—and an official known as a *punctator* checked each class, daily, to ensure that they had. But the competition was tough. In Sanctorius's day, the medical faculty of Padua comprised sixteen professors of medicine, including a second ordinary professor of medical theory, who would very likely have taught the exact same text as Sanctorius, at the same hour (Favaro 1888: 1060; Tomasini 1986: index, 291–330; Grendler

---

[21] As Bigotti has pointed out (Bigotti and Taylor 2017: 107 fn. 11), many early catalogues of medical books as well as many biographies of Sanctorius refer to an edition of the *Methodi vitandorum errorum* published in 1602 *Apud Societatem Venetam* (e.g., Castiglioni 1931: 750, Grmek 1975: 101, Eknoyan 1999: 229). Possibly these early scholars were mistaken; in any case, the edition seems to be no longer extant.

[22] Professors at the University of Padua were usually paid in Paduan florins (*fiorini*) instead of Venetian ducats. A Venetian ducat was worth 6 lire 4 soldi, whereas a Paduan florin equalled 5 lire (Grendler 2002: 22, fn. 55).

[23] Sanctorius's predecessor Orazio Augenio started his professorship on a salary of 900 florins (Tomasini 1986: 293). While, in the fifteenth century, the first ordinary professor of theoretical medicine was the most prestigious and best-paid member of the medical faculty, this changed during the sixteenth century, when the first ordinary professor of practical medicine first drew equal to and then surpassed the first ordinary professor of theoretical medicine both in prestige and salary (Grendler 2002: 352).

2002: 145, 161). Thus, Sanctorius's reluctance to accept the position is no surprise. In his inaugural lecture, he said:

> Therefore, I admit that before coming here I hesitated a lot, long undecided as to whether I should accept this position which was offered to me by the leaders of this academy, or whether I should rather refuse it (Capello 1750: XIX).[24]

In the end, according to his own words, he accepted because his leaders (*meorum Principum*) had chosen him, and because of the dignity the position conferred not only on him but also on his home country and his family (Capello 1750: XX).

The fact that Sanctorius kept company with Venetian high society and frequently visited the home of the Morosini, by then a meeting place of the most illustrious Venetian scholars and aristocrats, including Galileo Galilei (1564–1642) and Paolo Sarpi, may also have contributed to his appointment to the University of Padua. How important the so-called *Ridotto Morosini* circle was for Sanctorius, not only socially, but also intellectually, will be shown in the course of this book. The *Riformatori dello Studio*, elected by the Venetian Senate to oversee all aspects of the university, would hardly have left to chance one of the most prestigious university appointments.[25] The student rectors also played their part. They all wanted a star professor with an excellent reputation, who would attract students. Thus, they surely inquired beforehand how much it would take to convince Sanctorius and how receptive he would be to an offer (Grendler 2002: 160, 164).

They were not disappointed. In a letter of November 18, 1611, to the *Riformatori dello Studio*, following Sanctorius's inaugural lecture of November 17, the rectors congratulated themselves on their choice of "so famous a lecturer," who had already given a fine example of his worth and his intelligence; and they emphasized that the school was extraordinarily well attended (Castiglioni 1931: 738; Del Gaizo 1889: 56).[26] The university's international intake ensured, moreover, that Sanctorius's lectures were frequented by physicians and students not only from all over Italy, but also from Poland, England, and especially, Germany, to name but a few (Grendler 2002: 36 f.). As professor of *theoria*, he was obliged to interpret three classical books: Hippocrates's *Aphorisms* (ca. 450–ca. 380 BCE), Galen's *Ars medica*, *Ars parva*, *Tegni*, or *Microtechne* (The Art of Medicine; ca. 129–ca. 216 CE) and the first part of the first book of Avicenna's *Canon* (ca. 970–1037 CE). In fact, these three books are the basis of three of Sanctorius's six publications.[27]

---

[24] "Fateor equidem me priusquam huc accederem, diu multumque dubitasse, utrum Provinciam hanc a Supremis hujusce Academiae Moderatoribus mihi oblatam susciperem, an potius recusarem." Sanctorius's inaugural lecture was published posthumously in 1750 by Capello (Capello 1750: XIX–XXIV, cit. XIX). For an Italian translation, see: Ettari and Procopio 1968: 159–64.

[25] Sanctorius's close friend, Andrea Morosini, was *Riformatore dello Studio di Padova* in 1609, 1612, and 1616. See: Trebbi 2012.

[26] I was unable to consult the original letter due to its poor condition. It is in the *Archivio di Stato* in Venice. For a transcription of the letter, see: Del Gaizo 1889: 56.

[27] The publications are in order of appearance: *Commentaria in Artem medicinalem Galeni* (1612, 1630, 1631, 1632); *Commentaria in primam Fen primi libri Canonis Avicennae* (1625, 1626, 1646), *Commentaria in primam sectionem Aphorismorum Hippocratis* (1629). In the following I refer to these works as *Commentary on Galen*, *Commentary on Avicenna*, and *Commentary on Hippocrates*.

In the first period of his teaching career, in 1614, Sanctorius published his book *Ars de statica medicina* (The Art of Static Medicine), which immediately proved a great success.[28] It presented the results of the weighing procedures that Sanctorius had begun in 1584 or 1590. In addition to the weighing chair that he devised to this end, Sanctorius designed other precision instruments to supplement his research, and constructed apparatus for the improvement and alleviation of the sick. He published some of his findings in 1625, in his *Commentary on Avicenna*. In the preface he wrote:

> [...] since I hear that my pupils, coming from the most various parts of the world, instructed by me with the greatest disposition and with generous benevolence, attribute the invention of a lot of them [the instruments] to themselves: a ruthlessness that certainly may not be passed over in silence (Sanctorius 1625: Ad lectorem).[29]

Thus, when Sanctorius introduced the instruments into his commentary, he was acting under pressure, in response to those of his pupils in other countries who had published the results of his research under their own names.

2.4   The Collegio Veneto

On May 5, 1616, Sanctorius was named president of a new *Collegio* set up in Padua that year, which was later called the Collegio Veneto.[30] Strictly speaking, it was an examination board comprised of the first ordinary professors of the arts and medicine faculty of the University of Padua for the purpose of conferring doctorates. Officially, the Collegio granted doctorates only to poor students who were not in a position to pay the usual fee. But in fact, the Collegio was established to allow foreign, non-Catholic students to avoid making the profession of faith that Pius IV had imposed through the bull *In sacrosancta*.[31] For the first time, doctorates could be

---

[28] In the following I refer to this work as *De statica medicina*. For an enumeration of the numerous editions and translations of the *De statica medicina*, see: Appendix II.

[29] "… quia audio, discipulos meos in varias terrarum partes dispersos, quos summa caritate, & gratuita benevolentia docui, horum multorum sibi inventionem attribuere, quorum inhumanitas silentio certè non erat obvolvenda." See: Sanctorius 1625: Ad lectorem. For the Italian translation, see: Sanctorius and Ongaro 2001: 13 f.

[30] In the beginning it was called Collegium al Bo, Collegium universitatis, Collegium publicum, or Collegium auctoritate Veneta (Rossetti 1984: 374). For a list of the presidents of the Collegio Veneto, see: AAU 703: 1r.

[31] The correspondent decree used the following careful words to avoid conflicts with the Pope: "to give the insignia of the doctorate in the arts to poor *and other* students in accordance with the common ancient customs," [my emphasis]; see: ASVe-b: 340r, Rossetti 1984: 369.

conferred by the state directly, without ecclesiastical intervention.[32] This was of particular importance to the Venetian government, because it hoped to continue to attract international, often non-Catholic, students. Such students contributed not only to the Republic's economy, but also to the good reputation of its university in Padua. In the literature on Sanctorius, this episode is typically treated as an anecdote, greatly simplified, and often reduced to a single sentence. Other sources, however, reveal a fuller picture. I draw on them to expound in more detail this event in Sanctorius's life.

2.4.1 Quarrels with the Church

As was to be expected, the Collegio Veneto immediately provoked papal protest. As president of the institution, Sanctorius was in the thick of the disputes—but also in good company. His close friend Paolo Sarpi had been involved in the issue from the start, along with their mutual friend Nicolò Contarini (1552–1631).[33] What is more, Contarini was *Riformatore dello Studio*, at the time, as was Sarpi's friend Alvise Zorzi (1535–1616). Thus, Sanctorius had powerful support, when resisting the nuncio's demand that students graduating from the Collegio Veneto profess their Catholic faith. Paolo Sarpi tried to resolve the issue by emphasizing that the subjects examined in the Collegio Veneto, philosophy and medicine, were not directly connected to religious matters. "Saying that a heretic is a good physician is not prejudicial to the Catholic faith," he stated (Grendler 2002: 507).[34] Even though curial mistrust remained, the Collegio Veneto was able to continue its work and in 1635 it was officially extended to the law faculty (Rossetti 1984; Weigle 1965: 332 f.; De Bernardin 1983: 71 f.; Sarpi 1969: 562–71). However, Sanctorius's involvement in the matter left its mark on his career—and not only in the way one might expect.

---

[32] Until then, students who wanted to avoid the normal procedures could take their doctorates with count palatines. The count palatine degrees were cheaper and not granted on the basis of papal authorization, contrary to those conferred by the Sacred Colleges of doctors of law and arts. The count palatines did not insist on an oath of allegiance to Catholicism. In the late sixteenth century, however, most heterodox students chose this route, which was much to the dislike of the University. Thus, in 1612 the Venetian Senate deprived the count palatines of their privilege to confer doctorates in the Venetian state (Grendler 2002: 173 fn. 102, 183–6, Rossetti 1984: 366 ff.).

[33] The close relationship between Sanctorius and Nicolò Contarini is attested by Sanctorius's dedication of his work *De statica medicina* to him, in which he referred to their forty years of acquaintance (Sanctorius 1614: dedication).

[34] "… dicendo che un heretico sia un buon medico, non si pregiudica alla fede catholica." See: Grendler 2002: 507, fn. 119, Rossetti 1984: 373.

2.4.2   Quarrels with the German Nation of Artists

At first, things seemed to be going well. In 1617, after completing his six years at
the university, Sanctorius was reappointed by the Senate with a pay increase of 400
florins per annum (ASVe-b: 342v; ASVe-f).[35] But trouble soon raised its head.
Already in 1618, dissatisfaction arose because Sanctorius was absent from the doctoral degree award ceremony in the Collegio Veneto. The proceedings of the German
Nation of Artists, the association of philosophy, medicine, and theology students of
the University of Padua, reported that Cesare Cremonini (1550–1631) and Rodrigo
Fonseca (1550–1622), namely the other two first ordinary chairs of the arts and
medicine faculty, had granted three students their doctorate in the absence of
Sanctorius. The latter was not amused and stated: "Your doctorate is not worth
much; I, not Cremonini, am the president" (Rossetti 1967: 64).[36] Therefore, the
students had to present themselves again before the professors and members of the
Nation to publicly receive their doctoral degree from Sanctorius. This episode
shows, on the one hand, Sanctorius's insistence on executing his role as president
and, on the other, the disapproval that his behaviour provoked in the German Nation.
According to them, Sanctorius himself had decided not to take part in the initial
graduation ceremony, preferring instead to pursue lucrative business in Venice. His
subsequent complaints evoked little sympathy among the German students, who
then decided to stay away from his next lectures. This conflict should not be under-
estimated. In terms of their number, activity, and prestige, German students played
a preeminent role at the University of Padua. Moreover, many of them were
Protestant and pursued their degree at the Collegio Veneto (Rossetti 1967: IX, 63 ff.;
Grendler 2002: 193).

How important the presidency of the Collegio Veneto was for Sanctorius is
shown in another passage from the proceedings. In 1619, when his term of office as
president came to an end, he tried to extend it—and did not shy away from bringing
up the matter before the Venetian Senate. Even though he did not succeed and a new
president, Rodrigo Fonseca, was elected, Sanctorius was given a second chance.
Fonseca died in the spring of 1622 and Sanctorius was called upon to succeed him
provisionally, until the end of the period required in law for a presidential election

---

[35] Gaetano and Luisa Cozzi emphasized that twenty senators voted against Sanctorius's reappointment. They saw this as proof that some senators shared the preoccupations of the Holy See and of
the nuncio regarding Sanctorius's conferral of doctorates without the profession of faith (Sarpi
1969: 571, ASVe-f). What they did not take into account, however, is that ca. thirty voted against
Sanctorius's first appointment in 1611, which corresponded to ca. one third of the Senate, see:
ASVe-c, ASVe-b: f. 319v–320r. Thus, there is no proof that he lost support in the Senate due to his
presidency of the Collegio Veneto, just as it is not known whether the senators who voted against
his reappointment in 1617 did so in solidarity with the Holy See. On the contrary, it seems that
Sanctorius gained support in the Senate during his first period of teaching in Padua. Corrections in
the original senatorial document suggest that the pay raise was adjusted upwards from 300 to 400
florins. It is not known, however, whether this was the outcome of negotiations or simply a typo-
graphical mistake. See: ASVe-f.

[36] "Il tuo doctorato non val tanto, ego praeses sum, non Cremoninus, etc." See: Rossetti 1967: 64.

(Rossetti 1967: 79; 1984: 374 f.; AAU 703: 1r, 130r). A few months later, new allegations were made. Busy again with his medical practice in Venice, Sanctorius had
been unable to attend the graduation ceremony of a student, the librarian at the
German Nation of Artists' library,[37] and had himself proposed, this time, that the
doctorate be awarded in his absence. According to the students, this flew in the face
of opinion among the *Riformatori dello Studio* and, what is more, it reduced the
value of their doctorates. They complained first to Sanctorius, but when this did not
have the desired effect, they went a step further and reported the matter to the
Riformatori dello Studio. The *Riformatori* took the criticism seriously, but the students had to wait until 1624 before Sanctorius was replaced as president of the
Collegio Veneto by Giovanni Colle (1558–1631) (Rossetti 1967: 147–50, 173 f.;
1984: 375; AAU 703: 1r).

In that same year, 1624, Sanctorius's second term as first ordinary professor of
medical theory came to an end. Moreover, as Sanctorius's biographer Capello
claims, new allegations that Sanctorius was neglecting his office soon landed him in
court. However, he was acquitted on February 8. According to Capello, the records
of the case can be found in the proceedings of the Paduan Curia; but these seem now
to be lost.[38] The issue was possibly linked to the death of Sanctorius's nephew, as a
statement by Sanctorius later that same year attests: "I did not miss a single lesson
in recent years, except last year, owing to the death of my nephew and son" (Ettari
& Procopio 1968: 147).[39] Be this as it may, the recurrent complaints show, in my
opinion, that Sanctorius's travels to Venice and his medical practice there made him
neglect his professorial duties. But the administration of the University of Padua
carefully monitored the professors and paid heed to the students' opinions. Teachers
whom students considered unfavorably as not very diligent were not reappointed.
This was the fate of Sanctorius, too. But in his case, things are more complicated
than they seem (Capello 1750: XII, XII fn. c; Castiglioni 1931: 738 f.; Ettari &
Procopio 1968: 29, 39 n. 50).

---

[37] At the University of Padua, the German Nation of Artists had a library from 1586 onward,
whereas the university library was only established around 1631. See: Grendler 2002: 505 fn.
111, 506.

[38] It seems that Capello was the only biographer of Sanctorius who saw the records of the case,
because all later authors referred to his work.

[39] "… non ho in questi anni preterita alcuna letione se non che quest'ultimo anno per la morte di un
mio nepote et figlio …." See: Ettari and Procopio 1968: 147. I was unable to find the original in the
archives. *Figlio* (son) is used here affectionately; there is no evidence that Sanctorius had children
of his own.

2.5   Failed Reappointment and Resignation

On January 20, 1624, two of the three *Riformatori dello Studio*, Antonio Barbaro
(1565–1630) and the aforementioned Nicolò Contarini, proposed before the Senate
that Sanctorius be reappointed, and were full of praise for him. They further suggested a pay raise of 300 florins per annum. The Senate did not agree: ninety-four
members voted against the proposition, only thirty-five voted in its favour, and fifty-
seven abstained (ASVe-b: 372v; ASVe-d). Given the German students' many complaints about Sanctorius, it seems surprising that the *Riformatori dello Studio* were
so supportive of him. But not quite so surprising, when one considers who was on the
examination board at the time. The friendship with Nicolò Contarini may well have
played its part. Moreover, the students were not completely innocent either. Already
around 1615, Sanctorius had written in a letter to Contarini that the "audience did not
allow for ordinary lectures" (BMCVe-b: f. 193).[40] This touched on a broader problem. Since the late sixteenth century, the Italian universities had been struggling both
with the failure of matriculated students to attend classes and with an increase in
student violence (Grendler 2002: 477–508). Regardless of whether or not Sanctorius's
students actually showed up or disrupted his teaching, this episode shows that the
complaints against their professor could be made also about themselves. It must also
be recalled that it was his students' plagiarism which first drove Sanctorius to publish illustrations of his instruments. But there is still more to it than that.

A few months later, Sanctorius's successor was elected: Pompeo Caimo
(1568–1631). He was the personal physician of Alessandro Peretti (1571–1623),
then one of the most influential cardinals of the Curia in Rome. The Venetian ambassador to Rome, Pietro Contarini (1541–1613), recommended him to the University
of Padua. As a result, Antonio Barbaro and Giovanni Corner (1551–1629) proposed
him to the Senate. What united these men was that they were all closely aligned with
the Pope. Caimo was appointed to the professorship without a single dissenting
vote. Not even Nicolò Contarini objected. Yet, as Gaetano and Luisa Cozzi suggest,
he could easily have abstained from voting, just as his friends did (Sarpi 1969:
571 f.; ASVe-b: 373r–374r).[41] Did Sanctorius's anticlerical behavior as president of
the Collegio Veneto cost him his professorship? Did his friendship with Sarpi,
Contarini, and other Venetian patricians among the so-called *giovani* (youths) put
an end to his university career?[42] Or was it rather his neglect of his duties and the
displeasure of his students?

---

[40] "Heri die sabbati videlicet nostri auditores non permiserunt ordinarias lectiones, …." See:
BMCVe-b: f. 193. The letter bears no indication of the year, but as Sanctorius referred to a lecture
by Francesco Pola Veronese, who was appointed to the University of Padua in 1615 and died a year
later, it can be assumed that it was written in 1615 or 1616. See also: Del Gaizo 1889: 56.

[41] Pompeo Caimo was elected by ninety-one yea votes, with no dissenting votes or abstentions.
See: ASVe-b: 373r–374r.

[42] The so-called *giovani* were a politically motivated group, consisting of mostly young Venetian
patricians who distinguished themselves through their innovative ideas and their critical view of
the Church and the Pope. See: Cozzi 1979: 140 f. For more information on the topic, see also:
Cozzi 1958: ch. 1.

Interestingly, most of Sanctorius's biographers tell yet another tale, namely that it was Sanctorius's personal decision to give up teaching.[43] And indeed, there is some truth to this. In the preface to his *Commentary on Avicenna*, published in 1625, Sanctorius stated that he "requested the liberty on March 5, 1624, from the most excellent moderators, so that the not small trouble of those, who burdened [him] much because of this one affair, might be lifted ..." (Sanctorius 1625: Ad lectorem).[44] He explained that he would like to retire in Venice where, once freed from teaching, he would reissue his previously published books and complete and publish his unfinished works as well as new work of his own (Sanctorius 1625: Ad lectorem). "This one affair" may well refer to the pending prosecution of Sanctorius, which was to lead him, he tells us, to resign his professorship. In his letter of resignation, however, Sanctorius claims that the Senate's refusal to grant him the 300 florin pay raise is the cause (Ettari & Procopio 1968: 147 f.).[45] Possibly the two factors were connected. For although Sanctorius was publicly declared innocent, the allegations surely had an impact on his reputation and the esteem he enjoyed in the Senate. Under these circumstances, a pay raise might have been considered inappropriate.

While Sanctorius and many of his biographers emphasized that he personally decided to resign, it rather seems that he preempted the inevitable outcome. He tried to limit the damage.[46] In fact, the Senate voted in January 1624, not only against his pay raise—but against his reappointment, too.[47] Sanctorius argued that all of his predecessors had received a pay raise with each new reappointment. His colleague, Cesare Cremonini, and his rival (*concorrente*), Niccolò Trivisano (life dates unknown) had both recently received a wage increase. What is more, with his medical practice in Venice alone, he could earn as much as 3000 ducats per year. This, and the fact that his teaching was very popular and attracted scholars to the university,

---

[43] E.g., Castiglioni 1931: 739 f., Major 1938: 376, Premuda 1950: 119, Grmek 1975: 103, Sanctorius and Ongaro 2001: 13.

[44] "licentiam die quinta Martij 1624 petij ab Excellentissimis Moderatoribus, ut levatus non levi molestia illorum, qui mihi propter hoc onus negotium valde facescebant, ..." See: Sanctorius 1625: Ad lectorem.

[45] I was unable to find the original letter of resignation in the archives. For a transcription of the document, see: Ettari and Procopio 1968: 147 f.

[46] In a letter from April 1624, Johan Rode (ca. 1587–1659), member of the German Nation of Artists of the University of Padua, informed Caspar Hofmann (1572–1648), professor for theoretical medicine in Altdorf (Nuremberg), that Sanctorius declined the professorship of theoretical medicine to preempt a decision of the Senate. Interestingly, Rode wrote in the next sentence that Hofmann could take a look at a piece of writing testifying that Sanctorius was not rejected. Given the senatorial decree in January of the same year, one cannot but wonder which writing Rode was referring to. See: Rode to Hofmann 1624.

[47] As the proposition connected Sanctorius's reappointment with a pay raise of 300 florins, the outcome of the election was a refusal of both, the reappointment and the pay raise. There is no indication that a further vote took place on only one of the two issues.

made him unable to consider continuing his professorship without the pay increase
of 300 florins (ASVe-d; Ettari & Procopio 1968: 148). His reaction could not have
come as a surprise to the senators. To what extent their decision was shaped by the
displeasure of the students, the neglect of his duties, or his involvement in the
Collegio Veneto and Paolo Sarpi's circle remains an open question. Most likely it
was a combination of all of these.

In the light of the above, the idea usually advanced by Sanctorius's biographers,
that the Senate decreed to grant Sanctorius life-long tenure on a full stipend, must
be taken with a pinch of salt. It can be traced back to Niccolò Papadopoli
(1655–1740), an early historian of the University of Padua, whose work has, however, been proved to contain inaccuracies. On the title page of Sanctorius's first
publication after he left Padua, we read "once professor of theoretical medicine,"
which implies that he had had to give up his title.[48] In any case, his prosperous medical practice and the powerful connections that he still had among the Venetian patriciate surely allowed him to live without financial worries.[49] By now, Sanctorius's
name was famous throughout Europe and Capello claims that he received offers
from the Universities of Bologna, Pavia, and Messi, but did not accept them (Capello
1750: XIII; Burrow 1763).

2.6   Retirement in Venice: The Continuation of a Busy Life

Venice, a place Sanctorius had gravitated toward since his childhood, appears to
have become his second home. Besides his many friendships and acquaintances,
there was also his professional connection to the *Serenissima*. In June 1612, shortly
after Sanctorius had become professor in Padua, he became a member of the
Collegio dei Medici/ Fisici di Venezia (College of Physicians of Venice) (BNMVe:
f. 28v). This was a highly distinguished institution, because Venice attracted the
most competent physicians, owing to the high rewards of medical practice in the
city and the opportunities provided by the Venetian press. What is more, while
Colleges of Physicians elsewhere in Italy became increasingly exclusive when the
profession expanded in the sixteenth century, the Venetian College retained its cos-
mopolitan character and also attracted distinguished physicians from all over Italy.
Membership in the Venetian College was very common among leading professors
of medicine in Padua. However, the majority of the members were practising physicians. The College mainly fulfilled two functions: awarding degrees and defending
medical standards. Compared to its counterpart in Padua, the Sacro Collegio dei

---

[48] "Olim in Patavino Gymnasio Medicinae Theoricam Ordinar. Primo loco profitentis" See:
Sanctorius 1625: title page.

[49] Sanctorius's testament shows that he accumulated wealth during his lifetime. See: ASVe-g. For
a transcription of the testament, see: Ettari and Procopio 1968: 139–46; for an English translation,
see: Castiglioni 1931: 775–8. Castiglioni estimated that his fortune at the time of his death was
60–70,000 Venetian ducats (ibid.: 741).

Filosofi e Medici (Sacred College of Philosophers and Physicians), it awarded fewer degrees and was more concerned with regulating various aspects of medical practice. For example, it ensured that only doctors of arts and medicine could practice medicine in Venice (Palmer 1983: 8 ff., 13 f., 18).

Sanctorius was involved in the College's activities, but never presided over it, despite some of his biographers erroneously asserting that he did.[50] They may have confused it with the Collegio Veneto or with the Collegio dei Chirurgi di Venezia (College of Surgeons of Venice) with which the College of Physicians of Venice cooperated in arranging an annual public demonstration of anatomy in Venice. In 1613, the *Riformatori dello Studio* assumed responsibility for paying the *lector* and *incisor* at these anatomical events. The *lector* was responsible for a series of lectures on anatomy, whereas the *incisor* performed a separate series of anatomical demonstrations. Sanctorius was among those nominated for the position of *lector*, but he turned it down. The records of the College of Physicians show that Sanctorius participated in the institution's doctoral examinations (Fig. 2.2). In June 1626, he was named as *promotore* of Paulus Leonardus, who graduated in surgery.[51] This is not the only indication of Sanctorius's expertise in this medical field, a topic I return to

Fig. 2.2 Drawing of a doctoral examination in the College of Physicians of Venice (date and author unknown) (BUP, MS 318, 25r). (By kind permission of Ministero della Cultura)

---

[50] E.g., Capparoni 1925–1928: 56, Castiglioni 1931: 740, Major 1938: 379, Sanctorius and Lebàn 1950: 37, Grmek 1952: 11, Eknoyan 1999: 229 f.

[51] The *promotores* assisted the candidate during the doctoral examination. Usually, the candidate was entitled to choose three or four *promotores* from amongst the members of the College, and to have another four assigned by lot (Palmer 1983: 37).

in a later chapter (Sect. 4.2.1). In 1629, Sanctorius, together with the *protomedico* (chief physician) of Venice, Giovanni Battista Fuoli (life dates unknown), was charged with obtaining an amendment to a senatorial decree, in order that the College might elect its secretary without the *Riformatori*'s interference (BNMVe: 29v, 33r, 34v; Ettari & Procopio 1968: 30; Palmer 1983: 46 f., 50).

Besides his activities in the College of Physicians of Venice, Sanctorius took up the tasks that he had imposed on himself upon leaving the University of Padua: to publish and edit his works. As mentioned earlier, he published his *Commentary on Avicenna* in 1625, followed quickly by a second edition only one year later. In 1629 he published his *Commentary on Hippocrates*, along with his *De remediorum inventione* (On the Invention of Remedies).[52] Moreover, in 1630 he published revised editions of his books *Methodi vitandorum errorum* and the *Commentary on Galen*. However, one work announced several times by Sanctorius appears to have remained unpublished, the *Liber de instrumentis medicis* (Book on Medical Instruments). In his three commentaries he promised repeatedly to present in this book more written details of the construction and uses of his instruments, as well as more elaborate illustrations.[53] In 1624, Sanctorius requested the *privilegio*, a sort of copyright, not only for his *Commentary on Avicenna* but also for his "*De instrumentis medicis noviter inventis suo sanitate conservanda*" (On newly invented medical instruments to maintain one's health) (ASVe-e).[54] A work had to be published within twelve months of the *privilegio* being granted; otherwise the *privilegio* expired (Witcombe 2004: 41). Hence, Sanctorius must have intended to publish both books soon. Interestingly, five years later, in the *Commentary on Hippocrates,* it sounds as if the book on instruments had actually been published. Sanctorius wrote: "we show the contemplation mentioned here in the *Commentaries on Avicenna* and in the *Book on Instruments*" (Sanctorius 1629a: 51).[55] If this really was the case, all trace of the book has been lost.

In 1638 Johan van Beverwijck (Beverovicius, 1594–1647), a student of Sanctorius, published the work *De calculo renum & vesicae* (On kidney and bladder stones), which contains a *consilium* (word of advice) from Sanctorius and Hieronymus Thebaldus (life dates unknown). It is part of a longer piece on lithotomy, the surgical removal of bladder stones. The *consilium* and Beverwijck's statements show how experienced Sanctorius was in treating this affliction. As will be seen later, Sanctorius also designed surgical instruments, among them a special syringe to extract bladder stones (Sect. 4.2.1). The *consilium* also refers to his distinguished Venetian clientele, as he recounts the case of a Senator who suffered from a bladder stone. Furthermore, it hints at his friendship with Hieronymus

---

[52] In the following, I will refer to this work as *De remediorum inventione*.

[53] See: Sanctorius 1612b: 62, 136, 229, Sanctorius 1625: Ad lectorem, 12, 24, 78, 200, 303, 513, finis, Sanctorius 1629a: 51.

[54] For more information on copyright in the Renaissance, and in Venice specifically, see: Witcombe 2004.

[55] "… ostendimus in commentariis Avicennae, & in lib. de instrumentis huic contemplationi dicatis: …." See: Sanctorius 1629a: 51.

Thebaldus, a fellow Venetian physician, with whom he composed the advice. The two men were listed among the illustrious surgeons of Venice by Francesco Bernardi, in his account of surgery (Bernardi 1797: 49 f.).

2.7   Sanctorius's Role in the Treatment of the Plague

Thebaldus and Sanctorius were involved in treating the Venetian plague of 1630–1631, fighting, however, on opposite sides. The medical health officers (*Provveditori e Sopraprovveditori alla Sanità*) consulted the most famous physicians of the Republic to decide, after an examination of the sick, whether or not the latter were afflicted by plague. The opinions were conflicting and no conclusions were arrived at. Further discussions were held and, in August 1630, the Senate tried to solve the issue by organizing a plenary meeting of the physicians. The reports of the sessions illustrate the controversy. A group of physicians, including Sanctorius, persistently contested the existence of the contagion in the city. They were faced by another, smaller group of physicians, amongst them Fuoli and Thebaldus, who tried in vain to persuade the government of the reality of the plague.[56] Fierce disputes arose between the two parties and Fuoli, who had recognised the disease as plague from the beginning, faced public hostility and even death threats. Meanwhile, the epidemic spread. It was only toward the end of the year that the high mortality rate left no more room for doubt (Ettari & Procopio 1968: 80–3; Preto 1984: 382 f.).

Why did Sanctorius fail to realize the seriousness of the situation? What made him doubt that the plague was ravaging Venice? These questions seem all the more pressing given that there was a precedent to the ill-judged response dating back fifty years. Already in 1576, the Paduan professors Girolamo Mercuriale and Girolamo Capodivacca (died 1589) had mistaken the Venetian plague for other diseases. Strikingly, but maybe not surprisingly, they both taught Sanctorius.[57] Thus, medical education may have played a part here. Furthermore, economic and political factors must be taken into account. Confirming that there was plague in the city would have had immense social and economic consequences. Trade as well as public and private commerce would have stagnated, and the government feared for the freedom of Venice. Thus, the ruling patriciate struggled to ensure that their measures would not impinge on foreign political interests, and laid the groundwork for economic and social recovery. The denial of the existence of plague in the city was most welcome to them. It was, of course, also what the people wanted to hear. The fact that Sanctorius's friend, Nicolò Contarini, was the doge, at this time, surely increased the burden of liability on the physician's shoulders (Palmer 1978: 238–79; Preto 1984: 380–87).

---

[56] Two statements by Thebaldus, in which he insisted that the disease in Venice was plague, can be found in the following file: ASVe-a: f. 12r–13r, 31r–32r.

[57] In his *Commentary on Hippocrates*, Sanctorius referred to Capodivacca as his teacher: "Quarta opinio fuit Hieronymi capivacei praeceptoris nostri, ..." See: Sanctorius 1629a: 95.

In addition to signing joint statements that denied the existence of a plague in Venice, Sanctorius also gave his personal opinion, as requested by the authorities. In his assessment, he confirmed what he had previously claimed: there was no plague in Venice. Without going into the details of the document, it is notable that Sanctorius proposed that the sick be separated from the healthy and confined to the *lazaretto*. He warned that "what is not now may well still come about."[58] A sign of doubt? Or even fear? Either way, it was not enough to make him change his position. However, when reality proved him wrong, he fulfilled his duties and did not flee from the Black Death, as many of his colleagues did (ASVe-a: f. 47r–47v, 60r–61r; Ettari & Procopio 1968: 82 f.; Girardi 1830: 16; Dolfin 1843: 28).

2.8 Death and Legacy

A few years later, on February 25, 1636, Sanctorius died at his home in Venice.[59] In accordance with his wishes, he was buried in the Venetian Church *Santa Maria dei Servi*, where a bust was erected to his memory. His friend Paolo Sarpi, being a famous member of the Servite Order, had been buried in the same church. Sanctorius's connection to the Order is further illustrated by his testament, in which he specified that a certain sum of money be left to the Servite Church in Koper, in order that it might annually commemorate his death. Moreover, he bequeathed a sum to the College of Physicians of Venice, ten ducats of which were to be given every year to a doctor at the College on condition that he publicly commemorate his benefactor (ASVe-g; Cigogna 1824: 50 f., 91 ff.).[60] Hence, Sanctorius made sure that his name would not be soon forgotten.

And it was not only his name that was kept alive, but also his remains: the final rest eluded them. In 1812, the Venetian Servite Church was destroyed and the bust of Sanctorius was thereupon taken to the Ateneo Veneto in Venice, where it stands

---

[58] "… non bisogna però restare di usare le istesse diligenze, perché questo che non è potrebbe farsi: Ricordo pero alle E.V. Illme di far separare li sani, che hora stanno insieme con gl'Infetti, col'mandargli al Lazzaretto." See: ASVe-a: f. 6v. This file contains the joint judgments as well as Sanctorius's personal judgments regarding the plague of 1630–31. For transcriptions and paraphrases of the judgments, see: Dolfin 1843.

[59] Many biographers indicated the wrong date of death, e.g., Castiglioni 1931: 740, Ettari and Procopio 1968: 30, Grmek 1975: 101. The exact date results from the work of Emmanuele Antonio Cigogna, who referred to the epitaph and the record of Sanctorius's death. See: Cigogna 1824: 50 f., Cigogna 1827: 436 f. and Sanctorius and Ongaro 2001: 16, fn. 35.

[60] The list of the Sanctorian Orators (*Oratori Santoriani*) continues until 1774, according to the remaining notes compiled from the College records by Giuseppe Bolis, see: BNMVe: f. 85v–87v. Most of the original records of the College were destroyed by fire in 1800. Hence, the public commemorations for Sanctorius took place for more than one hundred years, almost until the closure of the College in 1806 (Palmer 1983: 52 f.).

to this day.[61] Francesco Aglietti (1757–1836), a Venetian physician and president of the Ateneo, collected Sanctorius's bones and kept them in a box in his library. Upon his death, the mummified body was found on top of the bookcase. It was then entrusted to Francesco Cortese (1802–1883), who had just become professor of anatomy in Padua. Except for the skull, he dispatched the bones to the cemetery, where they at last found their final resting place. He used the skull for his phreno-logical studies, until it was exhibited in the medical museum of the university. Later, it was displayed in the Hall of Medicine situated in the Palazzo Bo' of the University of Padua, and today it can be admired by visitors to Padua in the MUSME—the Museum of the History of Medicine (Fig. 2.3).[62]

Of course, Sanctorius's legacy comprises more than bodily relics and commemo-rations. The preceding paragraphs sped through seventy-five years of a life filled with intellectual vitality and community. It turned out that Sanctorius combined a prosperous medical practice with a successful university career that came, however, to an unfortunate end. At some points in his biography, it seems that his priorities lay in the practice of medicine, accepting the displeasure that this provoked on the part of his students. Nevertheless, he wrote three extensive commentaries on

Fig. 2.3 Skull of Sanctorius Sanctorius (MUSME Padova). (By kind permission of Università degli Studi di Padova)

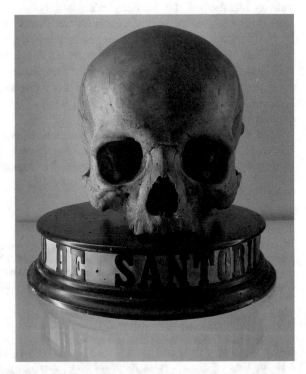

[61] The Ateneo Veneto di Scienze, Lettere ed. Arti is an institution for the promotion of the sciences, education, art, and culture. For more information, see: https://www.ateneoveneto.org/it

[62] Cigogna 1824: 51, Ettari and Procopio 1968: 30, Palmer 1983: 54, Sanctorius and Ongaro 2001: 16, Zanatta, et al. 2016.

traditional texts used on university courses, two of which he published only after resigning his professorship. From childhood on, he belonged to Venetian high society and kept company with highly influential men. In the course of his life, he was connected to several important institutions, which attests that he was held in high esteem; and, at the same time, this enabled him to further expand his fame and social network. The weighing experiments, for which he was most celebrated and which later earned him the title of founder of a new medical science, accompanied his practice quite early. It seems that he developed his quantitative approach to medicine in the period between his graduation and his appointment as professor at the University of Padua.

However, the Sanctorius who came to light in the course of this chapter differs greatly from the common image of him. There was another side to him, besides the brilliant, successful physician. As first president of the Collegio Veneto, he put himself in direct conflict with the Catholic Church, certainly aware of the major political agenda behind this—to free the Venetian Republic from papal power. In addition, there were the recurrent quarrels and tensions with his students and, finally, his fatal position on the Venetian plague. It turns out that his career was not as smooth as it is usually portrayed to be. This brief biographical survey of Sanctorius's social, institutional, and professional contexts thus paves the way for a comprehensive review of Sanctorius and his work. In the next chapter, I continue this review by scrutinizing the intellectual context in which Sanctorius produced his scientific output.

References

AAU 703: Collegio Veneto Serie dei Presidenti Informazioni. Dottorati dai Conti Palatini. Vertenza col Collegio di Venezia.

ASVe-a: Provveditori e Sopraprovveditori alla Sanità, Busta 562, Giudizi sulla peste.

ASVe-b: Riformatori allo Studio di Padova, "Registre delle parte prese in Senato", Reg./Filza I.

ASVe-c: Senato, Deliberazioni, Terra, Terra. Filze, Pezzo 200: 1611 sett./nov.

ASVe-d: Senato, Deliberazioni, Terra, Terra. Filze, Pezzo 265: 1624 gen./feb.

ASVe-e: Senato, Deliberazioni, Terra, Terra. Filze, Pezzo 266: 1624 mar./apr.

ASVe-f: Senato, Deliberazioni, Terra, Terra. Filze, Pezzo 226: 1617 sett./nov.

ASVe-g: Sezione Notarile Testamenti (Testamenti Crivelli), Busta 289, n. 537.

BMCVe-b: ms. Correr 1, 1377, Autografi O-Z, Correr 146–217.

BNMVe: Mss. Ital. VII 2342 (= 9695), Bolis, Notizie cavate dalli libri di Priori.

Anonym. 1882. Varietà: Documenti inediti sul Santorio medico Istriano. *La concordia: almanacco istriano coi varii ruoli di ciascun comune* 1: 90–92.

Baila, Ernesto. 1936. Santorio Santorio, il precursore della medicina sperimentale. *Gazzetta Sanitaria* n.a: 13–14.

Bernardi, Francesco. 1797. *Prospetto storico-critico del Collegio Medico-chirurgico, e dell'arte chirurgica in Venezia*. Venice: Constantini.

Bigotti, Fabrizio. 2016. Mathematica Medica. Santorio and the Quest for Certainty in Medicine. *Journal of Healthcare Communications* 1: 1–8.

Bigotti, Fabrizio, and David Taylor. 2017. The Pulsilogium of Santorio: New Light on Technology and Measurement in Early Modern Medicine. *Society and Politics* 11: 55–114.

Burrow, James. 1763. *A Few Anecdotes and Observations Relating to Oliver Cromwell and His Family; Serving to Rectify Several Errors Concerning Him, Published by Nicolaus Comnenus Papadopoli, in His Historia Gymnasii Patavini.* London: Printed for J. Worrall.

Bylebyl, Jerome J. 1979. The School of Padua: Humanistic Medicine in the Sixteenth Century. In *Health, Medicine and Mortality in the Sixteenth Century*, ed. Charles Webster, 335–370. Cambridge/London: Cambridge University Press.

Capello, Arcadio. 1750. *De Vita Cl. Viri Sanctorii Sanctorii … Accedit Oratio ab eodem Sanctorio habita in Gymnasio Patavino dum ipse primarium Theoricae Medicinae explicandae munus auspicaretur.* Venice: Apud Jacobum Thomasinum.

Capparoni, Pietro. 1925–1928. *Profili bio-bibliografici di medici e naturalisti celebri italiani: dal sec. XV al sec. XVIII.* Rome: Istituto naz. medico farmacologico.

Castiglioni, Arturo. 1931. The Life and Work of Santorio Santorio (1561–1636). *Medical Life* 135: 725–786.

Cigogna, Emmanuele Antonio. 1824. *Delle Iscrizioni Veneziane.* Venice: Giuseppe Orlandelli Editore.

———. 1827. *Delle Iscrizioni Veneziane.* Venice: Giuseppe Picotti.

Cozzi, Gaetano. 1958. *Il Doge Nicolò Contarini: ricerche sul patriziato veneziano agli inizi del Seicento.* Venice/Rome: Istituto per la collaborazione culturale.

———. 1979. *Paolo Sarpi tra Venezia e l'Europa.* Turin: Giulio Einaudi editore.

De Bernardin, Sandro. 1983. I Riformatori dello Studio: Indirizzi di politica culturale nell'Università di Padova. In *Storia della Cultura Veneta: Il Seicento* ed. Girolamo Arnaldi and Manlio Pastore Stocchi, 61–91. Vicenza: Neri Pozza Editore.

Del Gaizo, Modestino. 1889. *Ricerche Storiche intorno a Santorio Santorio ed alla Medicina Statica. Memoria letta nella R. Accademia Medico-Chirurgica di Napoli il dì 14 Aprile 1889.* Naples: A. Tocco.

———. 1936. Santorio Santorio nel terzo centenario della morte. *Il giardino di Esculapio* IX: 4–21.

Dolfin, Paolo Nob. 1843. *Della Peste: Opinioni dei Medici di Venezia nel 1630.* Padua: Tipografia Penada.

Eknoyan, Garabed. 1999. Santorio Sanctorius (1561–1636): Founding Father of Metabolic Balance Studies. *American Journal of Nephrology* 19: 226–233.

Encyclopaedia Britannica. 2018. *Pannonia.* https://www.britannica.com/place/Pannonia. Accessed 28 Sep 2018.

Ettari, Lieta Stella, and Mario Procopio. 1968. *Santorio Santorio: la vita e le opere.* Rome: Istituto nazionale della nutrizione.

Favaro, Antonio. 1888. Lo Studio di Padova e la Reppublica Veneta. *Atti del Reale Istituto Veneto di Scienze, Lettere ed Arti* 6: 1045–1069.

Gedeon, Andras. 2006. Science and Technology in Medicine. In *An Illustrated Account Based on Ninety-Nine Landmark Publications from Five Centuries.* Singapore: Springer.

Giordano, Davide, and Arturo Castiglioni. 1924. Centenarî e Commemorazioni: Santorio Santorio. *Rivista di Storia delle Scienze Mediche e Naturali* 15: 227–237.

Girardi, Giuseppe. 1830. *La peste di Venezia nel MDCXXX. Origine della erezione del Tempio a S. Maria della Salute.* Venice: Dalla tipografia di Alvisopoli.

Grandi, Giacomo. 1671. *De laudibus Sanctorii: Oratio Iacobi Grandii publicè Venetiis Anatomen profitentis.* Venice: Apud Ioannem Franciscum Valvasensem.

Grendler, Paul F. 2002. *The Universities of the Italian Renaissance.* Baltimore: The Johns Hopkins University Press.

Grmek, Mirko D. 1952. *Santorio Santorio i njegovi aparati i instrumenti.* Zagreb: Jugoslav. akad. znanosti i umjetnosti.

———. 1975. Santorio, Santorio. In *Dictionary of Scientific Biography*, ed. Charles Coulston Gillispie, 101–104. New York: C. Scribner's Sons.

Major, Ralph H. 1938. Santorio Santorio. *Annals of Medical History* 10: 369–381.

Mangeti, Joannis Jacobi. 1731. *Bibliotheca Scriptorum Medicorum Veterum et recentiorum.* Vol. II. Geneva: Sumptibus Perachon e Cramer.

Mattioli, Mario. 1985. *Grandi indagatori delle scienze mediche*. Naples: Idelson.

Palmer, Richard. 1978. *The Control of Plague in Venice and Northern Italy 1348–1600*. Phd diss., University of Kent at Canterbury.

———. 1983. *The Studio of Venice and its Graduates in the Sixteenth Century*. Trieste: Lint.

Premuda, Loris. 1950. Santorio Santorio. *Pagine istriane* 1: 117–124.

Preto, Paolo. 1984. La società veneta e le grandi epidemie di peste. In *Storia della Cultura Veneta: Il Seicento*, ed. Girolamo Arnaldi and Manlio Pastore Stocchi, 377–406. Vicenza: Neri Pozza Editore.

Pusterla, Gedeone. 1891. *I Rettori di Egida "Giustinopoli Capo d'Istria"*. Koper: Tipografia Cobol & Priora.

Renauldin, Léopold Joseph. 1825. Sanctorius (Sanctorius). In *Biographie Universelle, Ancienne et Moderne. Saint L – SAX*, ed. L.G. Michaud. Paris: Imprimerie d'Everat.

Rode to Hofmann. 1624. Letter from Johan Rode to Caspar Hofmann, Padua, 18.04.1624 Erlangen, UB, Trew Rhodius Nr. 3; Regest [Ulrich Schlegelmilch]. www.aerztebriefe.de/id/00000912. Accessed 8 Jan 2019.

Rossetti, Lucia. 1967. *Acta Nationis Germanicae Artistarum (1616–1636)*. Padua: Editrice Antenore.

———. 1984. I Collegi per i dottorati "Auctoritate Veneta". In *Viridarium Floridum. Studi di Storia Veneta Offerti dagli Allievi a Paolo Sambin*, ed. Maria Chiara Billanovich, Giorgio Cracco, and Antonio Rigon, 365–386. Padua: Editrice Antenore.

Sanctorius, Sanctorius. 1603. *Methodi vitandorum errorum omnium, qui in arte medica contingunt, libri quindecim*. Venice: Apud Franciscum Barilettum.

———. 1612a. *Commentaria in Artem medicinalem Galeni*. Vol. I. Venice: Apud Franciscum Somascum.

———. 1612b. *Commentaria in Artem medicinalem Galeni*. Vol. II. Venice: Apud Franciscum Somascum.

———. 1614. *Ars Sanctorii Sanctorii Iustinopolitani de statica medicina, aphorismorum sectionibus septem comprehensa*. Venice: Apud Nicolaum Polum.

———. 1625. *Commentaria in primam Fen primi libri Canonis Avicennae*. Venice: Apud Iacobum Sarcinam.

———. 1629. *Commentaria in primam sectionem Aphorismorum Hippocratis, &c. … De remediorum inventione*. Venice: Apud Marcum Antonium Brogiollum.

———. 1902 [1615]. Santorio Santorio a Galilei Galileo, 9 febbraio 1615. In *Le Opere di Galileo Galilei*, ed. Galileo Galilei, 140–142. Florence: Barbera. http://teca.bncf.firenze.sbn.it/ImageViewer/servlet/ImageViewer?idr=BNCF0003605126#page/1/mode/2up. Accessed 5 Nov 2015.

Sanctorius, Sanctorius, and Evaristo Lebàn. 1950. *De statica medicina: con un saggio introduttivo di Evaristo Lebàn*. Florence: Santoriana, A. Vallecchi.

Sanctorius, Sanctorius, and Giuseppe Ongaro. 2001. *La medicina statica*. Florence: Giunti.

Sarpi, Paolo. 1969. *Opere, a cura di Gaetano e Luisa Cozzi*. Milan/Naples: Riccardo Ricciardi Editore.

Schmitt, Charles B. 1985. Aristotle Among the Physicians. In *The Medical Renaissance of the Sixteenth Century*, ed. A. Wear, R.K. French, and Iain M. Lonie, 1–15. Cambridge/London: Cambridge University Press.

Siraisi, Nancy. 1987. *Avicenna in Renaissance Italy: The Canon and Medical Teaching in Italian Universities After 1500*. Princeton: Princeton University Press.

Stancovich, Pietro. 1829. *Biografia degli uomini distinti dell'Istria*. Trieste: Gio. Marenigh Tipografo.

Tomasini, Giacomo Filippo. 1986. *Gymnasium Patavinum (Rist. anastatica [of the ed.] Utini, 1654)*. Sala Bolognese: Forni Editore.

Trebbi, Giuseppe. 2012. Morosini, Andrea. In *Dizionario biografico degli italiani*. Vol 77. Rome: Istituto dell'Enciclopedia Italiana. http://www.treccani.it/enciclopedia/andrea-morosini_%28Dizionario-Biografico%29/. Accessed 5 Sep 2019.

Valdera, Marc'Antonio. 1604. *L'Epistole d'Ovidio*. Venice: Appresso Francesco Bariletto.

Vedrani, Alberto. 1920. Santorio Santorio da Capo d'Istria: (1561–1636). *Illustrazione medica italiana* 2: 26–29.

Vida, Hieronimo. 1621. *De' cento dubbi amorosi*. Padua: Gasparo Crivellari.

Weigle, Fritz. 1965. Die deutschen Doktorpromotionen in Philosophie und Medizin an der Universität Padua 1616–1663. *Quellen und Forschungen aus italienischen Archiven und Bibliotheken* 45: 325–384.

Witcombe, Christopher L.C.E. 2004. *Copyright in the Renaissance: Prints and the Privilegio in Sixteenth-Century Venice and Rome*. Leiden: Brill.

Zanatta, Alberto, Giuliano Scattolin, Gaetano Thiene, and Fabio Zampieri. 2016. Phrenology Between Anthropology and Neurology in a Nineteenth-Century Collection of Skulls. *History of Psychiatry* 27: 482–492.

Ziliotto, Baccio. 1944. *Accademie ed Accademici di Capodistria*. Trieste: L. Smolars & Nipote.

Chapter 3
Sanctorius's Galenism

Abstract The chapter deals with Sanctorius's intellectual background and places his book *De statica medicina* within the framework of contemporary Galenic medicine. Usually, the book is celebrated for its innovative quantitative approach to medicine yet read in isolation from its broader context. However, as I will show, an analysis of this context is crucial to understanding how Sanctorius developed his novel ideas and revised the medical knowledge of his day. Of particular importance in this regard are the dietetic doctrine of the "six non-natural things" and the concept of insensible perspiration, an invisible excretion of the human body. Potential relations of Sanctorius's notions to the doctrine of the ancient medical school of the Methodists and to corpuscular ideas are also scrutinized. The chapter concludes with an analysis of the *De statica medicina* itself, focusing on the conceptual backdrop against which Sanctorius developed his weighing procedures, the results of which he presented in the book. References to Sanctorius's other publications help set his ideas in the broader context of his endeavors and contribute to an understanding of the theoretical context in which the *De statica medicina* emerged.

Keywords Dietetics · Galenic medicine · Humoral theory · Perspiration

If one thinks of medicine and medical practitioners in the early modern period, very diverse images may spring to mind: the apothecary amidst bottles and jars full of different tinctures and remedies, the woman healing her family members and other sick people at home, with poultices and herbal infusions, the town physician examining his patient's urine, the surgeon setting broken bones, or the charlatan trying to cure with dubious remedies. There is some truth to all of them, and many more characters could be added to the list. The European medical world was highly diverse, comprising different areas of knowledge, various intellectual interests, and a broad range of commitments within a variety of institutions, occupations, skill sets, and activities. However, there was a large body of shared knowledge, too, and the boundaries between the different actors were often blurred. Medicine was a craft, a profession, and a scholarly activity. It is within this context that Sanctorius and his work must be considered.

© The Author(s) 2023   35
T. Hollerbach, *Sanctorius Sanctorius and the Origins of Health Measurement*,
https://doi.org/10.1007/978-3-031-30118-6_3

With his medical university education, Sanctorius belonged to a privileged group and enjoyed a special status at the top of the hierarchy of medical practitioners. As soon as he became university professor of theoretical medicine, he climbed even further up the ladder, personally teaching a new generation of physicians. Theoretical medicine (*theoria*) at the time included the teaching of the nature of medical science, the position of medicine in the hierarchy of arts and sciences, and the proper relationship of medicine and philosophy, as well as the basic principles of physiology, pathology, and regimen (Siraisi 1987: 10; 2012: 492–514).[1] All of this was taught against the backdrop of the medical tradition—Galenic medicine.

Galenic Medicine as the Leading Authority For more than thirteen centuries, the medical system known as *Galenism* prevailed in Western and Arabic medical thought. Its influence began to slowly decline during Sanctorius's lifetime, but was still substantial, especially in the universities. It goes back to the Greek physician Galen of Pergamon (ca. 129–ca. 216 CE), who practiced mainly in Rome. He was one of the most prolific writers of Western antiquity and many of his works survive. Galenism refers to the school of thought that emerged from Galen's work. This differentiation is important, so as not to confuse the historical figure and his original works and doctrines with the transformations that the latter underwent over time. There are many "Galens," namely reshaped and updated versions of the ancient original, and thus they are by no means all identical. So which Galen did Sanctorius encounter in Padua? And what kind of Galenism did he later teach his students? These are not easy questions and it would probably take at least another monograph to answer them in full. In the following account, therefore, I mainly rely on secondary literature to outline the intellectual framework in which Sanctorius was trained at the University of Padua (Temkin 1973; Salmón 1997; Arrizabalaga et al. 2002; Singer 2016). To this end, I address Galen's scientific output insofar as this enhances understanding of the Renaissance teaching derived from it. Rather than studying the phenomenon of Galenism in Padua at the turn of the seventeenth century in its own right, this chapter aims to analyze Sanctorius and his work against the backdrop of this medical tradition. Sanctorius is my point of departure in this endeavor to add another piece to the enormous puzzle of Renaissance Galenism.

Sanctorius's publications are all deeply informed by Galenic medicine. This is no surprise, especially in the case of his three commentaries—on Galen's *Ars Medica*, Avicenna's *Canon,* and Hippocrates's *Aphorisms*. These reflect his teaching as a professor of *theoria* in Padua and therefore necessarily refer to the medical tradition. But his other works—the *Methodi vitandorum errorum,* the *De statica medicina,* and the *De remediorum inventione*—likewise pursue this same theoretical thrust. This is well worth emphasizing given that the image of Sanctorius as an innovator who promulgated a new medical science *at the expense of* Galenic medicine still haunts the literature. The *De statica medicina,* Sanctorius's most famous work, is usually celebrated for its innovative quantitative approach to medicine, in isolation

---

[1] For the distinction between theoretical medicine (*theoria*) and practical medicine (*practica*), see Sect. 2.1, fn. 8.

from its broader context. However, the book's very structure is modelled on an ancient concept that originated in Galen's work and is fundamental to the Galenic tradition. It serves insofar as an introduction to the intricate world of Galenism.

3.1 The "Six Non-Natural Things"

The *De statica medicina* is divided into seven sections: *De ponderatione insensibilis perspirationis* (weighing of insensible perspiration), *De Aere & aquis* (air and water), *De Cibo & potu* (food and drink), *De Somno & vigilia* (sleep and wake), *De Exercitio & quiete* (exercise and rest), *De Venere* (coitus), *De Animi affectibus* (affections of the mind) (Fig. 3.1) (Sanctorius 1614: index).[2] While this may not ring a bell with the modern reader it surely did among his contemporaries; for sections II to VII correspond to the list of the so-called six *res non-naturales*, albeit in slightly altered fashion. These six non-natural things were of great importance in traditional dietetic medicine, as they were considered to be the main determinants of health and disease. They are categories of factors to which human beings are unavoidably exposed in the course of daily life and that influence health or disease, depending on the circumstances of their use or abuse. Generally, they are classified as follows: (1) air, (2) food and drink, (3) sleep and wake (or: wakefulness), (4) motion and rest, (5) evacuation and repletion, (6) passions of the mind. Management of the patient's regimen (that is, of these six sets of factors) was for centuries the physician's most important task (Rather 1968: 337; Jarcho 1970: 374). Thus, Sanctorius used a common concept of dietetic medicine to structure his work with the weighing chair.[3] However, as the first section implies, he shifted the focus to the *perspiratio insensibilis*, an insensible perspiration of the human body, and to how its excretion is affected by the non-naturals. More will be said about this later.

3.1.1 The Origin of the "Six Non-Natural Things"

The expression "six non-natural things" was so familiar to scholars until the end of the eighteenth century, and so embedded in the Galenic medical tradition, that it was usually not explained in more detail, nor were clues given as to its origin. Therefore,

---

[2] As a response to a harsh critique by Ippolito Obizzi, a physician and philosopher of Ferrara, who attacked the *De statica medicina* violently in his work *Staticomastix sive staticae medicinae demolitio* (The Scourge of Statics, or the Demolition of Static medicine) (Obizzi 1615), Sanctorius added an eighth section to his book, called *Ad Staticomasticem* (To the Scourge of Statics). It was often reprinted as a supplement to the original work. The earliest edition I could find of the *De statica medicina* with the additional eighth section dates from 1634 (Sanctorius 1634). However, a statement by Sanctorius in the *Commentary on Avicenna* implies that he had published his defense against the *Staticomastix* before 1625 (Sanctorius 1625: 81). See also Sect. 5.3.2.

[3] In the following, I refer to the "six non-natural things" as a "concept," or a "doctrine," but it is important to note that they comprise normative, practical, and theoretical aspects.

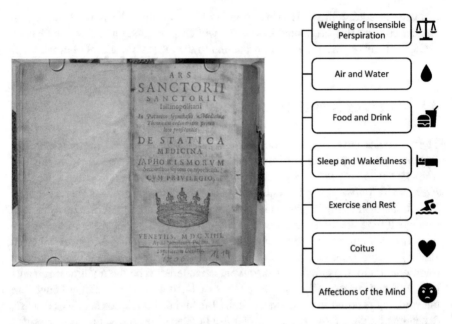

Fig. 3.1 The subjects and their order, as presented in the *De statica medicina* (Sanctorius 1614). Image of the frontispiece courtesy of Universitätsbibliothek Erlangen-Nürnberg, TREW.Xx 400

once the term had been largely forgotten, it was very difficult for historians to trace where it originally came from. In 1970, Saul Jarcho set out to do so and finally found the decisive clue in the writings of the famous anatomist Giovanni Battista Morgagni (1682–1771) (Jarcho 1970). In his posthumously published lectures on Galen's *Ars medica*, Morgagni discussed, amongst other things, the six non-naturals. Following up on this clue, Jarcho continued his research in the Galenic corpus and was successful.[4] The relevant passage reads:

> Accordingly, some of the changes of the body are necessary and some are not. I call 'necessary' those which it is impossible for a body not to be associated with and 'not necessary' the rest. Thus constant contact with the ambient air, eating and drinking, waking and sleeping are necessary to it whereas constant contact with swords and wild animals are not necessary, whence the art devoted to the body resides in the first class of causes whereas the

---

[4] In 1968—two years before Jarcho, but unbeknown to him—L.J. Rather had traced the source of the doctrine of the six non-naturals to Galen's *Ars medica*. See: Rather 1968: 341. In 1988, Galen's authorship of the *Ars medica* was called into doubt for the first time in the history of medicine by Jutta Kollesch (Kollesch 1988). A few years later, García-Ballester supported Kollesch's hypothesis in his study *On the origin of the "six non-natural things" in Galen*, which was first published in 1993 (García-Ballester 2002: 114 f.). Boudon-Millot examined the matter of authenticity again in 1996 and came to the conclusion that there was no reason to doubt Galen's authorship (Boudon-Millot 1996). In the introduction to the recent edition and English translation of the *Ars medica*, Johnston makes no mention of any uncertainty regarding the authenticity of the work (Galen and Johnston 2016: 137–55). In the present account, I follow Boudon-Millot and Johnston in assuming that Galen was the author of the *Ars medica*.

second doesn't apply any more [i.e., in the first class of causes but not in the second there is an art devoted to the protection of the body]. And so, if we distinguish all those changes of the body which are necessary, we shall discover, in respect of each of them, some specific class of causes of health. There is, then, one from association with the ambient air, another from movement and rest of the whole body and its parts, a third from sleeping and waking, a fourth from those things taken in, a fifth from those things excreted or released, and a sixth from the affections of the soul (Jarcho 1970: 376; Galen and Johnston 2016: 247 ff.).

Interestingly, Morgagni's commentary on this passage leads to another famous scholar who had been appointed professor at the University of Padua exactly a century before him: Sanctorius Sanctorius. A look at Sanctorius's *Commentary on Galen* shows that Morgagni copied this and other comments almost word for word (Morgagni 1965: esp. 83–100)—and thus, that a professor at the University of Padua one hundred years later, in 1712, still found Sanctorius's thoughts on the subject so relevant that he did not care even to revise them. Of course, one could argue that Morgagni relied so heavily on Sanctorius's work in his lectures on the *Ars medica* because he felt that teaching Galen's classic was a mere formality, a statutory obligation inherited from the past, but barely worth any effort. Nancy Siraisi has shown, however, that Morgagni supplemented his lectures on another traditional textbook, Avicenna's *Canon*, with lengthy descriptions of contemporary physiological ideas and thus evidently was prepared to introduce new material into his teaching on established subjects. In fact, Morgagni's lectures on the *Canon* dealt only cursorily with traditional ideas, and it was for these precisely that he relied heavily on Sanctorius's *Commentary on Avicenna*. This further suggests that Morgagni considered parts of Sanctorius's work useful to eighteenth-century students. In fact, of the many commentaries on the *Ars medica* that existed in Morgagni's time, Morgagni advised students to choose only three—among them, Sanctorius's *Commentary on Galen*—and "to keep them day and night within arm's reach" (Morgagni 1965: 23; Siraisi 1987).[5]

With respect to the six non-naturals, it is striking that Morgagni, who is regarded as the founder of anatomical pathology and follower of the new quantitative approach introduced by Sanctorius, referred neither to insensible perspiration nor to the *De statica medicina*.[6] What this implies for the reception of Sanctorius's thoughts will be scrutinized later. However, it already hints at the problem of applying the

---

[5] "Quamobrem hos tres ultimos Enarratores, ex omnibus electos Vobis propono quos nocturna diuturnaque manu prae caeteris versetis." See: Morgagni 1965: 23 and also 18, 30. The two other commentaries on Galen's *Ars medica* that Morgagni suggested to his students were the commentaries of Francisco Vallés (1524–1592) and Luca Tozzi (1638–1717) published in 1567 and 1703 respectively.

[6] In the two volumes of Morgagni's lectures on Galen's *Ars medica,* edited by Adalberto Pazzini, Morgagni referred only rarely to the *De statica medicina* and not in the context of the doctrine of the six non-natural things (ibid.: 343, 351, 355, 359, Morgagni 1966: 568, 674, 688 f., 710 f., 751). The respective passages show that Morgagni accepted Sanctorius's static doctrine yet did not discuss it at length. In the second volume, Morgagni mentioned Sanctorius's *pulsilogium*, but it seems that he did not use a similar instrument himself. For more information on Giovanni Battista Morgagni, see: Ongaro 2012.

categories of tradition and innovation retrospectively to the knowledge and work of historical figures. Did Morgagni differentiate between Sanctorius, the Galenist, and Sanctorius, the pioneer? Why did he rely so heavily on Sanctorius's commentaries precisely for the "traditional" thoughts that refer to Galenic medicine? Why did he not focus instead solely on the more "innovative" thoughts? The discussion of these and similar questions is postponed until the end of this book (Chap. 8). At this point, a closer look needs to be taken at what Sanctorius himself had to say on the concept of the non-natural things.

As is apparent from the citation above, Galen did not use the expression "non-natural" nor the phrase "six non-naturals" in the *Ars medica*. Instead, he discussed these factors in terms of "necessary" and "non-necessary" causes. Sanctorius explained that it was the Arabs (*Arabes*) who introduced the term. Galen, he wrote, had used the expression "non-natural" only in his work *De pulsibus ad tirones* (On the Pulse for Beginners), in reference to the causes of alteration in the pulse. The non-naturals, so Sanctorius, were explained by such an indefinite denomination, because their proper name was unknown; for they were factors which produced not only health but also disease, depending on their use respectively their abuse (Sanctorius 1612b: 19; 1625: 59). Indeed, later medical historical research confirmed that Arabic Galenism connected the necessary causes mentioned in the *Ars medica* with the non-natural causes referred to in the *De pulsibus ad tirones*. However, according to L. J. Rather, it seems unlikely that the Arabic authors used a term equivalent to "non-natural" with reference to Galen's six necessary causes. More probably, the term was introduced into the Western European medical vocabulary in Latin translations of Arabic works largely based on Galen (Rather 1968: 341).[7]

Sanctorius also referred to other passages in Galen's works, in which the latter expressed similar ideas to those in the *Ars medica*. In Galen's treatises *De sanitate tuenda* (Hygiene) and *Thrasybulus,* a similar group of factors was mentioned but they were divided into four groups instead of six and they were called neither "non-naturals" nor "necessary." The classification is as follows: (1) things administered (food, drink, drugs), (2) things evacuated (the bodily secretions and excretions), (3) things done (exercises, wake, insomnia, sleep, sexual activity, anger, anxiety, bathing), and (4) things befalling a person externally (air, water, seawater, olive oil, etc.) (Galen & Johnston 2018b). Sanctorius explained that Galen defined here the "non-natural" factors more broadly, and pursued a different aim than with the sixfold division. The fourfold classification was the most universal, per Sanctorius, because it comprised more causes that effect health or disease than any other classification

---

[7] The designation "non-natural" has often led to discussion, because the factors it describes seem among the most "natural" things in our experience. One explanation for the term was that the six things are non-natural in the sense that they can be manipulated by humans for the purpose of prophylaxis or cure (Strohmaier 1996: 172 f.). For a more comprehensive analysis of the development of the term "non-natural" and the phrase "six non-natural things," see: Rather 1968, Niebyl 1971.

scheme and included every conceivable non-natural thing (Sanctorius 1603: 98v–99r; 1612b: 20; 1625: 60).

As Sanctorius indicated, Galen had introduced these two categories in different contexts. While in the *Ars medica* (citation above), he highlighted the role of the non-naturals in pathology as inevitable causal factors, he focused in the *De tuenda sanitate* (Hygiene) and the *Thrasybulus* on their therapeutic role. In the latter case, they are understood as regulators of human life for the preservation of health, as aspects of diet and regimen demanding special medical attention (Rather 1968: 341; García-Ballester 2002: 106). With regard to pathology, Sanctorius stressed the occasional character of the six non-naturals as causes of disease. In his *Commentary on Galen*, he wrote:

> From the six non-naturals nothing certain can be obtained, because they do not necessarily cooperate in the production of internal affections. Sometimes we see men slip into cachexia and anasarca in the summer and after using strong wine and aromatics. In the winter some old men occasionally develop ardent fever after taking cold liquids. For this reason Hippocrates and Galen did not want to call these non-natural things causes in any way, but προφάσεις, i.e., occasions [in the sense of *a juncture of circumstances*] (Sanctorius 1612a: 173).[8]

Thus, Sanctorius warned that our involvement with these factors was purely fortuitous. Even though they played a substantial role in the causal system of Galenic pathology, there were other aspects to be considered when searching for the causes of a disease; and so Sanctorius reminded his students to not treat them in isolation (Sanctorius 1603: 99v; 1612a: 173; 1625: 47).

In the above citation, Sanctorius also pointed to the possible source of the non-naturals in Galen's works, namely Hippocrates.[9] In his last work, *De remediorum inventione*, Sanctorius wrote that Hippocrates had dealt with the six non-natural things in the *Libri epidemiorum* (Books on Epidemics) (Sanctorius 1629b: 144). What is more, when discussing issues related to the non-naturals in his *Commentary on Hippocrates* and his *Commentary on Avicenna*, Sanctorius referred to Galen's commentaries on *Epidemics* and on two other Hippocratic works—*De victus ratione in morbis acutis* (On Regimen in Acute Diseases) and *De natura humana* (On the Nature of Man) (Sanctorius 1625: 59 f.; 1629a: 100, 389). The assumption that

---

[8] "… ex rebus non naturalibus nihil certi colligi potest, quia hae in affectuum internorum productionem non necessario conspirant: videmus enim aliquando homines tempore aestivo post unum [sic] generosi visi [sic], & aromatum in cachexiam, & anasarcam praeterlabi: & tempore hyberno senes aliquos post usum frigidorum aliquando in ardentem febrem incidere: Quo fit ut Hippocrates, & Galenus noluerint has res non naturales ullo modo appellare causas, sed προφασεις, idest occasiones: …." See: Sanctorius 1612a: 173. In this edition, there is an error in pagination; the correct page number would be 169. The English translation was made on the basis of Jarcho 1970: 376, who refers to Morgagni's comment, which is, however, nearly identical to the passage in Sanctorius 1612a, see: Morgagni 1965: 85.

[9] Around 60 medical treatises attributed to Hippocrates have been handed down to us, compiled in the so-called *Corpus Hippocraticum*. It is difficult to determine exactly which works of the *Corpus* are his, but it has been proved that not all of the treatises were written by the same author (Jouanna 1996: 38 f.).

Galen developed concepts on the basis of Hippocratic ideas that were subsequently systematized as the "six non-naturals" is further supported by recent historical research. Luis Garcìa-Ballester revealed that Galen considered the contents of the doctrine of the non-natural things in commentaries on various works by Hippocrates: Epidemics I and VI, *De aere, aquis et locis* (On Airs, Waters, and Places), and *De natura hominis* (On the Nature of Man) (García-Ballester 2002: 108). An analysis of the respective passages in the Hippocratic works in connection with Galen's commentaries still remains to be done.

But whatever the influence of Hippocrates's thoughts on Galen may have been in this regard, it was through a study of Galen's works that later generations established the doctrine of the six non-natural things. Sanctorius's statements illustrate how Galen's thoughts on the matter, scattered throughout various works, were collected, interpreted, and further developed. Although Galen's apparently imprecise and unsystematic treatment of these concepts gave rise to discussion—for example, as in the different listings (fourfold and sixfold) and their respective functions in pathology and therapy—the doctrine of the six non-naturals remained intact for centuries and was dealt with under both headings, pathology and therapy. And, as mentioned above, Morgagni held the exact same lecture on the subject as Sanctorius one hundred years before him (García-Ballester 2002: 105, 115; Rather 1968: 341).[10]

The relevance of the doctrine of the six non-naturals for Sanctorius is evident from his decision to structure the results of his weighing procedures around it. In doing so he wittingly or unwittingly tied in with the literary genre of *Regimina sanitatis*—a medieval tradition of rules of health, which followed the organizational criterion of the six non-natural things. In what way the *De statica medicina* resembles these writings on hygiene will be outlined in Sect. 4.1.2. Here, it is important to note that even though Sanctorius's use of the six non-naturals as a structural element in his work was not unique, the fact that he considered this of all concepts suitable for the presentation of his new quantitative findings is of interest. What is more, the six non-naturals may even have played a crucial part in the preparation and conduct of his weighing procedures. Hence, this is a striking example of the way in which Sanctorius integrated innovative ideas into the traditional framework of Galenic medicine. To understand how Sanctorius organized his static medicine around the doctrine of the six non-naturals and what this implies, one has to dig deeper and scrutinize the contents of the doctrine. Therefore, the next sections analyze the effect of the six non-natural things on the body and how they restored health and produced disease.

---

[10] For more information on the origins and the development of the doctrine of the "six non-natural things," see: García-Ballester 2002, Ottosson 1984: 253–70, Bylebyl 1971, Niebyl 1971, Jarcho 1970, Rather 1968.

3.1.2 The Role of the Non-Naturals in Pathology

In line with the well-known Hippocratic tradition, according to which moderation was praised as a key to good health, every excess in the non-natural things was thought to harm the body. This view was embedded in Galenic humoral theory, which was likewise based on Hippocratic ideas. Here, it is important to remind ourselves that Galen's views of Hippocrates and Hippocratic medicine were very influential in the Renaissance. Scholars trusted that he followed the teachings of Hippocrates accurately, that he understood the works of the Corpus, and knew which were authentic and which were not. Hence, whenever Sanctorius referred to Hippocrates or the Hippocratic teachings, it can be assumed that he was guided by Galenism. This is not to say that he did not have the Hippocratic Corpus at hand, but rather that he read these works through Galenic lenses. Thus, for example, although Galen attributed the four humors theory to Hippocrates, it is clear to us today that this famous theory was expounded in fact by Polybos, a student of Hippocrates (Sanctorius 1612b: 37 f.; Smith 1979: 13; Jouanna 1996: 38 f.).

In fact, the idea that human bodies contain fluids which affect their physiology and their state of health can be found in various Hippocratic treatises; yet these diverged regarding the number of humors contained in the body. As already indicated, Galen identified the four main kinds of humor, blood, phlegm, yellow bile, and black bile, as the Hippocratic humors. Ever since, the four humors theory has been the standard form of humoral theory. The various schemes included in this theory and addressed in the following paragraphs were shared by different physicians and medical schools, and are to be found not only in the Hippocratic Corpus. Thus, the humoral theory that Galen presented was rather eclectic and it is very difficult to pinpoint Galen's particular contributions. Galen himself specifically identified Hippocrates, Plato (ca. 429–347 BCE), and Aristotle (384–322 BCE) as his precursors in adopting this concept (Temkin 1973: 18 ff.; Siraisi 1990: 104 f.; Galen & Johnston 2016: xxviii). But I will not dwell on these issues any further here. In the following, some basic features of the theory will be outlined, also in the context of Sanctorius's understanding and adoption of it. The origins of certain ideas will be touched on only where Sanctorius's statements demand this clarification.

According to Galenic humoral theory, the four humors were each related to a vital organ: blood to the heart, phlegm to the brain, yellow bile to the liver, and black bile to the spleen. The humors were also linked to the primary qualities of hot, cold, moist, and dry, which in turn characterized the four elements of the macrocosm: fire, air, water, and earth (Fig. 3.2). Health was thought to consist of a balanced mixture of the four humors (*eucrasia*), whereas an imbalance of the humors (*dyscrasia*), caused for example by an excess or deficit of one or more of the humors, was thought to be the direct cause of all disease. The qualities of the humors influenced the nature of the diseases they caused. Hence, balance and moderation were crucial to maintaining health. The non-naturals could change the balance of the

**DIAGRAM OF
GALENIC HUMORAL THEORY**

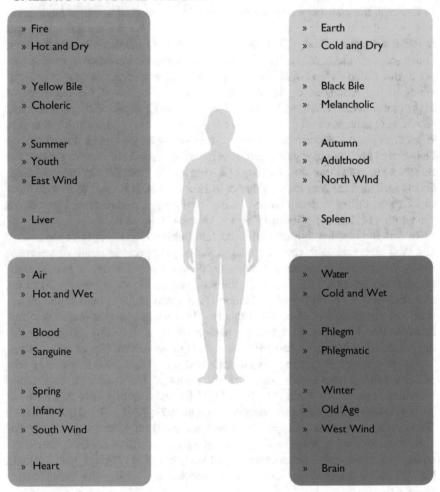

» Fire
» Hot and Dry

» Yellow Bile
» Choleric

» Summer
» Youth
» East Wind

» Liver

» Earth
» Cold and Dry

» Black Bile
» Melancholic

» Autumn
» Adulthood
» North WInd

» Spleen

» Air
» Hot and Wet

» Blood
» Sanguine

» Spring
» Infancy
» South Wind

» Heart

» Water
» Cold and Wet

» Phlegm
» Phlegmatic

» Winter
» Old Age
» West Wind

» Brain

Fig. 3.2 Diagram of Galenic humoral theory: elements, humors, qualities, temperaments, seasons, ages, winds, and organs. Diagram made with resources from Freepik.com, designed by macrovector/Freepik

primary qualities and thus influence the character of the humors and the state of the humoral balance (Rather 1968: 339; Temkin 1973: 17 f., 103; Gourevitch 1996: 141).

Therefore, it was important that they were used in moderation, which means in due quantity and quality, as Sanctorius further specified. Referring to Galen, he explained that food was healthy for those who fasted, or whose body had just evacuated, whereas it was unhealthy for replete bodies. In the same way, moderate exercise was beneficial, but became harmful if done to excess (Sanctorius 1612b: 49). With regard to quality, Sanctorius wrote in the *De statica medicina*:

Cold air and cold baths heat up robust bodies and render them lighter by removing [whatever is] superfluous. They cool weak bodies and render them heavier by prevailing over their heat (Sanctorius 1614: 20r).[11]

Hence, cold air and cold baths could have both wholesome and harmful effects, depending on the physical constitution, i.e., on the bodies' individual balance of the primary qualities. In Galenic medicine, this balance was called complexion (*complexio*), or temperament, and was believed to result from the mixture of the elements in the human body.[12] Every individual had their own innate complexion, acquired at the moment of conception and persisting throughout life. Accordingly, some people were hotter than others, and this characterization would apply to them their whole life long. Moreover, there was a second type of complexion, which Sanctorius called adventitious constitution. According to Sanctorius, this complexion could be attained by using the six non-natural things and was the one that physicians should use to assess every affection of the body. While the innate constitution could hardly be changed, the adventitious constitution was permanently influenced by the use of the six non-naturals and the habits associated with them. A body with a healthy innate complexion could still be affected by a disease that was introduced through an error committed in the six non-naturals. Hence, the adventitious constitution represented a person's current state of health (Sanctorius 1603: Book IV, esp. 81v, 82r, 97r; 1612a: 117 f.).

The well-balanced complexion, which is to say, a good mix of the four humors, was vital for good health. If the complexion was out of balance, meaning that it was too hot, cold, moist, or dry, weakness occurred. But the boundaries between a "balanced" and an "imbalanced" complexion were vague; no absolute measure of the healthy complexion existed. Instead, there was a spectrum of health, ranging from the ideal condition to that where the functions of the body were disturbed such that one could definitely speak of disease. In between there was thought to be a neutral state (Temkin 1973: 18; Grendler 2002: 315).

Sanctorius's statement shows how the non-natural pair, air and water, had a different influence on a body with a strong and very healthy complexion than on a body with a weaker complexion that was further removed from the ideal constitution.[13] In his *Commentary on Galen*, Sanctorius explained that, according to Galen, those who knew best how much (*quantum*) and in what way (*quomodo*) the six

---

[11] "Aer frigidus, & lavacra frigida corpora robusta calefaciunt, eaque; auferendo superfluum reddunt leviora. Debilia refrigerant, eaque; vincendo calorem ponderosiora efficiunt." See: Sanctorius 1614: 20r.

[12] Danielle Jacquart introduced a distinction between *complexio* (complexion), connected with the doctrine of qualities, and *temperamentum* (temperament), based on humoral theory (Jacquart 1984, see also García-Ballester 1992: 129, n. 19). For the sake of simplicity, I use these terms interchangeably in the present work.

[13] Sanctorius added water to the first category, "air," of the traditional list of the six non-naturals. In this, he may have been inspired by the *Isagoge Johannitii*, a standard introductory textbook at medical university faculties, where the list of the non-naturals included as special categories also "coitus" and "bath" (Ottosson 1984: 254). For more information on the *Isagoge Johannitii*, see: Temkin 1973: 104–8.

non-natural things heated, cooled, moistened, and dried the body, knew how to preserve the health of bodies that were out of balance as well as how to return them to
a better condition. Hence, the correct quantitative and/or qualitative management of
each of the six non-natural things was virtually a guarantee of maintaining a suitable
standard of health. Conversely, incorrect management of these factors—quantitative
and/or qualitative, here too—led to a pathological state. In this context, Galenic
physiology distinguished the non-natural things from the natural things (for example, humors, complexions, or members) as well as from the contra-natural (*praeternatural*) things, which were pathological conditions of all kinds (Sanctorius 1612b:
19, 111).[14]

Sanctorius, still expounding Galen's teachings in the *Ars medica*, pointed out
that bodies with an optimal complexion could autonomously prescribe themselves
the proper quantity and quality of all the non-natural things as well as their proper
timing.[15] Thus, they needed neither a supervisor nor doctor to monitor the management of the six non-naturals, per Sanctorius, as they were able to do this perfectly
well on their own. But this optimal complexion was an ideal that could only be
approximated and probably never reached. As a result, all people needed support in
managing their health, i.e., in regulating their lifestyle in line with the concept of the
non-naturals (Sanctorius 1612b: 79; Siraisi 1990: 101–23; García-Ballester
2002: 105).

3.1.3 The Role of the Non-Naturals in Therapy

According to Galenic medicine, therapeutics were divided into surgery, drug lore,
and dietetics.[16] While the non-naturals were rather insignificant for surgery, they
were all the more important for drug lore and dietetics. It is the latter category that
I will focus on first. In his *Commentary on Hippocrates*, Sanctorius pointed to the
double meaning of the word "diet" (*victus* or *diaeta*) in the works of Hippocrates
and Galen. On the one hand, diet was understood in the context of dietetics and
included the six non-naturals. On the other hand, diet simply meant food (Sanctorius
1629a: 100).[17] The first meaning reflects the integration of nutrition into a broader

---

[14] This division can be traced back to Galen's work *De pulsibus ad tirones*, in which Galen used the
expression "non-natural" as an intermediate category between "natural" causes and *praeternatural*
causes that change the pulse. See: Galen. 1997b: 462–73, Sanctorius 1612b: 19, García-Ballester
2002: 106 f.

[15] Time appears here as a third category, alongside quantity and quality. Sanctorius explained that
this third category had been introduced by Galen only for teaching purposes, to simplify his doctrine. In fact, so Sanctorius, the opportune time was necessarily integral to the other two factors,
quantity and quality, because only if these coincided with the opportune time (i.e., if their timing
was right) could they be said to occur in an appropriate way (Sanctorius 1612b: 51 f.).

[16] Sanctorius mentioned this tripartition in his *Commentary on Avicenna*, see: Sanctorius 1625: 4.

[17] See also Sanctorius 1629a: 389.

concept of a healthy lifestyle that included, among other things, the influence of environmental factors, like climate and weather (represented by the first non-natural pair, air and water). The six non-naturals provided Hippocratic dietetics with a doctrinal framework that guided patient and doctor in their pursuit of a healthy regimen. Moreover, the concept integrated these aspects of Hippocratic medicine into Galenic complexional theory as a system of explanation providing the rational link between disease and therapy. Thus, besides the treatment of disease, the preservation of health through a preventive health regime was the main task of the physician.

The physician had to tailor the use of the non-naturals to every individual patient so as to maintain him or her with the optimum complexion. He needed to identify how much and what kind of food, exercise, sleep, etc. was beneficial or harmful to the respective person and would accordingly have a positive or negative effect on the qualities of his or her complexion. Contrariwise, the effect of the non-naturals revealed to the physician the complexion of his patient. Identifying the general complexion was particularly complex because each organ of the human body was considered to have its own complexion. Adding to the complexity, bodily parts each had their own predominant complexional quality. Hence, the heart was hotter than the brain, the brain was colder than the heart, and so on. Medical textbooks helped the physician not lose track by providing long lists of body organs and their predominant qualities. Moreover, some general rules could be applied to different groups of people. Young people were thought to have a warm and moist complexion that, over time, gradually turned into a cold and dry complexion in old age. Women were thought to be colder than men and complexion varied also among geographical regions (Sanctorius 1612a: 107, 531; 1625: 382–6; 1629a: 293 f.).

If the physician detected a complexional imbalance, i.e., ill-health, he tried to restore the balance by changes in the six non-naturals according to the theory of cure by contraries. A body that was too hot had to be cooled down. A body that was too dry had to be moistened. With regard to the non-natural pair of exercise and rest, Sanctorius wrote:

> If whoever lies in bed for a long time suffers from pain in the feet, walking will cure them: if those who travel suffer, the remedy is rest (Sanctorius 1614: 63v–64r).[18]

The principle that every cure is effected by contraries and every conservation by similarities was fundamental to Galenic medicine (e.g., Sanctorius 1612a: 606). However, the physician had to be careful when changing the lifestyle of his patient. Sanctorius, referring to Hippocrates, explained that a faulty but habitual regimen was less harmful to one's health than a suddenly switch to a better regimen. A body had to be slowly accustomed to changes in the non-natural things, as for example to more or less exercise, warm or cold food, longer or shorter sleep, and so on (Sanctorius 1629a: 413).

As reflected in the double meaning of the word "diet" that was addressed at the beginning of this section, the third non-natural thing, food and drink, was a special

---

[18] "Si diu iacenti dolores pedum superveniant, remedio est ambulatio: si iter facienti, quies." See: Sanctorius 1614: 63v–64r.

category and one often considered individually. It is closely connected to the second form of therapy of interest here, namely drug lore. Sanctorius wrote:

> Or the name diet is used in its meaning of food, just as Hippocrates does in this place. In his commentary, Galen divides this food according to its differences, without a distinct knowledge of which sick people, suffering from an acute disease, cannot be managed and healed (Sanctorius 1629a: 100).[19]

Thus, foodstuffs were used to heal diseases, just as drugs were. Both were *complexionate*, meaning that they were characterized by the same qualities (hot, cold, wet, dry) as the four humors of the body. Moreover, drugs were closely related to food by their mode of administration, i.e., ingestion. In this way, the actions of drugs and food were integrated into the Galenic theory of digestion, affecting the complexion of the person who ingested them. Spices and various vegetables were sometimes counted as food, sometimes as drugs. In the *Commentary on Avicenna*, Sanctorius explained that food (*alimentum*) could be either considered in the strict sense, according to which it nourished the body by increasing its substance, or with regard to its ability to change the body; and in this latter sense, it qualified not as food, but as a drug. The close connection between food and drugs is further emphasized by the fact that Galen put both into the same category in his quadruple classification of the non-naturals in his work *Thrasybulus*. Yet there were differences too. The decisive criterion was the direction of their action. As suggested by Sanctorius in the aforementioned statement, while the body acted upon foodstuffs by digesting, i.e., assimilating them, drugs acted upon the body, and their respective impact indicated their place on the broad spectrum from food-like drugs to detrimental poisons. Dietetic treatment, conceptualized in the doctrine of the six non-naturals, mainly sought to preserve health. Drugs, on the contrary, were used to counteract the noxious impact of an illness (Sanctorius 1612b: 42; 1625: 63; Siraisi 1990: 100–23; Touwaide 1996: 289 f.; Vogt 2008: 304, 306 f.; Galen and Johnston 2018b: 295).

With this overview of the role of the non-naturals in Galenic pathology and therapy at hand, one can more readily tackle Sanctorius's special use of the doctrine of the non-naturals, namely the shift in focus to the *perspiratio insensibilis*. By means of his weighing chair, Sanctorius claimed to be able to measure this physiological process and to argue, on a quantitative basis, for its central role in health and disease. His new findings made him reconsider the doctrine of the six non-naturals and readjust the rules for a healthy lifestyle. Innovative ideas thereby met long-established concepts and instead of displacing each other, they intermingled and started evolving into something new: static medicine. To understand this process, a closer look into the development and content of the concept of *perspiratio insensibilis* is necessary. The theoretical backdrop against which Sanctorius developed his concept of *perspiratio insensibilis* and the way he presents it in his works have to be analyzed.

---

[19] "Vel sumitur nomen diaetae, pro ut significat cibum, sicuti in hoc loco sumitur ab Hippocrate quem cibum Galenus in comm. Dividit in suas differentias, sine quarum distincta cognitione aegri acuto morbo laborantes regi, & sanari non possunt." See: Sanctorius 1629a: 100.

3.2 The Concept of *perspiratio insensibilis*

According to Sanctorius, the constant supervision of bodily discharges was essential for the preservation of health. In keeping with Galenic humoral theory, he conceived of health as an ideal balance between ingestion and excretion, meaning that the quantity of substances consumed by the organism should be proportionate to the amount of substances rejected by it. In Galenic medicine, this equilibrium was thought to be an expression of the balance of the humors. The measurements Sanctorius is said to have conducted with the weighing chair demonstrated that a large part of excretion takes place invisibly through the skin and lungs. He wrote: "Insensible perspiration alone is usually much more abundant than all sensible evacuations taken together" (Sanctorius 1614: 2r).[20] Thus, in Sanctorius's view, the monitoring of the *perspiratio insensibilis* by means of systematic weighing was fundamental to the preservation of health; a very strong claim indeed.

3.2.1 Early Ideas on perspiratio insensibilis

The conception of an insensible perspiration of the body—*perspiratio insensibilis*—dates back to ancient times. Mystical and religious beliefs have always linked the life principle to air, breath, and breathing. Here may lie the origin of the early conviction that not only the lungs but the whole body breathed in and out. Hence, expressions related to the Latin term *respiratio* (breathing), such as *transpiratio, exhalatio,* or *perspiratio,* were used for this activity of the body. And, perhaps to emphasize its invisible nature, it was sometimes referred to as *insensibilis (occulta) transpiratio, exhalatio*, or *perspiratio*.[21]

In the Hippocratic writings, numerous references to an imperceptible, vaporous excretion of the body attest that Hippocrates and his followers had knowledge of this phenomenon. The following passage is taken from the treatise *De alimento* (Nutriment):

> Porosity of a body for transpiration is healthy for those from whom more is taken; denseness of body for transpiration is unhealthy for those from whom less is taken. Those who transpire freely are weaker, healthier, and recover easily; those who transpire hardly are stronger before they are sick, but on falling sick they make difficult recovery (Hippocrates and Jones 1923: 353).

From citations and references in the Galenic corpus, it is known that in the third century BCE, ideas of *perspiratio insensibilis* also existed in the Alexandrian medical school. Erasistratus demonstrated material losses by weighing fowls and their food and excreta, and explained them by the existence of an insensible perspiration

---

[20] "Perspiratio insensibilis sola solet esse longe plenior, quam omnes sensibiles simul unitae." See: Sanctorius 1614: 2r.

[21] The focus in this section is on Latin terminology, as this is the language Sanctorius wrote in.

in animals. Further ideas on the matter were expressed by Theophrastus (ca. 370–
ca. 297 BCE), Aristotle, and Aretaeus of Cappadocia (second century CE), to name
but a few. Galen finally systematized the scattered notions and integrated them into
his physiology. But it was not until more than a millennium later that the concept of
perspiratio insensibilis gained considerable attention (Renbourn 1960: 135–39).[22]

3.2.2 Sources of Sanctorius's Concept *of* perspiratio insensibilis

In a letter Sanctorius sent together with a copy of his *De statica medicina* to Galileo
Galilei, he explained that his work was based on two principles: first, Hippocrates's
view that medicine is essentially the addition of what is lacking and the removal of
what is superfluous; and second, experience (Sanctorius 1902).[23] In contrast to most
studies on Sanctorius, I focus in the following first on the medical tradition to which
Sanctorius referred, an aspect that has hitherto been analyzed only marginally; and
in a later chapter treat the second principle, experience, which has already gained
the attention of many scholars (Sect. 7.5).

In the first aphorism of the *De statica medicina*, Sanctorius wrote:

> If there is daily an addition of what is wanting and a removal of what abounds, in the
> required quantity and quality, lost health will be restored and the present [health] always
> preserved (Sanctorius 1614: 1r).[24]

Sanctorius further explained in the letter to Galileo:

> That this art, by me invented, should be important is clear, because I am able accurately to
> measure insensible transpiration, which if altered or impeded, according to the opinion of
> Hippocrates and Galen, is the origin of nearly all ills; …. That this art is alluded to by Galen
> is clear in many places, and especially in the sixth [book] of *De tuenda sanitate* cap. 6,
> where may be read these words: *Whenever those things dispersed in vapor from the body*

---

[22] For more information on ancient concepts of *perspiratio insensibilis*, especially on Galen's, see:
Debru 1996: 178–210. I thank Caroline Petit for drawing my attention to this work. A comprehensive historical study on the medical concept of *perspiratio insensibilis* still remains to be written.
For a short historical survey of the topic, see: Renbourn 1959 and Renbourn 1960. Weyrich 1862
prefaced his physiological study on insensible perspiration with a historical overview that contains
a more detailed analysis of the concept in Galen's works and in Sanctorius's *De statica medicina*.
The most recent study is Stolberg 2012, which focuses on early modern meanings of sweating and
transpiration and the theories and practices surrounding them.

[23] "L'opera è ridotta in afforismi, i quali nascono da due principii certissimi. Il primo è la diffinition
della medicina, proposta da Hippocrate nel libro *De flatibus*, dove dice: *Medicina est additio et
ablatio: additio eorum quae deficiunt, et ablatio eorum quae excedunt* …. Il secondo principio di
quest'arte è l'esperienza …." See: Sanctorius 1902. For a transcription, see: Sanctorius and Ongaro
2001: 34–8 and for an English translation, see: Castiglioni 1931: 773 f.

[24] "Si quanta, & qualis oporteat, quotidie fieret additio eorum quae deficiunt, & ablatio eorum quae
excedunt, sanitas amissa recuperaretur, & praesens semper conservaretur." See: Sanctorius
1614: 1r.

are less than those things taken in, the plethoric diseases arise. What must be preserved, then, is the balance between foods and drinks, on the one hand, and those things evacuated, on the other. There will be balance when we give consideration to the quantities in each (Sanctorius 1902).[25]

Thus, without any doubt, Sanctorius learned from the teachings of Hippocrates and Galen about the *perspiratio insensibilis* and its effects on health and disease. However, in the preface to the *De statica medicina*, he also pointed out the novelty of his work: the exact weighing of insensible perspiration (Sanctorius 1614: Ad lectorem).[26] The sheer volume of fluid that the body excreted insensibly everyday showed the outstanding importance of insensible perspiration and made Sanctorius claim that it needed particular attention and care. Outshone by the quantitative method, Sanctorius's clearly articulated adherence to the Galenic conception of *perspiratio insensibilis* took a back seat in the reception of the *De statica medicina*. If mentioned at all, it was usually subject to criticism.[27] Contrary to this, I think it is crucial to include exactly these aspects that are often dismissed as old-fashioned Galenism in the analysis of Sanctorius's works, in order to understand their content and the scientific legacy of Sanctorius. This is the aim of the following sections.

3.2.3 *Sanctorius's Conception of* perspiratio insensibilis

To the *perspiratio insensibilis* Sanctorius gave different synonymous expressions: *perspirabile, perspirabilis* (matter of perspiration), *perspirantia, perspiratio insensibilis* or *transpiratio insensibilis, halitus invisibilis, insensibilia excrementa* or *insensibilis excretio, evacuatio insensibilis, exhalatio, difflatio, occulta perspiratio,* or simply *perspiratio* and *transpiratio.*[28] The variability in nomenclature makes it difficult to grasp Sanctorius's understanding of insensible perspiration. Are the

---

[25] "Che quest'arte, da me inventata, veramente sii importantissima, è cosa chiara, perchè può distintamente mesurar l'insensibile transpiratione, che, alterata o impedita, secondo l'opinion d'Hippocrate et Galeno, è origine quasi de tutti i mali; ... Che quest'arte sii accennata da Galeno, è cosa chiara in molti luoghi, et spetialmente nel sesto *De tuenda sanitate*, cap. 6°, dove si leggono queste parole: *Ubi quod ex corpore exhalat minus est iis quae accepit, redundantiae oriri morbi solent; ergo prospiciendum est, ut eorum quae eduntur ac bibuntur, respectu eorum quae expelluntur, conveniens mediocritas servetur. Sane is modus servabitur, si ponderabitur a nobis in utrisque quantitas.*" See: Sanctorius 1902. For a transcription, see: Sanctorius and Ongaro 2001: 34–8 and for the English translation, see: Castiglioni 1931: 773 f. For the English translation of the passage quoted here by Sanctorius from Galen's *Hygiene*, see: Galen and Johnston 2018b: 153. Original emphasis.

[26] "Novum atque inauditum est in medicina posse quēpiā ad exactam perspirationis insensibilis ponderationem pervenire ..." See: Sanctorius 1614: Ad lectorem.

[27] An important exception is Paolo Farina's paper on the influence of Sanctorius on his disciple Henricus Regius (1598–1679), in which the author repeatedly points out Sanctorius's strong adherence to Galenic medicine (Farina 1975).

[28] This is not a comprehensive list and I still find some other variations in Sanctorius's works.

different expressions interchangeable? Or are they connected to different aspects of insensible perspiration? When Sanctorius omitted adjectives like *insensibilis, invisibilis* etc., he often still referred to *insensible* perspiration. However, the exact meaning has to be deduced from the context. This can be said also with regard to the interchangeability of the different expressions. Generally, Sanctorius used them synonymously, but caution is still needed, as there are always exceptions to the rule. When he wrote, for example, of *meatus insensibilis*, he referred to the insensible channels or pathways in the body through which humors, vapors, and insensible perspiration passed (e.g., Sanctorius 1629a: 82, 472). However, *meatus* can also be translated with "a going" and therefore describe perspiration itself.

3.2.4 The Dual Origin of perspiratio insensibilis

In the beginning of the first section of the *De statica medicina,* Sanctorius explained the dual origin of insensible perspiration:

> Insensible perspiration either occurs through the pores of the body, which is completely transpirable and covered by the skin like a net; or it occurs by means of respiration that is made through the mouth, which usually amounts to about half a pound during one day; the drops on a mirror placed in front of the mouth actually indicate this (Sanctorius 1614: 2r).[29]

Thus, according to Sanctorius, insensible perspiration was generated either through the pores of the skin, or through the mouth. In the quoted aphorism, he even noted the quantity of daily respiration. This suggests that he differentiated between the two different forms of *perspiratio insensibilis* in his weighing experiments.[30] However, when explaining the difference between sensible and insensible evacuations in his *Commentary on Avicenna*, Sanctorius wrote quite plainly: "… but it is insensibly [evacuated] through the pores of the skin" (Sanctorius 1625: 60). Hence, from a conceptual point of view, Sanctorius seems here to be somewhat inconsistent. What this implies for his measurements will be explored in a later chapter (Sect. 7.5.5).

The dual origin of *perspiratio insensibilis* mentioned by Sanctorius hints at the Galenic conception that insensible perspiration resulted from the respiratory and digestive activities of the body. These were the physiological processes responsible for ingestion and excretion and therefore crucial to keep the balance between the substances ingested by the body and those excreted by it (Weyrich 1862: 5). As the precondition of health was, above all, a proper and regular evacuation of the

---

[29] "Perspiratio insensibilis vel fit per poros corporis, quod est totum transpirabile, & cutem tanquam nassam circumpositam habet, vel fit per respirationem per os factam, quae unica die ad selibram circiter ascendere solet; hoc enim indicant guttae in speculo, si ori apponatur." See: Sanctorius 1614: 2r. The measurements used by Sanctorius will be discussed in Sect. 5.4.2, fn. 39.

[30] The method Sanctorius used to measure respiration as distinct from insensible perspiration of the skin is far from clear and will be analyzed in Sect. 7.5.5.

consumed material, it is worth considering the processes of digestion and respiration in more detail.

3.2.5   Digestion

In line with Hippocratic ideas, digestion was understood as a cooking by means of heat and subsequent refinement, for use by the body. Hence, the Latin terms *coctio* (coction) or *concoctio* (concoction) and the verb *concoquere* (to concoct) were used to describe this process.[31] Fundamental to this concept was the idea that every living being is the product of heat and moisture. In this context, the human body was often compared to an oil lamp—a metaphor that Sanctorius employed as well. At birth, every living being acquired a certain amount of radical moisture, corresponding to the oil in a lamp. Throughout life this moisture was consumed by an inborn, or innate heat (*calor nativus*), just as is the oil in a burning lamp. With age, the radical moisture and innate heat decreased and the body naturally became colder and drier. Food was needed in order to maintain the heat by replenishing the substance of the body that had been consumed. During digestion, food was transformed into the body, becoming flesh itself (e.g., Sanctorius 1612a: 313, 348, 610; 1625: 351, 357; 1629a: 290).

However, there were always elements that withstood incorporation. Hence, on the one hand, food contained the nutriments needed to replace the natural deterioration of the body. But on the other hand, it also contained superfluities, which could harm or destroy the body. The evacuations helped the body to get rid of the superfluities and waste and to keep the blood pure. Sanctorius explained in the *Commentary on Galen* that digestion fulfilled three purposes: to transform or convert nutritive food into body substance, to separate useful material from useless material, and to expel those excrements which were useless. The digestive process, Sanctorius continued, took place in three steps while each step produced different excreta. After chewing, the food entered the stomach, where it was concocted by means of heat. This first step was crucial, as a bad concoction could never be corrected later. What was more, the digestive process could only continue after the food was fully concocted. In the process, food was transformed into *chyle* and solid waste was produced and expelled in the form of stools.[32] Then, the *chyle* was directed to the liver,

---

[31] Sanctorius differentiated between "digestion" and "concoction." While the former described the transmission of nutrition from the stomach to the liver, to the guts, or to the skin and then into the ambient air, the latter referred to the transmutation of substance, of food into *chyle* and *chyle* into blood (Sanctorius 1612a: 611, Sanctorius 1629a: 305, 312). As Ken Albala has argued, this distinction was common in Renaissance nutritional theory, even though Roman authors used the term "digestion" in its broader sense, applying it to the whole process (Albala 2002: 54). Sanctorius, however, did not consistently make the distinction and sometimes referred to the whole digestive process as *coctio* or *concoction* (see e.g., Sanctorius 1612b: 84, Sanctorius 1625: 589 f.).

[32] *Chylus* from the Greek *chylos* was the synonym for the masticated food turned into a fluid state. See: Orland 2012: 465.

where in the second stage of digestion it was converted into blood. The residual matter was excreted via the urinary tract. The final step in the digestive process took place in the organs and at the bodily periphery, and its excreta were insensible perspiration and filth (*sordes*), or sweat. The blood, generated in the liver, was now distributed throughout the body via the venous system. At this stage the blood was, however, impure, still containing in it the other unrefined humors. As Sanctorius explained in the *De remediorum inventione*, it was refined in three organs close to the liver: the gallbladder, the kidneys, and the spleen. Accordingly, the spleen, for example, generated in, or purged from, the blood, melancholy, i.e., black bile. Apart from entering the organs, the blood could also make its way from the liver to other parts of the body, including the heart. It was in the former that assimilation took place and the nutrients were converted into flesh (Sanctorius 1612b: 70, 84; 1625: 465, 589; 1629a: 276; 1629b: 45 f.; Albala 2002: 17–64; Kuriyama 2008: 430; Stolberg 2012: 505).

3.2.6   Respiration

In addition to food, the body continuously takes in air. In line with the teachings of Galen, the main functions of breathing were, so Sanctorius, to cool the heart, to nourish the vital spirits, and to cleanse the body from smoky vapors. During inspiration, air was drawn from the lungs into the left ventricle, where vital spirits were generated.[33] This happened simultaneously to the diastole of the heart and the distension of the arteries. During the formation of the vital spirits, smoky vapors were produced. Throughout systole, i.e., the compression of the arteries, which coincided in Galenic medicine with expiration, the smoky vapors were expelled. Spirit (*spiritus*, the Greek *pneuma*) was thought to be an exhalation (*halitus*) itself, a very fine vapor essential for maintaining life. The vital spirits were carried by the blood through the arterial system and reached via the carotid arteries the *retiform plexus*, a network of fine arteries at the base of the brain. Here, they were prepared to become animal spirits, which were finally generated in the ventricles of the brain from the vital spirits, from inhaled air, and from the surrounding substance of the brain.[34] According to the Galenic teachings, the brain itself was able to "breath" and

---

[33] Rudolph E. Siegel argued that according to Galen, air as a substance could not be absorbed by the body. Thus, only an invisible quality of heat, which Galen considered to be the predominant component of air, was absorbed from the inhaled air (Siegel 1968: 151, 155, 158). Julius Rocca did not refer to an invisible quality of heat, but explained that, in the opinion of Galen, inspired air was altered in the lungs into a "pneuma-like" substance (Rocca 2012: 637). Sanctorius simply explained that the vital spirit was created in the left ventricle by the inhaled air and the pure blood of the right ventricle (Sanctorius 1625: 367).

[34] In the sixteenth century, the existence of the retiform plexus (*rete mirabile*) was challenged, because anatomists could not observe it in the human brain. However, as Andrew Wear has argued, the existence of the animal spirits that were produced in the retiform plexus according to the Galenic teachings was not denied. Wear also analyzed Sanctorius's thoughts on the issue, who had

thus the anterior ventricles of the brain performed the actions of inspiration and expiration. During inspiration, the brain attracted the outside air, necessary for the generation of animal spirits. In the process, smoky vapors were produced and expelled by the diastole, i.e., the expiration of the brain, just as happened in the heart, when vital spirits were generated. Sanctorius described that the air, necessary for the formation of the animal spirits, was inhaled by the brain through the mamillary processes (*processus mamillares*). Accordingly, he stated that the smoky vapors, produced during the formation of the animal spirits, were expelled by means of the mamillary processes. Different from today's meaning of mamillary process, Sanctorius understood by *processus mamillares* the olfactory tracts located directly above the ethmoid bone. Thus, air was drawn into the brain via the nasal passages and the residual vapors were expelled the same way through the nose (Sanctorius 1612a: 258, 261, 356, 443, 447 f.; 1612b: 58; 1625: 209 f., 319 f., 367; 1629a: 362).[35]

The generation of the spirits was thought to be analogous to the notion of the concoction of nutriment. It therefore was connected to the concept of combustion, which explains the formation of *smoky* vapors as residual matter of the processes of the formation of the two spirits.[36] The animal spirits resided in the ventricles of the brain and spread through the nerves and the spine. They provoked sensation and voluntary motion, whereas the vital spirits served to nourish the animal spirits and to heat the body (Sanctorius 1612a: 422; 1625: 298, 319 f.; 1629a: 362; Rocca 2003: 65, 211–27).[37]

Sanctorius did not describe in detail the process of respiration via the skin pores. In the *Methodi vitandorum errorum*, he plainly explained that "the whole body is transpirable," just as he referred in the citation, quoted above (Sect. 3.2.4), to "the body, which is completely transpirable."[38] In the *De statica medicina*, Sanctorius was a little more explicit when stating that the external air passed through the arteries into the body. Thus, here again, Sanctorius seems to be true to the teachings of Galen, according to which, during diastole, the arteries attracted some air from the

---

stated that the retiform plexus was conspicuous. See: Wear 1981: 233–7; 251 ff. and Sanctorius 1612a: 260, Sanctorius 1629: 363 as well as Sect. 4.2.1. Julius Rocca pointed to the controversy and confusion that the doctrine of the retiform plexus caused for later physicians and gave a survey of some of these later Galenic accounts. See: Rocca 2003: appendix two.

[35] Sanctorius's anatomical knowledge and experience will be treated later in Sect. 4.2.1.

[36] In his commentaries, Sanctorius denied the existence of a third *natural* spirit (Sanctorius 1612a: 257–61, Sanctorius 1625: 51, Sanctorius 1629a: 360–5). This is exceptional, as Galenic pneumatology was usually interpreted as a tripartite system and much of the secondary literature follows this assumption. In fact, Owsei Temkin, Rudolph E. Siegel and Julius Rocca have shown that there is no reason to postulate the existence of a natural spirit in Galen's physiology (Temkin 1951, Siegel 1968: 186, Rocca 2012).

[37] According to Sanctorius, it was not the animal spirits themselves which provided sensation and voluntary motion, but an incorporeal radiation emanated by them. Similarly, he thought that incorporeal radiation of the vital spirits produced the faculties responsible for the systole and diastole in the arteries. See: Sanctorius 1612a: 255 ff.; 424 ff., Sanctorius 1625: 93 f., 298, 650 f., 749 ff., Sanctorius 1629a: 364 f.

[38] "totum enim corpus est transpirabile …." See: Sanctorius 1603: 31r.

outside through the pores of the skin, in a manner similar to the thoracic movements which caused air to enter the blood through the pores of the terminal bronchial tubes. Throughout systole, residual vapors, like the smoky vapors produced during the formation of the two spirits, were expelled not only into the lungs, but also through the skin pores (Sanctorius 1612a: 257; 1614: 20v; Renbourn 1960: 136; Siegel 1968: 103).

In this light, Sanctorius's comparison of the skin to a net (Sect. 3.2.4) is instructive, as it provides some insight into his understanding of skin. The analogy might reflect the influence of the Italian physician Girolamo Mercuriale, who, drawing on Plato's *Timaeus*, defined the skin as a fisherman's net (*nassulae piscatoriae*). Just as a net, he thought, the skin was a common bond holding together the separate body parts. According to Mercuriale, the only function of the skin was to receive waste materials. Sanctorius's reference to the analogy of a net implies that he shared Mercuriale's conception of the skin as an inherently porous layer of interchange between body and environment (Te Hennepe 2012: 526). This is further reinforced by the fact that Sanctorius organized the *De statica medicina* according to the six non-naturals, which, as was already shown, included also environmental and meteorological aspects that served Sanctorius to examine how the skin and its excretion of *perspiratio insensibilis* were affected, for example, by the climate in which a person lived and the weighing took place.

To put it in a nutshell, according to Sanctorius, insensible perspiration resulted from the respiratory and digestive activities of the body. It expelled the residual matter of both, respiration and digestion, thereby cleansing the body of superfluous matter. The distinction between the two different forms of *perspiratio insensibilis*, through the mouth and through the pores of the skin, will be of interest again in a later chapter, when it comes to the question of how to quantify them (Sect. 7.5).[39]

3.2.7 Perspiratio insensibilis *and Sweat*

Besides the two origins of insensible perspiration (the skin and the mouth), Sanctorius referred to two different kinds of *transpiratio insensibilis*. One was generated during sleep, when the body concocted, and it increased strength. The other was generated while awake and arose from a crude (unconcocted) humor through violent motion, which was why it decreased strength. This illustrates the close connection between insensible perspiration and digestion.[40] For Sanctorius, the

---

[39] In her analysis of Galen's concept of perspiration, Armelle Debru differentiated between *perspiratio insensibilis* and cutaneous respiration, the latter of which fulfilled the same functions as oral respiration (Debru 1996: 178–210). I do not follow this distinction here, because Sanctorius did not explicitly refer to cutaneous respiration. Therefore, I subsume any perspiration that occurs via the skin under *perspiratio insensibilis*.

[40] In the context of sleep, Sanctorius pointed to the difference between digestion and concoction in several passages in his works (see also Sect. 3.2.5, fn. 31). He explained that concoction, per

differentiation between insensible perspiration during sleep and during wake was linked to the rest or motion of the body. In the *Commentary on Galen*, he explained that sleep was a type (*species*) of rest, while wake corresponded to movement. While the body rested during sleep, digestion was carried out undisturbed. As soon as a person woke up, movement occurred and, with the movement, violence. The greater the movement, the greater the violence with which digestion took place. It was due to this violence that crude material was expelled from the body (Sanctorius 1612b: 38; 1614: 5r; 51v–52r). Sanctorius wrote:

> That which is evacuated through the pores during violent movement is sweat and occult perspirable matter; but being violent, it is for the most part raised by uncooked juices. For there is seldom collected in the body as much cooked perspirable matter as is evacuated by means of violence (Sanctorius 1614: 61r–61v).[41]

With violent motion, another excretion occurred: sweat. In the works of Galen, the relation between *perspiratio insensibilis* and sweat (*sudor*) is far from clear. Galen sometimes put forward the view that sweat simply came through the skin pores in the form of small drops of liquid. On other occasions, he insisted that sweat arose from the insensible perspiration caused by a thickened skin or through the condensing effect of a cold air. In his commentary on the Hippocratic Aphorisms, Galen referred to the claim of Diocles of Carystos (ca. 375–ca. 300 BCE), that liquid sweat was pathological (*preternatural*), a diagnostic sign of excess fluid in the whole body. Only if it occurred due to violent movement, hot baths, or summer heat was it healthy. In normal circumstances, the innate heat was strong enough to transform the superfluous humors into such fine, subtle parts that they escaped notice. Sweat was produced only under conditions of great external heat, such as body heat increased by violent exertion or fever, or arose from considerable weakness in the expelling force. As Michael Stolberg has pointed out in his article, Galen did not argue against this idea, but he still had his doubts (Stolberg 2012: 506).[42]

According to Sanctorius, sweat always originated from a violent cause and could impede the insensible excretion of concocted perspirable matter (*perspirabilium*). Due to the violence, the three stages of digestion could not be concluded and the body expelled crude, unconcocted matter in the form of insensible perspiration and, above all, sweat. In the *Commentary on Galen*, Sanctorius agreed with Diocles that sweat was always pathological (*praeter naturam*), because it only emerged with

---

Galen, occurred during sleep, while digestion occurred during wake (Sanctorius 1612a: 513, Sanctorius 1612b: 39, 76, Sanctorius 1614: 149, Sanctorius 1629a: 305). This implies that during sleep only the transmutation of substances took place, while the transmission of nutrition was carried out only while awake. Following this argument, *perspiratio insensibilis* was expelled only in a waking state. This, however, is contrary to Sanctorius's statement that the body perspired insensibly during sleep twice as much as while awake (Sanctorius 1612b: 40, Sanctorius 1614: 52r).

[41] "Quod in motu violento per poros evacuatur, est sudor & perspirabile occultum: sed ut violentum magna ex parte elevatur ex incoctis succis: raro enim tantum cocti perspirabilis in corpore colligitur, quantum per violentiam evacuatur." See: Sanctorius 1614: 61r–61v.

[42] An overview of ancient notions of sweat has been given by Armelle Debru, who, however, only briefly described Galen's concept of sweat. See: Debru 1996: 187–90.

violence. However, Sanctorius also referred to the beneficial effects of sweat—the evacuation of potentially harmful matter. Especially during the *crisis*, the decisive phase of a disease, a "critical sweat" could free the patient of the morbid matter.[43] Thus, Sanctorius's conception of sweat and sweating was ambivalent, probably reflecting the ambiguity of the issue found in Galen's works (Sanctorius 1612b: 62; 1614: 61v; 1629a: 285).

Insensible perspiration and sweat ultimately originated from the same matter as did urine. The three evacuations only differed in their refinement. Urine was the least refined, whereas insensible perspiration excreted the finest and more volatile parts of serum, which heat had resolved into vapors. The evacuations could also substitute for each other. Sanctorius wrote in the *De statica medicina* that people, who urinated more than they drank, perspired less or not at all. Moreover, abundant perspiration could not occur simultaneously to abundant sensible evacuations (Sanctorius 1614: 3v–4r; Renbourn 1959: 206; 1960: 136; Stolberg 2012: 504–7).

3.2.8 The Composition of perspiratio insensibilis

According to Sanctorius, *perspiratio insensibilis* always consisted of heavier parts and lighter parts. If the heavier parts accumulated, they could give rise to creatures such as bugs and lice, or even to contagious infections. Sanctorius thought that the lighter parts "flew away," whereas the heavier parts stayed and vitiated the body. There was also a connection between the emotions of a person and the two parts of perspiration. In sadness and fear the lighter parts of the perspiration were evacuated, but the heavier parts remained. And correspondingly, the heavier perspirable matter that was excessively retained brought about sadness and fear. Thus, the subtler the perspiration, the healthier it was. Sanctorius also differentiated between thick (*crassus*) and fine (*tenuis*) parts of *perspiratio insensibilis*. They seem, however, to correspond to the heavy and light parts of perspiration (Sanctorius 1614: 6r, 18r, 75r–76r, 79r) (Sect. 3.3.6).

In his article, Michael Stolberg has explained that Galen, and early modern physicians with him, described sweating as an excretion of thin serous humors. This suggests that sweat and insensible perspiration were closely related to the bodily humor serum. Even though I could not find this description in Sanctorius's works, a closer look at the idea might help to understand his concept of the composition of *perspiratio insensibilis*. The meaning of the Latin term *serum* is "whey," the watery residue from making cheese. In early modern medical writing, it was commonly used to describe the thinner, more watery parts of the blood. As Stolberg has pointed out in his article, the pores of the skin were thought to act like the kidneys, as a

---

[43] The so-called *crisis* was thought to be the turning point of an illness, leading toward recovery or death. It usually took the form of a sudden excretion of "bad humors" like a heavy sweat, vomiting, diarrhea, or the onset of menstruation (Siraisi 1990: 135). Sanctorius dealt with the topic in his *Commentary on Hippocrates*, see: Sanctorius 1629a: 195, 263 f., 438–47.

"sieve" for the serum. Hence, only the very watery, fine parts of the blood passed through the narrow pores, while the coarser parts were retained. However, depending on the width of the pores, the quality of the blood, and the strength of the expelling forces, sweat and insensible perspiration could sometimes also contain larger or thicker parts. This understanding of perspiration as a process of "sieving" the blood and of separating the serum also pointed to the danger of a defective skin function resulting in the accumulation of heavier, thicker material that could pollute the body. These residues could lead to the obstruction of the pores or invisible channels through which humors, vapors, and insensible perspiration passed (Stolberg 2012: 504–8).

3.2.9   Perspiratio impedita

Since antiquity, hindered or blocked perspiration (*perspiratio impedita*) had been identified as a major cause of illness and death. Accordingly, the concept of *perspiratio insensibilis* was an important factor in Galen's humoral pathology and therapy. Sanctorius, too, repeatedly warned of the effects of impeded perspiration. In the *De statica medicina* he wrote:

> If nature is hindered in the function of perspiration, it immediately begins to fall short of many things (Sanctorius 1614: 9v).[44]

> The complete hindrance of insensible perspiration, not only of the principal parts, but also of one single lower part, takes away life. With regard to the principal parts, it produces an apoplexy in the brain, palpitation in the heart, polyemia in the liver, suffocation in the womb; in the lower parts it produces a gangrene (Sanctorius 1634: 13r).[45]

Hence, it was crucial that during the third stage of digestion (Sect. 3.2.5), insensible perspiration was properly produced, not only in the organs, but also in the parts of the body. The skin pores needed to be open, as insensible perspiration and sweat provided one of the principal pathways through which morbid matter was evacuated, and prevented harmful substances from accumulating. But there were more things to be considered. In the *De remediorum inventione* Sanctorius criticized the view, "wandering through the schools," that putrefaction caused by hindered perspiration could always be reduced to obstruction. In fact, Sanctorius explained, a contraction of the narrow passages in the body through which the perspirable matter

---

[44] "Natura, dum in perspirandi officio est impedita, incipit statim in multis deficere." See: Sanctorius 1614: 9v.

[45] "Perspiratio insensibilis non solum principum, sed unius partis infimae omnino vetita vitam tollit. Principum dum in cerebro fit apoplexia, in corde palpitatio, in iecore polyaemia, & in utero praefocatio: Infimae partis gangraena." See: Sanctorius 1634: 13r. I refer here not to the first edition of the *De statica medicina*, because Sanctorius added the quoted aphorism, along with further 107 aphorisms, only to later editions of the work. My citation is from the earliest edition of the *De statica medicina* that I could find containing the added aphorisms (ibid.). Whenever I refer to this later edition of the *De statica medicina*, I allude to the added aphorisms unless otherwise indicated.

passed, their compression, coalescence, subsidence, or occlusion, could also impede perspiration. Accordingly, the physician had to know that a pathological tumor, compressing the passages, just as the outside cold contracts the skin, could be the cause of a hindered perspiration in his patient. Hence, he had to be careful in choosing the right remedy, knowing exactly the variety and specifics of the affection (Sanctorius 1629b: 75–81; see also Stolberg 2012: 511–5).

3.2.10 The Doctrine of Sympathy

To make things more complicated, "sympathy" or "consent" between different parts of the body also had to be taken into account. Along traditional Galenic lines, Sanctorius thought that a relationship exists between the organs and their secretions in health and in disease. Thus, harmful substances that accumulated due to defective evacuations could be directed from one part to another, affecting it as well. Sanctorius wrote in the *De statica medicina*:

> No cause more frequently disturbs sleep than a corruption of food: this happens because of the sympathy that exists between the stomach and the brain (Sanctorius 1614: 56v).[46]

In the *Commentary on Galen*, Sanctorius discussed as many as 60 possible sympathies, for example between the spleen and the stomach, the scalp and the neck, or between the septum, intercostal muscles, and nerves. Referring to Hippocrates and Galen, he described four forms of "consent" in his work *De remediorum inventione*. The first consent emerged from the continuous parts by contact, the second from vapors. The third arouse from a transfer of the humors brought about by "insensible channels" (*meatus insensibiles*). The fourth emerged from humors that flowed from one part to another due to the insensible channels of the vessels. In another passage, he warned that if the channels were open (*meatus apertos*), the humors that were discharged would thereby accumulate and cause disease (Sanctorius 1612a: 724–46; 1629b: 37 f., 42).[47]

Without going deeper into the details of the doctrine, the idea of sympathy explains that the fear of blocked pores or channels was connected to the fear that either absorbed miasmas or retained morbid putrescent matter could produce affections, like fevers and inflammations that might be directed to other parts of the body and affect those parts as well. Blocked substances might be transferred to the lungs with a cough and inflammation, to the nose with a catarrh or nose bleed, or to the stomach with a disturbed appetite or vomiting (Renbourn 1960: 138).

---

[46] "Nulla causa saepiùs somnum interturbat quàm ciborum corruptela: id efficit quae est inter stomachum & cerebrum sympathia." See: Sanctorius 1614: 56v.

[47] Sanctorius dealt extensively with the concept of "sympathy" in the second book of *Methodi vitandorum errorum*, see: Sanctorius 1603: 28r–58v, esp. 31r–31v. For an account of the doctrine in Galen's works and its further development, see: Siegel 1968: 360–82.

3.2.11 The Influence of Medical Methodism

In the context of blocked perspiration (*perspiratio impedita*), one repeatedly comes across the doctrine of *strictum et laxum* (tightening or loosening of atoms, corpuscles, pores, or ducts) as a cause of health and disease. It was developed by the medical school of the Methodists, which was founded in the first century BCE by the Greek physician Themison of Laodicea. His successors, Thessalus of Tralles (first century CE) and Soranus of Ephesus (ca. 98–138 CE) refined the doctrine. Methodism was a dominant medical school in Rome for over three hundred years. It represented one of the three ancient sects, together with the dogmatic, or rational, and the empirical sect. At the basis of the methodic doctrine lay the assertion that illnesses were ultimately forms of three different conditions: constriction (*strictum*), laxity (*laxum*), or a mix of these. The Methodists adopted these categories from Asclepiades of Bithynia (124–ca. 40 BCE), who argued that all diseases derived from blockages and flows of corpuscles in the invisible passageways of the body. Thus according to the methodic theory, the states of *strictum* and *laxum* refer to the tightening or loosening of atoms, corpuscles, pores, or ducts. Accordingly, fevers or inflammations were thought to arise from pores being too wide or narrow, blocked by cold air, or by excretions too abundant or too thick to pass through the pores (Renbourn 1960: 136; Webster 2015: 658 ff.).

A lot of information about the methodic sect was passed down by Galen, who, however, was mostly skeptical of their doctrine. Still, he adopted some of their ideas. In Book III of *Hygiene*, he wrote:

> I term in this way [*stegnosis*] damage of the pores due to which the superfluities are prevented from being dispersed. This arises through blockages or constriction (condensation) which people also call occlusion of the pores. Blockage arises from viscid or thick superfluities when they come to be overly collected together in the skin, while constriction arises due to astringents and cooling agents (Galen & Johnston 2018a: 319).

In the sixteenth century, discussions of corpuscular ideas arose in medical circles in Italy, when Girolamo Fracastoro (1470–1553) published his book *De contagione* (On Contagion, 1546).[48] In this work, Fracastoro defined attraction and sympathy, interpreted in quasi-mechanistic and atomistic terms, as a basic phenomenon in nature.[49] In the same decade, the French physician Jean Fernel (1497–1558), whose own concept of the elements was unconventional but not atomist, drew attention to the ideas of Democritus (ca. 460–ca. 370 BCE) and the corpuscularism of the

---

[48] For an overview of late medieval and early modern corpuscular matter theories, see: Lüthy et al. 2001.

[49] Fracastoro treated the issue of contagion as one of a larger class of sympathies and antipathies. In doing so he tried to remove it from the realm of magic, given that contagion was often conceived as an occult force at the time. Contrary to this view and with reference to the atomism of Lucretius, Fracastoro explained sympathy as a mechanical attraction resulting from a flow of particles between objects. According to him, contagion was carried by especially fine *seminaria* or seed particles with the ability to cover great distances and to penetrate the bodies they struck. See: Copenhaver and Schmitt 1992: 305 f.

ancient medical Methodists. He indicated that the number of adherents to the school of the Methodists (which he confused with the atomists) was considerable in his time. Even though Fernel did not follow them, he did not escape their influence, as he sometimes explained phenomena by means of pores. Shortly after, there was an actual revival of Empedoclean corpuscular theory among physicians. At the beginning of the seventeenth century, in the very year Sanctorius began teaching *theoria* in Padua, a colleague of his, Prospero Alpini (1553–1617), a lecturer on simples, and prefect of the botanical garden in Padua, published his treatise *De medicina methodica* (1611).[50] Interestingly, Alpini was friends with a family to which Sanctorius, too, had a special connection: the Venetian Morosini.[51] Hence, the two of them might have encountered each other in the *Ridotto Morosini* (Sect. 2.3), potentially discussing ideas connected to Alpini's publication. Directed to the Methodists, it marked a comeback of the ancient solidist and therefore anti-humoralistic and anti-Hippocratic medicine. It must be noted, though, that besides the Methodist doctrine, Alpini always referred to the Hippocratic-Galenic conceptions as well, and aimed for a certain reconcilability that he finally achieved (Hooykaas 1949: 74; Rothschuh 1978: 227; Siraisi 1987: 242; 2012: 513; Sanctorius and Ongaro 2001: 40; Garber 2006: 33 f.). Sanctorius was well aware of Alpini's treatise and referred to the *De medicina methodica* in his *Commentary on Galen*. In the first part of the commentary, when Sanctorius wrote about the medical sects, he stated:

> Lately, Prosperus Alpinus published a most sophisticated book about this sect, in which the principles of this sect are most completely declared (Sanctorius 1612a: 52).[52]

With "this sect," Sanctorius alluded to the *thessalici*, meaning the disciples of Thessalus and hence to the Methodic sect. E. T. Renbourn has argued in his article that Sanctorius's concept of the *perspiratio insensibilis* was also influenced by ideas attributed to the medical Methodists. He went as far as to identify Sanctorius's medical doctrine as "New Methodism" and a "resuscitated Methodic doctrine" (Renbourn 1960: 139, 142). In the *Commentary on Galen*, Sanctorius wrote:

> Thessalus reduces the whole art to laxity [*laxum*], and constriction [*densum*], or mixed, and from these he collects three remedies, which he says are sufficient to remove all pathological affections. But they are most vain [statements] and those, who would like to penetrate

---

[50] The lectures on simples (*lettura dei semplici*) were part of the teaching of the medical faculty at Italian universities from the sixteenth century onward. This teaching was an intermediate between what we call today botany, pharmacognosy, and pharmacology. See: Treccani enciclopedia on line 2019.

[51] Alpini dedicated his work *De medicina aegyptiorum* to Antonio Morosini (Alpini 1591: dedication). A collection of copied letters by Andrea Morosini, containing many letters to Prospero Alpini, can be found in the library of the Museo Correr in Venice, see: BMCVe-a.

[52] "Edidit his diebus de hac secta Prosperus Alpinus librum eruditissimum, ibique huius sectae principia plenissimè declarantur." See: Sanctorius 1612a: 52.

the vanities of the grand Thessalus, shall read Books 1 & 2 of *Methodus medendi*, where Galen carefully refutes him (Sanctorius 1612b: 318).[53]

This statement makes clear that Sanctorius did anything but identify with the methodic doctrine. However, just as in the case of Galen, this does not mean that he did not adopt some of their ideas. From the previous analysis it has become clear that Sanctorius shared the idea that the tightening or loosening of pores or ducts was a cause of health and disease. Moreover, in his view, the observation and regulation of bodily evacuations contributed more to the preservation of health and cure of diseases than any other means. But Sanctorius strongly adhered to Hippocratic-Galenic conceptions and integrated his new finding, the large quantity of insensible perspiration, into this theoretical framework. Only very few references to corpuscles, atoms, and small parts, or particles (*minima partes, minimae particulae*) can be found in his works.[54] The *De statica medicina*, his major publication on the causes and effects of *perspiratio insensibilis*, does not contain a single mention of corpuscles, particles, and the like. Hence, to classify Sanctorius's medical doctrine as "New Methodism" is far-fetched.[55]

3.3   The Non-Naturals Reconsidered

In the preceding account of *perspiratio insensibilis*, one repeatedly comes across the non-natural factors. This is consequential, as the subject was dealt with through the perspective of Sanctorius, who was original in examining the non-naturals with regard to their effect on insensible perspiration. When considering that, in Galenic

---

[53] "Thessalus totam artem ad laxum, & densum, vel mixtum ex his referebat, & ex his colligebat tantum tria remedia, qua dicebat sufficere pro auferendis universis affectibus praeter naturam: Caeterum haec vanissima sunt, & qui vult vanitatem Thessali altius penetrare, legat lib. 1. & 2. methodi medendi, ubi Galenus exactissime illum conuincit." See: Sanctorius 1612b: 318.

[54] Sanctorius used the term *corpusculum* in his works only when he discussed the doctrine of Asclepiades, see ibid.: 36, 399, Sanctorius 1629b: 9. With regard to his employment of the term *atomus*, there is only one passage in his works, in which Sanctorius did not connect it to a refutation of ancient atomism, but used it in a discussion on the transparency of the air: "Nec mirum est, quod vitreus ex se transparens in oculi profunditate gerat vicem cubiculi umbrosi: quia etiam aer ipse in sua immensitate diminuit transparentiam, vel fiat hoc propter atomos, vel propter alias causas." See: Sanctorius 1625: 762 and Bigotti 2017: 10, fn. 19. Sanctorius used more often the terms *particula minima, pars minima,* or simply *minima* (Sanctorius 1603: 158v, 218v, Sanctorius 1625: 167–70, 176, 186, 218, 385, 426, 430, 455, 466, 472, 476, 561, 690 f., 728).

[55] The same applies to Fabrizio Bigotti's argument that Sanctorius adopted a corpuscular theory that pre-empts both Galileo Galilei and Daniel Sennert (1572–1637) (Bigotti 2017). There is no hint that Sanctorius connected his quantitative approach to corpuscular ideas or that such influenced his static medicine and his new approach to the six non-natural things. Nowhere in his works did Sanctorius connect his concept of *perspiratio insensibilis* to corpuscular notions and the *De statica medicina* shows no trace of corpuscular ideas. In 1975, Paolo Farina also argued against a corpuscular theory of Sanctorius, highlighting instead his strong adherence to Galenic medicine (Farina 1975: esp. 369–74, 377).

dietetics, evacuations were closely connected to the processes of digestion and respiration, i.e., to air, food, and drinks, and were influenced also by the motion or rest of the body, the examination of insensible perspiration in the context of the six non-naturals does not come much as a surprise. What is more, according to the traditional list, "evacuation and repletion" constituted a non-natural thing itself. As Ken Albala has explained, the details of evacuation were usually not included in treatises on the digestive process, but considered rather in the frame of the doctrine of the six non-natural things, as a process believed to be controllable by external factors (Albala 2002: 60). From this, it is easy to understand why Sanctorius chose this of all concepts to structure the results of his weighing experiments. And from this, it is also clear that evacuations had a distinct place in the medical literature, even in the time before the *De statica medicina* was published.

In his study of sweat, Stolberg argued that early modern medical literature dealt considerably with sweat, but not especially prominently (Stolberg 2012: 504). To accurately identify the role that *perspiratio insensibilis* played in early modern medicine, especially in the times before and contemporary to Sanctorius, further research is needed. Based on Stolberg's study, it can be assumed that the phenomenon was usually treated within the context of sweat, or, as the analysis of the doctrine of the six non-naturals has shown, in the context of the fifth non-natural factor "evacuation and repletion." Moreover, a treatise on sweat, interestingly published by a Neapolitan physician in the same year as the *De statica medicina*, gives some insight into the topicality of *perspiratio insensibilis* at this time (Baricellus 1614).

Without deeply analyzing the more than 400-page work, a look into the table of contents shows, on the one hand, that the non-natural things, like motion, food, and environmental factors, were discussed, and on the other hand, that insensible perspiration—*transpiratio insensibilis* appeared only twice. The author, Julius Caesar Baricellus, explained in the preface that he set out to write a treatise on sweat, because physicians did write nothing or only deceiving and unnecessary things on the topic. But when carrying out the task, he soon discovered that many of the wisest men dealt with *sudorific matter* and much more was contributed to the topic than he had expected. This is very much in line with Stolberg's assessment of the non-prominent place of sweat in early modern medical writing. At the beginning of the fourth book, Baricellus wrote that the physicians of his time rarely used sweats in their treatments, while this had been commonly done in antiquity. This implies that he was not acquainted with Sanctorius's weighing procedures. The fact that Baricellus mentioned Sanctorius's work *Methodi vitandorum errorum* in the context of the mixtures of the humors and of tastes reinforces this impression. As the *De statica medicina* was published in the same year as Baricellus's book on sweat, it can be assumed that the two works were conceived independently of each other (Baricellus 1614: index, 2 f., 165, 358). Hence, apparently there was a general awareness at the time of the important effects that sweat and perspiration had on health and disease. By focusing on insensible perspiration specifically and, even more so, by adding a quantitative dimension to the study of the phenomenon, Sanctorius gave new relevance to the topic.

The next paragraphs consider these new approaches from a conceptual point of view. The focus is on the way in which Sanctorius connected the traditional doctrine of the non-natural things with insensible perspiration. Rather than fully analyzing the content of the *De statica medicina*, I will focus on those factors that are relevant to understanding the conceptual backdrop against which Sanctorius developed his weighing procedures. References to Sanctorius's other books will help set his ideas in the larger context of his endeavors. In doing so, I seek to contribute to an understanding of the theoretical context in which the *De statica medicina* emerged, and of how a new medical idea, the quantification of insensible perspiration, was integrated into a well-established Galenic doctrine. In my opinion, this understanding is fundamental to comprehending the practical and material dimensions of static medicine, which are inextricably interwoven with theoretical medical knowledge. Therefore, before analyzing the weighing of *perspiratio insensibilis*, I will pay attention to this hitherto neglected aspect of Sanctorius's work: his reinterpretation of the six non-natural factors.[56]

3.3.1 Air and Water

After the first section of the *De statica medicina,* which deals with the weighing of insensible perspiration and will be considered in later chapters, Sanctorius proceeded to the non-natural pair, air and water (Fig. 3.1). It is striking that these two non-natural factors represented also two of the four elements and, as such, not only formed part of the complexion of human bodies, but were the unifying explanatory model for all nature. However, insofar as air and water changed the body, preserved health, and led to diseases, they were compiled among the non-natural things, as Sanctorius explained in his *Commentary on Avicenna*. While air was the common first thing of the non-naturals, water was usually not contained in the list. This can be explained by the fact that the traditional concept of "air" included considerations of pollution, seasonal variations, climate, and region—and might be described in modern terms as "environment." In pointing explicitly to water, Sanctorius might have been inspired by the *Isagoge Johannitii*, a standard introductory textbook at medical university faculties, where the list of the non-naturals included also "bath" as a special category (Sanctorius 1625: 70).[57]

Air In the account of the respiration process (Sect. 3.2.6), the importance of air for the human body has already become clear. Indeed, medieval and Renaissance dieticians agreed that it is the most important particular factor among the six non-naturals. In his *Commentary on Galen* Sanctorius too concluded that, compared to the other non-natural things, air changed the body the most (Sanctorius 1612b: 58).

---

[56] In the following sub-chapters, my references to the traditional Galenic teaching of the six non-natural things are mainly based on Sotres 1998 and Albala 2002.

[57] For more information on the *Isagoge Johannitii*, see: Temkin 1973: 104–8, Ottosson 1984: 254.

Hence, in the *De statica medicina* he showed how the quality of the air influenced body weight and the amount of *perspiratio insensibilis*:

> The external air, which passes through the arteries into the depths of the body, may render the body lighter and heavier. Lighter, if it is fine and warm; heavier, if it is thick and moist (Sanctorius 1614: 20v).[58]

> External cold, by concentrating heat, renders the nature so much more robust, that it can carry, in addition to the usual weight, also around two pounds of repressed perspirable matter (Sanctorius 1614: 30v).[59]

The first citation shows the different effects of fine and warm air in contrast to thick and moist air. Just as Sanctorius described liquid excretions as heavier than solid ones, and liquid food as heavier than solid food elsewhere in the *De statica medicina*, he thought that moist air made the body heavier. Therefore, dry weather was healthier than continuous rain, making the bodies lighter. The second citation might explain why balanced bodies were, according to Sanctorius, around three pounds lighter in summer than in winter. Due to cold air, the skin contracted and correspondingly the pores narrowed, which made it difficult for the perspirable matter to leave the body. In summer, on the contrary, when the pores widened because of the warm air, perspirable matter could be excreted more easily. However, cold air did not only affect the pores of the skin, but also the internal heat of the body. As has been explained, inspired (inhaled) air served to cool the heart and the blood and hence, the colder the air that entered the body, the more it concentrated the latter's heat. The concentration of heat was directly related to the robustness of a body, and Sanctorius stated that external coldness impeded perspiration in weak people, because it dissipated their heat. In robust people, on the contrary, cold air increased perspiration, as their heat was drawn back deep into the body, where it doubled, the nature of the body strengthened, and shortly after, the weight of the repressed perspirable matter was consumed and the body became and felt lighter. This is why insensible perspiration was, in robust bodies, greater in winter than in summer (Sanctorius 1614: 6r–6v, 16r–16v, 24v–25r; 1629a: 382).

These examples illustrate that the qualities of the air could affect the body not only *per se*, insofar as warm air heated the body, but also indirectly (*per accidens*) when for example the humidity of the air obstructed insensible perspiration and generated putrefaction. By the same token, warm air sometimes dissipated internal heat and cooled the body, and cold air sometimes warmed the body by concentrating or compressing heat (Sanctorius 1603: 9r; 1612b: 25, 27). This feature applied to the other non-natural things, too.

Changes in the Air   In his *Commentary on Galen*, Sanctorius named three causes that changed the air: region, time of the year, and the constitution of the heavens. With regard to the latter, Sanctorius was very critical of astrology and thus allowed

---

[58] "Aer externus per arterias in profundum corporis penetrans, potest reddere corpus levius, & gravius: levius si tenuis, & calidus; gravius, si crassus, & humidus sit." See: Sanctorius 1614: 20v.

[59] "Externum frigus concentrando calorem reddit naturam tantò robustiorem, quanto ultra solitum pondus ferre quoque possit duas libras circiter retenti perspirabilis." See: ibid.: 30v.

air to serve as a medium only for the celestial influences of light and movement. In the *Commentary on Avicenna*, he explained that air contained so much light that it always dried, even though its nature was actually exceedingly humid. It has already been indicated above how the seasons and the weather affected insensible perspiration and body weight. Accordingly, perspiration decreased daily by around one pound from the autumnal equinox to winter solstice, whereupon it began to increase until the vernal equinox. Time of day also played its part. Robust bodies perspired more during the day in the summer, while in winter they perspired more during the night. In Galenic medicine, each season corresponded to a complexion. Spring air was hot and moist, and blood dominated. Summer was connected to yellow bile, with the air being hot and dry. Fall was the season of black bile, with the predominating qualities of dry and cold. In winter, cold and moist phlegm prevailed (Fig. 3.2) (Sanctorius 1603: 137v; 1612b: 25; 1614: 27r, 28v; 1625: 61).

According to Sanctorius, all philosophers and physicians agreed that spring air was the most temperate of all the seasons, but they argued about the most temperate climate or region. This discussion was, however, fruitless, because there was not one absolute temperate climate; rather, each climate had its own temperate climate, depending on the complexion of its inhabitants. Thus, there was the idea that a population adapted to the region and climate it lived in. The complexion of someone living in the German territories was totally different from the complexion of someone living on the African continent. In this context, geographical differences like proximity to the sea or the mountains were important, as they influenced the quality of the air. They also affected the movement of the air and the wind a population was exposed to. And so Sanctorius explained that windy air might harm one person, but benefit another. Wind blowing from the north was healthy, whereas south wind was harmful. With regard to insensible perspiration, he concluded that wind which was colder than the skin always blocked and harmed, especially the head, because this body part was the most exposed (Sanctorius 1603: 137v, 138v; 1612b: 62; 1614: 24v; 1625: 212, 225, 245 f.). This hints at the importance of clothing. As a means to protect oneself from the immediate environment, from bad weather, from heat, and from cold, clothes also affected the excretion of insensible perspiration:

> Because of cold air that follows on heat, those who take off clothes usually perspire in the course of one day about less than two pounds without noticing any trouble (Sanctorius 1614: 22v).[60]

> Those body parts which are covered perspire healthier. But if they are discovered bare after sleep, their pores are compressed by very warm air, too (Sanctorius 1634: 28v).[61]

Hence, clothing promoted insensible perspiration during the day and night, which was why Sanctorius suggested covering the body also during sleep.

---

[60] "Ob aerem frigidum supervenientem calori, vestibus denudatus, minùs duabus libris circiter diei cursu perspirare solet, nulla ab ipso animadversa molestia." See: ibid.: 22v.

[61] "Corporis partes tectae salubriter perspirant: Si verò à somno detectae inveniantur, etiam ab aere calidissimo eorum pori condensantur." See: Sanctorius 1634: 28v.

Pure Air and Plague Another important aspect was the idea of pure or clean air. In line with the teachings of Avicenna, Sanctorius conceived of air as pure when it was not mixed with any extraneous vapors, any type of smoke, or any harmful substance. Accordingly, muddy air impeded insensible perspiration and the retained matter harmed the body. Therefore, country air was preferable to thick city air.[62] Being exposed to impure air could have serious consequences, as plague and other diseases were thought to stem from miasmas or foul vapors contaminating the air (Sanctorius 1612b: 27; 1614: 16r, 22r; 1625: 64; 1634: 28v). Interestingly, Sanctorius added to the 1634 edition of the *De statica medicina* a subsection with the title *De peste* (On plague) (Sanctorius 1634: 17v–20r).[63] Published 3 years after the plague raged in Venice, this most likely reflects Sanctorius's experiences during the epidemic (Sect. 2.7). Contrary to what one might expect, the added aphorisms were not included in the discussion of air, but were printed as a sort of appendix to the first section that deals with the weighing of insensible perspiration. Even more curious is that none of the 15 aphorisms relates to either weighing or insensible perspiration. Instead, there is a clear connection to air:

> The plague is conveyed not by contact, but by inhalation of pestilential air or by the vapor of furniture. It happens like this: the vital spirit is infected by the air, from the infected spirit the blood coagulates, which, pushed to the external parts, produces plague spots [*carbones*], black papules, and buboes; if it remains inside, it brings about death; if everything is thrust out, we survive (Sanctorius 1634: 18r).[64]

Thus, Sanctorius opposed the view that plague was a contagious disease.[65] According to him, plague resulted from bad air that first corrupted the vital spirit, then the blood. People with a *loose* lung were more prone to infection than people with a *compact* lung, and wind was a means of spreading the pestilential rays, which were similar to light rays (Sanctorius 1634: 19r). The closest one comes to any idea of

---

[62] In his analysis of *Regimens of Health*, Pedro Gil Sotres argued that medieval physicians considered the city a far more hygienic place to live than the unhealthy countryside (Sotres 1998: 302 f.). Ken Albala drew a different picture in his study on Renaissance dietary works and referred to practices of purifying city air (Albala 2002: 116–20). Hence, the conception of the beneficial and harmful effects of city and country air seems to have changed during this period.

[63] The following statements are partly based on a talk given by Vivian Nutton at the international conference *Humours, Mixtures, Corpuscles* and the ensuing discussion (Nutton 2017).

[64] "Peste non tactu, sed inspiratu aeris pestiferi, vel halitus supelectilium inficimur: Sic fit: spiritus vitalis ex aere inficitur ab infecto spiritu congelatur sanguis, qui extrapulsus carbones, nigras papulas, & Bubones, si manet intus, mortem: si totus pellitur ad extra, evadimus." See: Sanctorius 1634: 18r.

[65] In his study on the perceptions and reactions of university medical practitioners with regard to the Black Death, Jon Arrizabalaga argued that air spread and contagion had not been contradictory views of the diffusion of pestilence, but rather referred to two different and successive stages of its dissemination (Arrizabalaga 1994: 259 f., 287). In this context, it is interesting that, even though Sanctorius denied that plague was transmitted by contact, in another of the plague aphorisms he blamed the authorities for not shutting down the poultry market, as handling of chickens by infected persons transmitted the disease (Sanctorius 1634: 19v–20r).

quantification, or rather mechanization, is Sanctorius's comparison of the course of the plague with the movement of a clock:

> Things that are infected with plague corrupt as long as the remote causes persist. If, however, only one of the causes is missing, the venom diminishes, like the movement of a clock, which stops when a single tooth of the cogwheel breaks down (Sanctorius 1634: 17v–18r).[66]

Without reading too much into the analogy that Sanctorius employed elsewhere in his works to the body and its physiology (Chap. 8), it further confuses matters. And even more so, when Sanctorius wrote in another aphorism that there was no cure for the plague; one could only evade it (Sanctorius 1634: 19v). This remarkably pessimistic attitude leaves one to wonder whether Sanctorius's plague aphorisms reflect his devastating experiences and the general sense of helplessness and resigned fatalism that crept into the Venetians in view of the epidemic, or whether they are related to the fact that Sanctorius was already an old man, sensing probably that he was facing the end of his life, or whether they truly reveal his notions of the plague.[67] One should not forget here that Sanctorius was among those physicians, who persistently denied the existence of the plague in Venice, until the many deaths proved them to be wrong (Sect. 2.7). Maybe for this reason he did not want to remain silent on the topic later. But why then, one is inclined to ask, did he choose the *De statica medicina* to present his thoughts on the disease? Given the fact that he did not refer to *perspiratio insensibilis* or the weighing procedures, this choice seems rather peculiar. It might have been practical reasons, the aphoristic form, or the popularity of the *De statica medicina*, that led him to this decision. This, however, is pure speculation. In fact, the consideration of plague in the frame of the six non-natural things was fairly common. In the fourteenth century, in line with the medieval rules of health (Sect. 4.1.2), even a new literary genre was created—that of the plague *regimina*.[68] Nevertheless, Sanctorius did not add his plague aphorisms to the section on air, nor did he consider the disease in the context of the other non-natural things. Thus, his true motives behind the plague aphorisms remain a puzzle yet to be solved.

Water In the preceding section, it has already become clear how water, insofar as it affected the climate, had an important influence on the complexion of the body and the excretion of insensible perspiration. Lakes, rivers, and the sea determined the quality of the air and shaped the weather in different regions. But water could also directly act on the body through baths and swimming. In the *De statica medicina* Sanctorius explained that hot baths, just as hot air, promoted perspiration. But he also warned that perspiration provoked by the force of hot air or water was harmful, except when its damage was compensated by a much greater benefit. Relating

---

[66] "Res peste infectae inficiunt quousque durant proximae, & remotae causae: unica tamen deficiente ceßat virus ad instar motus horologij, dum rotarum unico irrito dente quiescit." See: Sanctorius 1634: 17v–18r.

[67] For more information on the Venetian plague of 1630–31 and the trauma of the Venetian population, see: Preto 1984: 379, 384. See also Sect. 2.7.

[68] For more information on plague *regimina*, see: Zitelli and Palmer 1979: 24–37, García-Ballester 1992: 120, Arrizabalaga 1994: 273.

to the non-natural pair of exercise and rest, he advised against a swim in cold water after a violent exercise, for this was "most pleasant, but lethal," because these were two opposed movements (Sanctorius 1614: 20r–20v, 23r–23v, 27r). Overall, Sanctorius treated water only very briefly in the second section of the *De statica medicina*, while he dealt more extensively with air. In addition to the possible influence of the *Isagoge Johannitii*, the close connection between water and air with regard to climate and region, as well as Sanctorius's interest in the weight not only of air, but also of water, which will be considered later (Sect. 5.3.2), might have made him include water in the list of the non-natural things, pairing it with air.

In conclusion, Sanctorius followed traditional Galenic notions in his account of air and water, but reconsidered their influence on the human body by focusing on body weight and insensible perspiration. The fact that he paired water with the first non-natural thing, air, is unusual and indicates that Sanctorius considered it important with regard to the quantity of *perspiratio insensibilis*.

3.3.2   Food and Drink

As a product of the digestive process, insensible perspiration was necessarily closely connected to the food and drink ingested by the body. The quantity of *perspiratio insensibilis* depended on the digestive power, which therefore had to be taken into account when prescribing food and drink. This power in turn was determined by an individual's innate heat—the hotter it was, the greater the power to concoct and digest more food (Sanctorius 1629a: 300). Referring to Galen, Sanctorius wrote: "nourishment must be proportionate to difflation" (Sanctorius 1629a: 382) and in the *De statica medicina* he advised: "One should only ingest such a quantity of food that nature can concoct, digest, and perspire" (Sanctorius 1614: 39r).[69] Hence, the more a body perspired, the more food it needed. But it was not as simple as that. In keeping with the doctrine of the non-natural things, multiple factors continuously influenced the body and its digestive power.

Meals and Mealtimes In summer, for example, when the stomach was thought to be colder and the forces weaker, one should not eat an abundant meal, but rather several sparse meals. In winter, on the contrary, when the stomach was thought to be hotter and the forces stronger, bigger but few meals were recommended. The same applied to regions. In hot regions, one should eat little and often, whereas in cold regions, one should eat a lot, but rarely. According to the Hippocratic teachings, the four seasons corresponded to the four regions, to the four ages of man, to the four complexions, to the four humors, to the four elements, and to the four times of day, which was why, so Sanctorius, it was enough for the physician to know what

---

[69] "… alimentum debet proportionari difflationi." See: Sanctorius 1629a: 382. "Illa cibi copia est ingerenda, quam natura potest coquere, digerere, & perspirare." See: Sanctorius 1614: 39r.

arrangement of meals applied to one of these factors, because from this, he could infer what he had to recommend with regard to the others (Fig. 3.2) (Sanctorius 1629a: 382, 410, 423).

Generally, Sanctorius thought that several meals a day were healthier than only one. A body, for example, was weighted down more by eight pounds of food that were eaten in one meal, than by ten pounds that were eaten in three separate meals. And as the fullness of the stomach took the insensible evacuation away, a meal of around four pounds was harmful when it was taken all at once, but beneficial when divided into two or three meals. Moreover, the amount of food for each meal had to be equally divided. Instead of eating six pounds at lunch (*prandium*) and two pounds at supper (*caena*), it was healthier to eat four pounds at each, both lunch and supper. This, however, contradicts Sanctorius's statement in the *Commentary on Galen*, according to which one should eat more for supper than for lunch. Referring to the ongoing discussions on the topic among medieval and Renaissance dieticians, Sanctorius adopted the Galenic position that the digestive power was stronger at night, which was why a larger supper was recommended.[70] This might imply that his experiences with the weighing chair made him change his mind. Sanctorius was thus ready to revise traditional knowledge on the basis of his novel quantitative observations. But only to a certain extent, as will be shown later (Sanctorius 1612b: 76 ff.; 1614: 38v, 40r; 1634: 40r).

In the context of the number and size of meals, mealtimes were important, as one had to make sure that the previous meal was already digested before taking in new food and drink. Perspiration occurred least when the stomach was full, Sanctorius explained. The moment to eat, according to him, was when the body, shortly before ingesting the first food of the day, had returned to the same healthy weight as the previous day. In the *Commentary on Galen*, he recommended 9 to 10 h between lunch and supper and, correspondingly, 14 to 15 h between supper and lunch. It is important to remember in this context that the digestive process was conceived of as proceeding in distinct stages. This means, knowing how long each of them took also disclosed the ideal time to eat. By continuously weighing the body and monitoring its excretions, Sanctorius attempted to gain exactly this knowledge (Sanctorius 1612b: 78; 1614: 41v, 46v–47r; 1634: 42r).

Quality of Food Besides the questions of what time to eat and how much, one wonders, of course: What to eat? According to Sanctorius, food could be nutritive, abundant, raw, vaporous, scarce, fat, dry, liquid, or solid, and most importantly, it might also have the ability to perspire. If not, obstructions, corruptions, lassitude, sadness, and heaviness of the body would loom (Sanctorius 1614: 32r–48r; 1634:

---

[70] The opinion that the digestive power was stronger during night was connected to the differentiation between digestion and concoction. While digestion was thought to occur during waking, concoction was thought to occur during sleep. See also Sects. 3.2.5, fn. 31 and 3.2.7, fn. 40. For more information on the controversy whether the midday meal or evening meal should be larger, see: Albala 2002: 112 f.

39v–42r). But fortunately, some foodstuffs could enhance the perspirability of others:

> Onions, garlic, mutton, pheasants, but most of all, *succus cyrenaicus* help the perspiration of hardly perspirable foodstuffs (Sanctorius 1614: 48r).[71]

Unfortunately, this worked also the other way around:

> The use of pork and of porcini is bad, both because they do not perspire, and because they do not permit that the other foodstuffs, ingested at the same time, perspire (Sanctorius 1614: 36r).[72]

During Sanctorius's times, mushrooms were commonly thought to be unhealthy, and pork was usually recognized as difficult to digest. It is thus not surprising that Sanctorius categorized them as bad also with regard to their effect on insensible perspiration. When Sanctorius explained in another passage of the *De statica medicina* that melons perspired poorly and therefore repressed perspiration, he also followed the fashion of his time, as Renaissance dieticians never tired of launching into tirades against melons. With regard to onions, garlic, mutton, and pheasants, the picture is, however, different. All of them were usually not considered healthy by the Italian dieticians of the time. But in the *Methodi vitandorum errorum*, Sanctorius hinted at his sources, when he wrote that Galen had counted garlic among warm and dry things that dissipate flatus. Moreover, in the work *De alimentorum facultatibus* (On the Properties of Foodstuffs), Galen wrote that garlic had the power to digest and to open obstructions. Notwithstanding that Sanctorius did not cite this exact passage, he frequently referred to the treatise in his books (e.g., Sanctorius 1603: 169v; 1612a: 196, 513). From this, it can be inferred that Galen's ideas on food and diet probably played a part in Sanctorius's qualification of nutrition according to its ability to perspire (Sanctorius 1603: 137r; 1614: 36r–36v; Galen. 1997a: 658 f.).

In this context, Sanctorius's mention of *succus cyrenaicus* deserves brief consideration. It refers to the resin of silphium, a plant that was an important commodity of the ancient North African city of Cyrene. In fact, it became a legendary spice, praised by many Greek and Roman physicians for its digestive qualities. However, in the first century it disappeared. Why? The reasons are uncertain (Dalby 2000: 17 ff.). What is certain is that Sanctorius did not have the spice at his disposal. Instead of being based on his own experiences with the weighing chair, Sanctorius's praise of silphium thus seems to rather depend on the well-known Greek physician again. Galen wrote in one of his books: "Indeed, *succus cyrenaicus* surpasses all [simple drugs] in heat and in fineness and therefore also dissipates the most through vapor, ..." (Galen. 1997c: 90 f.).[73] Hence, it was Galen, who already pointed to the perspirability of silphium. This confirms the suspicion mentioned above that Galen

---

[71] "Caepae, allium, caro vervecina, phasiani, sed maximè omnium succus cyrenaicus iuvant perspirationem eduliorum aegrè perspirabilium." See: Sanctorius 1614: 48r.

[72] "Usus carnis suillae, & boletorum malus, tum quia haec non perspirant, tum quia non permittunt caetera edulia simul ingesta perspirare." See: ibid.: 36r.

[73] "Succus Cyrenaicus quidem omnes et caliditate et tenuitate exuperat, ac proinde etiam omnium maxime per halitum discutit, ..." See: Galen. 1997c: 90 f.

provided the basis not only for the quoted aphorism, but also, on a more general level, for Sanctorius's qualification of foodstuffs with regard to their effect on insensible perspiration.[74] The familiarity of Sanctorius with Galen's respective treatise, *De simplicium medicamentorum temperamentis et facultatibus* (On the Mixtures and Powers of Simple Drugs), is confirmed by the frequent citations that can be found in Sanctorius's publications (e.g., Sanctorius 1612a: 494; 1612b: 397; 1629a: 68; 1629b: 70).

Another interesting aspect in this regard is that Renaissance dieticians considered the nutritive value of food, amongst other things, with regard to the proportion of the food expelled as excrement. The material that passed through had obviously not been assimilated and foods that produced abundant excrement were therefore considered less nutritious. If they were thoroughly processed, they left behind only a small amount of waste. Thus, in a certain way, the quality of food was here too connected to its perspirability. Following this line of argument, Sanctorius recommended food that was little nutritious (*pauci nutrimenti*), as to him it was fundamental that a body perspired sufficiently. As peculiar as this might seem to the modern eye, it was not unusual for contemporary physicians to not always recommend the most nutritious substances. What they feared most was an overburdening of the system and there was no equivalent as yet to the modern concept of minimum daily requirements or even of energy supplied by nutrients. Moreover, the nutritive value was not the most important criterion for a choosing foods (Sanctorius 1614: 32v, 40v, 41r, 42v).

In addition to general recommendations for certain foodstuffs, Sanctorius also took account of the fact that individual bodies reacted differently to nutrition, depending on their complexion. "Honey," he wrote, "is good for cold bodies, because it nourishes them and perspires, whereas it is harmful for warm bodies, as in them it is turned into bile" (Sanctorius 1634: 42r).[75] By the same token, fasting was not for everyone. In the *Commentary on Hippocrates*, Sanctorius reminded the reader that a distinction had to be made with regard to time, age, region, and habits to decide whether a person should, or should not do fasting.[76] In the *De statica medicina*, he emphasized the importance of weighing in order to decide whether fasting is healthy or not. Following the important premise mentioned above, according to which nourishment must be proportionate to perspiration, fasting was only beneficial, if there was still food left in the stomach to digest from the previous day. Being careful to eat the right food with the qualities that matched the individual needs of a body, and to abstain from eating under certain circumstances, was, however, not enough. A too great a variety of food also needed to be avoided, as

---

[74] Thomas Secker (1693–1768), archbishop of Canterbury from 1758 until his death, critically mentioned Sanctorius's adherence to Galen with regard to *succus cyrenaicus* in his medical doctoral dissertation, see: Secker 1721: 10.

[75] "In frigido corpore mel iuvat, quia nutrit, & perspirat, in calido nocet quia bilescit." See: Sanctorius 1634: 42r.

[76] In a later part of the *Commentary on Hippocrates*, Sanctorius discussed the suitability of bodies to fasting at length in a separate question (*quaestio*), see: Sanctorius 1629a: 293 ff.

Renaissance physicians commonly agreed. Sanctorius, too, warned of three harms that resulted from the variety of food: eating too much, concocting too little, and perspiring too little (Sanctorius 1614: 37v, 41v; 1629a: 102).

There is much more to say about Sanctorius's ideas on nourishment and eating. It is a very important topic in the *Commentary on Hippocrates*, in which he considered nutrition with regard to sick bodies, too (Sanctorius 1629a: e.g., 100–3). In the *De statica medicina*, in contrast, the focus is on prevention and the preservation of health. While food and drink played an important role in Sanctorius's quantitative study of insensible perspiration, this close connection had a less prominent place in his other works. Even though Sanctorius repeatedly pointed to his static observations, there are also passages in which he examined the influence of nourishment on the body detached from considerations of insensible perspiration (e.g., Sanctorius 1629a: 183–6). How this may relate to the importance of food and drink in Sanctorius's quantitative approach to physiology will be elucidated in later chapters. Here, I have limited myself to analyzing the topic in the context of the doctrine of the non-natural things and Sanctorius's revision of it.

3.3.3 Sleep and Wake

The relevance of sleep and wake with regard to the digestive process and to the production of *perspiratio insensibilis* was already mentioned above (Sects. 3.2.7 and 3.3.2, fn. 70). The body concocted during sleep and digested during wake. The digestive power was thought to be stronger at night and Sanctorius repeatedly stated that the body perspired insensibly twice as much while asleep as while awake (Sanctorius 1612b: 40; 1614: 52r). In this context, Sanctorius also referred to two different kinds of *transpiratio insensibilis*—one was generated during sleep, the other during wake. This differentiation was connected to the rest or the movement of the body. In order to gain a better understanding of these ideas, I will next present Sanctorius's physiological understanding of sleep and waking hours, and refer to some characteristics that I consider important with regard to his medical doctrine of static medicine.

The Physiological Concept of Sleep In his descriptions of the physiological processes that occurred during sleep and wake, Sanctorius followed Galen, contrasting his views with those of Aristoteles. All evidence put together, Sanctorius's explanation of sleep can be summarized as follows: sleep arose when the heat, dispersed over the sensory organs (*sensiteria*), withdrew into the inner parts of the body. Due to the activity of the sensory organs during waking hours, this heat was dried up and exhausted and needed to be moistened and restored. When it drew back into the body, the influent heat merged with the innate heat of the internal organs, which was why the overall heat within the body doubled. As a result, the digestive power increased and enabled the body to transmute substances, i.e., food into *chyle* and *chyle* into blood. In short, the doubled heat allowed the body to perform concoction.

In this process, the heart was heated and the brain moistened. The moisture in the brain prevented the animal spirits from flowing to the sensory organs. Moist vapors filled up the brain and the animal spirits were overwhelmed by these vapors. As the animal spirits were the active agent in all the sensory organs, spreading through the nerves and the spine, their retention led not only to their repose, but also to the repose of the five external senses. What is more, the vital spirits, from which the animal spirits arose, rested as well. Thus, sleep was needed in order to perform concoction, to restore heat, and to regenerate the spirits. As soon as this was accomplished, heat returned to the outside of the body and the animal spirits continued to flow again, thereby provoking sensation and voluntary motion, and the body woke up (Sanctorius 1612a: 358 ff., 364; 1612b: 39, 81; 1625: 362, 726; 1629a: 305).[77]

So far, so clear. Yet there was a problem when Sanctorius connected this conception of sleep and wake with insensible perspiration and the observations he made during his weighing procedures. According to his measurements, the insensible perspiration of the body was twice as great in sleep as during waking hours (Sanctorius 1612b: 40; 1614: 48v–49r, 52r; 1625: 68) (Sect. 3.2.7, fn. 40). But how can this be possible given that concoction occurred during sleep, and digestion during waking hours? According to Sanctorius's differentiation, concoction described the transmutation of substance and referred to the first two stages in the digestive process. Digestion, on the contrary, was, according to him, the transmission of nutrition; and it referred to the third and final step in the digestive process, the step during which *perspiratio insensibilis* was excreted. Why then did the body perspire insensibly during sleep at all? Why even twice as much as when awake? The only conclusion that I can draw is that Sanctorius was simply not able to conclusively integrate his novel observations with the weighing chair into the Galenic concept of sleep. The merging of old and new ideas was certainly not always easy and the way in which Sanctorius coped with the problem is telling for an understanding of his works. Hence, a closer look into the *De statica medicina* will reveal more about how Sanctorius connected insensible perspiration with sleep and waking hours.

According to Sanctorius, sleep and perspiration were interdependent. He wrote: "The things which impede sleep, also impede the perspiration of cooked perspirable matter" (Sanctorius 1614: 50r).[78] By the same token, short sleep was produced by the acrimony of retained perspirable matter, and a minor perspiration announced restless sleep and a tiresome night. Ultimately, both, good sleep and healthy insensible perspiration were determined by undisturbed concoction. As was mentioned

---

[77] In the 2001 edition of the *De statica medicina*, Giuseppe Ongaro erroneously referred to natural spirits in his translation of aphorisms XLVII and XLVIII (in the 1614 edition aphorisms XLVIII and XLIX), which describe the physiological processes during sleep and waking. With the knowledge of Sanctorius's commentaries and his denial of the existence of a natural spirit (see Sect. 3.2.6, fn. 36), I argue that Sanctorius referred here rather to the three faculties (*facultates*), or their respective virtues (*virtutes*) than to spirits. See: Sanctorius and Ongaro 2001: 144 f., Sanctorius 1614: 58r.

[78] "Quae impediunt somnum, impediunt quoque perspirationem cocti perspirabilis." See: Sanctorius 1614: 50r.

above, Sanctorius not only stated that insensible perspiration was more abundant during sleep, but also that it was of a different kind than insensible perspiration that occurred during waking hours (Sect. 3.2.7). This differentiation of *perspiratio insensibilis* seems to correspond to Sanctorius's differentiation between concoction and digestion along the same lines as between the rest or the movement of the body. While the perspirable matter that was produced during sleep was *cooked*, Sanctorius described perspiration that was expelled during waking hours as *crude*. What is more, the cooked perspiration that occurred during sleep seems to have been more beneficial. Contrary to its counterpart, it increased strength (Sanctorius 1614: 5r, 50r–50v, 53r; 1634: 50r). Sanctorius's solution to reconcile his discovery of the high amount of *perspiratio insensibilis*, which was expelled during sleep, with the Galenic differentiation between concoction and digestion was, it seems, the introduction of two different kinds of *perspiratio insensibilis*. A detailed explanation of the weighing procedures will be given below to further elucidate how Sanctorius connected his quantitative findings with Galenic physiology (Sect. 7.5).

Sleeping Times and Duration of Sleep Following the functions of sleep described above, namely the performance of concoction, the restoration of heat, and the regeneration of the spirits, the duration of these processes also determined the healthy length of sleep. This duration, however, was influenced by the individual complexion of the body. And so Sanctorius explained that warm bodies needed short sleep, while cold bodies needed longer sleep. In a warm brain, the moisture, which caused sleep, was quickly dissipated and the flow of animal spirits could not be hindered for long, as moving, spreading out, and diffusing from the center to the outside was characteristic of heat. But in a cold brain, the animal spirits could be hindered longer, as cold things naturally rested and withdrew heat for the sake of concoction (Sanctorius 1612a: 358, 364; 1612b: 79).

In the *Commentary on Avicenna*, Sanctorius explained that, according to Galen, the quantity of sleep varied, depending on which food was consumed beforehand. If lettuce was eaten for dinner, long sleep followed; if spices were eaten, short sleep followed. Hence, suitable sleeping times could vary between 7, 8, or 9 h.[79] This parallels the discussion mentioned above, as to whether the midday meal or evening meal should be larger (Sect. 3.3.2). As the duration of sleep was connected to the duration of concoction, it was only consequential that food also determined the duration of sleep. This instance illustrates well the interconnection between the different non-natural things. Food and drink, just like sleep and waking hours, had a major influence on the physiology of the body, especially on the digestive process, and had to be harmonized with each other in order to preserve health. They could not be treated in isolation. In the *De statica medicina*, Sanctorius wrote: "From food comes sleep, from sleep concoction, and from concoction a good transpiration"

---

[79] According to Ken Albala, Renaissance dieticians often recommended lettuce to combat sleeplessness, following Galen. See: Albala 2002: 137.

(Sanctorius 1634: 50r).[80] Accordingly, if one went to bed with an empty stomach, the amount of insensible perspiration was three times less than usual (Sanctorius 1614: 52v; 1625: 69).

Just as knowledge of the duration of the distinct stages in the digestive process disclosed the ideal time to eat, it also indicated the ideal sleeping time. Sleep should stop as soon as the first two stages of digestion had been completed. Sanctorius connected these stages with a certain kind of beneficial *perspiratio insensibilis* and it seems that he based his recommendation for a general duration of sleep on the amount of this perspiration excreted. In the *De statica medicina*, he specifically and positively referred to 7 h of sleep, which implies that this is the time by which concoction was completed and a healthy amount of *perspiratio insensibilis* excreted (Sanctorius 1614: 48v–49r, 52r, 59v).

But again, things are not that easy. Sanctorius wrote that it was very beneficial to sleep around 4 h after eating, as the body during this time was less occupied with the first concoction and better able to restore what was lost. Moreover, this favored perspiration. Why did he not recommend sleeping directly after eating, as sleep was the time when the first two stages of concoction took place? In the *Commentary on Avicenna*, Sanctorius explained that sleep should not, per Avicenna, begin directly after supper, because the food should arrive first at the bottom of the stomach in order to be concocted during sleep. This was why Galen recommended a brief walk before going to bed. However, 4 h seems a long time for the food to arrive at the bottom of the stomach. And what is more, Sanctorius implied that the first stage in the digestive process started before sleep began. The solution to this riddle lies in the following citation, taken from the *De statica medicina*: "He who concocts and digests well every day will really have a long life: concoction occurs during sleep and rest; digestion during waking hours and exercise" (Sanctorius 1634: 50v).[81] Hence, it was possible that concoction occurred while awake, as long as the body rested. In turn, digestion could also take place during sleep, if the body moved. However, while Sanctorius pointed to the beneficial effects of yawning and stretching of the limbs immediately *after* sleep for the expulsion of insensible perspiration, he did not say a word about the effects of movement *during* sleep on insensible perspiration.[82] At least, not in this section. The analysis of the next non-natural pair, exercise and rest, may bring more to light. But first, some other aspects of sleep and wake must be mentioned (Sanctorius 1614: 53v–55v; 1625: 69).

---

[80] "A cibo somnus, à somno coctio, à coctione utilis transpiratio." See: Sanctorius 1634: 50r.

[81] "Ille verè longaevus, qui quotidie bene concoquit, & digerit: coctio fit somno, & quiete: digestio vigilia & exercitio." See: ibid.: 50v.

[82] Following Sanctorius's differentiation between concoction and digestion it makes total sense that slight movements that occurred directly after sleep promoted the excretion of insensible perspiration. The perspirable matter was concocted and refined during sleep and was then ready to be expelled by the body during the third and final step of the digestive process: digestion, which occurred during waking and exercise. In another passage of the *De statica medicina*, Sanctorius wrote that the hour of the best perspiration was usually in the period of two hours after sleep (Sanctorius 1614: 55r–55v). It remains the question why insensible perspiration was twice as big during sleep than during waking.

Whatever the duration of sleep and its timing with regard to eating, sleep and wake had also to be arranged according to the time of day. Daytime sleep, Sanctorius wrote, was not as beneficial as nighttime sleep. He explained that the animal spirits were luminous (*lucidi*), and therefore were diverted by the sunlight in the daytime, when being drawn back into the inside of the brain during sleep. As a result, daytime sleep occurred with violence, because the animal spirits could be retained in the interior parts of the brain only with violence. By contrast, at nighttime the air was colder and innate heat withdrew easily into the inner parts of the body and was retained without violence there, which was why nighttime sleep was more quiet and more pleasant.[83] With regard to insensible perspiration, a midday sleep could, however, also be useful. In the *De statica medicina*, Sanctorius explained that it served to excrete perspiration that had been retained the day before (Sanctorius 1612a: 80; 1614: 55v–56r; 1625: 68 f.).

With regard to the seasons, Sanctorius was of the opinion that sleep was longer in winter and spring, and shorter in autumn and summer. In this context he suggested that the length of sleep was derived not from the weakness of the spirits, but from their concentration inside of the body. During winter, because of the external cold, the concentration was stronger and sleep therefore longer. Moreover, the external cold made the influent heat, which came from the sensory organs, withdraw more easily inside, into the body. In spring, he wrote, sleep was longer than in summer and autumn, because of the predominance of blood (Sect. 3.3.1). As blood was a sweet humor and without acrimony, sleep was longer.[84] Contrariwise, acrimonious, bilious, and melancholic humors encouraged wakefulness. According to Sanctorius, sleep in winter was more salutary than in summer, not because of the length of sleep, but because the bodies grew warmer before dawn in winter and as a result perspired very much, whereas they were colder in summer and perspired less (Sanctorius 1612b: 82; 1614: 56v; 1629a: 376 f.).[85]

One last remark has to be made about Sanctorius's conception of sleep and wake. In the *De statica medicina*, he wrote that the exhalation of sleeping bodies was so abundant that not only the sick who slept with the healthy, but also the healthy among themselves mutually communicated good and bad dispositions (Sanctorius 1614: 60v). This implies that Sanctorius thought that vapors, most likely insensible perspiration, could be transmitted from body to body and might affect their health and disease. It is, however, difficult to make sense of this statement, as Sanctorius did not pursue the idea any further. Interestingly, he did not connect it with his

---

[83] Sanctorius's reference to the luminosity of animal spirits fits his suggestion that animal spirits emanated an incorporeal radiation that caused sensation and voluntary motion (see Sect. 3.2.6, fn. 37).

[84] In Galenic humoral theory, the distinction between the humors was also by taste: blood was sweet, yellow bile was bitter, black bile was sour, and phlegm was salty. See: Jouanna 2012: 339, fn. 20.

[85] In this context, see also the discussion of the effects of cold and warm air on the body and on its excretion of insensible perspiration in 3.3.1.

notions of plague, or air. Another case, in which there are more questions than answers.

To sum up, Sanctorius generally followed the Galenic conception of sleep and wake. However, his new focus on *perspiratio insensibilis* required a reworking of the traditional notions of sleep and wake, as was the case for all the other six non-naturals, too. However, what makes his treatment of sleep and wake so interesting is that inconsistencies occur. Seemingly, Sanctorius was not able to coherently integrate his novel finding, the high amount of insensible perspiration which was expelled during sleep, into the traditional medical framework. Therefore, this section provides a valuable insight into the way in which Sanctorius struggled to compromise between innovation and tradition, between new and old ideas. It reveals how he handled this struggle and presented it to his readers and pupils. It is observations like these that contribute to an understanding of how Sanctorius developed and stabilized his static doctrine.

3.3.4 Exercise and Rest

Playing ball, walking, swimming, horseback riding, jumping, and dancing—the list of exercises Sanctorius mentioned in the *De statica medicina* reads like the program of a modern sports center. However, on closer examination, one finds also activities like tossing and turning in bed, frictions, being treated with cupping glasses, and travelling in a boat, palanquin, or carriage, as well as exercises of the mind that can certainly not be counted among current leisure activities, or be included in present-day workout plans. In line with Galen, Sanctorius described exercise as a movement during which a change happened and breathing altered. These aspects distinguished exercises (*exercitia*) from other movements (*motus*). Closely connected to the topic of exercise were periods of relaxation, as the proper alternation between activity and rest was thought to be crucial to a healthy life. Hence, in Galenic medicine the concept of exercise was somewhat broader than our modern notion of exercise as an activity chosen in moments of free time (Sanctorius 1612b: 64; 1614: 61r–67v).

The suitable quantity of exercise for a temperate body was, so Sanctorius, until the body started to tire. Referring to Hippocratic-Galenic teachings, the physician pointed out the importance of a body not continuing to exercise upon reaching this point, as only the first signs of fatigue could be easily and immediately remedied by rest. If a body experienced real fatigue, exercise was unhealthy. Avicenna had suggested, as Sanctorius explained, that exercises should be done until vaporous sweat occurred. If sweat turned fluid, further exercise should be avoided (Sanctorius 1612b: 63 ff.). This relates to the above discussion of the connection between movement and sweat (Sect. 3.2.7). Given Sanctorius's ambiguous attitude toward sweat and his conviction that this excretion only emerged with violence, it is likely that he interpreted Avicenna's vaporous sweat as insensible perspiration. Understood in this way, exercise was healthy as long as insensible perspiration was expelled; and it became harmful once the body started to sweat. This is also in line with the beneficial effects that Sanctorius connected with exercise.

Beneficial Effects of Exercise Referring to Hippocrates, Galen, and Avicenna as authorities on the matter, Sanctorius described three benefits of exercise. First, it rendered the muscles hard and very robust, and hence less susceptible to fatigue and pain during exercise. Secondly, it disposed the body to resolve excrements via insensible perspiration. And thirdly, it carried nourishment to the parts that needed to be nourished. As movement was the cause of heat, the internal heat was increased during exercise, which led to an augmentation of the attractive and the distributive faculties. Therefore, the digestive process and nutrition were better performed. Moreover, the spirits became finer and faster and therefore readier to act. Due to these physiological processes, bodies became lighter with exercise, as Sanctorius explained in the *De statica medicina*. While, today, losing weight is one of the main motivations for exercising, in Sanctorius's day, it was not necessarily considered beneficial. A median body size was the medical and cultural ideal, so dietetics did not put great emphasis on keeping the body slim. Too much exercise was even seen as positively harmful. According to Sanctorius, violent exercise speeded up the aging process and increased the risk of premature death (Sanctorius 1612b: 64 f.; 1614: 62v–63r, 65r; 1625: 64, 96, 369; 1629a: 385 f.).

Time of Exercise In connection with the functions and beneficial effects of exercise, its timing was determined by a number of conditions. Sanctorius's comments and suggestions on this topic were again based on Galenic and Galenist dietetics. Physical exercise should be done before meals, when the first two stages of digestion were completed. It was most healthy, Sanctorius argued in the *De statica medicina*, if the body returned to its usual weight two times a day, before eating. Moreover, before exercising, the body had to be free of superfluities, most importantly, of crude humors. During exercise, crude humors would be distributed throughout the body and produce adiapneustia, defective perspiration. Immediately after exercise, the body should rest and under no circumstances eat. Food would not restore the body's exhausted virtue, but rather, overwhelm it. Aggravated by the food, the body would also perspire less. Therefore, one should eat only when the heat, produced during exercise, has dissipated (Sanctorius 1612b: 65, 68; 1614: 62r, 66r; 1629b: 144).

Based on these considerations, Sanctorius recommended a moderate lunch and a substantial supper. To make sure that the foodstuffs ingested for lunch were concocted and healthy, they should be easy to digest; also, exercise could be done before supper. Supper, on the contrary, had to be of more copious and solid foodstuffs, as sleep would follow, and a longer time period in which exercise might be done before the next meal. Following his advice on mealtimes, according to which 9 to 10 h must be scheduled between lunch and supper, and 14 to 15 h between supper and lunch (Sect. 3.3.2), Sanctorius wrote that 1 h exercise in the period from 7 to 12 h after eating produced more insensible perspiration than 3 h exercise at any other time (Sanctorius 1612b: 68; 1614: 62r–62v). Considering that Sanctorius suggested elsewhere that one should sleep around 4 h after eating (Sect. 3.3.3), living according to his schedule turns out to be a complicated task. It was already mentioned that in his treatment of the non-natural pair, sleep and wake, there are inconsistencies which, together with the complex timing of the various non-natural things, highlight a tension between his recommendations and their individual implementation. Sanctorius not only struggled to reconcile his novel quantitative findings with traditional medical theory, but also to accommodate his newly formulated rules to their practical application, and vice versa. This raises the question of the feasibility of the static aphorisms, which will be addressed later, when analyzing the practical and material aspects of Sanctorius's work.

Exercise and Sleep In the previous section on sleep and wake, I pointed to the issue of reconciling the Galenic differentiation between concoction and digestion with Sanctorius's observation that insensible perspiration was twice as great during sleep as during wake (Sect. 3.3.3). Digestion, the third stage in the digestive process and the one that produced insensible perspiration, occurred, according to Sanctorius, during wake and exercise. Did Sanctorius reveal more on the effect on insensible perspiration of movement *during* sleep, in the present section? Does this explain why insensible perspiration and, hence, digestion also took place during sleep? Not quite. In the *De statica medicina*, Sanctorius wrote: "A body perspires much more when resting in bed than when tossing and turning with frequent and repeated agitation" (Sanctorius 1614: 61v–62r).[86] Accordingly, if the body moved at night, insensible perspiration was less and, hence, digestion was neccessarily hindered.[87] Thus, exercise at nighttime had different effects than exercise in the daytime. This, however, fits with the suggestion that exercise should be done only after the first two stages in the digestive process were complete, which is to say, concoction, which occurred during sleep when the body rested. How digestion and, consequently, the excretion of insensible perspiration happened during sleep remains an open question. Be that as it may, Sanctorius, drawing on his quantitative findings, explained that the body perspired less during exercise than during sleep. Around 10 h after supper, provided that the body had rested in bed during this time, insensible perspiration was optimal (Sanctorius 1614: 61r, 63r–63v, 65v).[88]

---

[86] "Longe magis perspirat corpus in lecto quiescens, quam in lecto frequenti & crebra agitatione circumvolutum." See: Sanctorius 1614: 61v–62r.

[87] One may argue that tossing and turning in bed was rather considered as movement than as exercise and hence did not have the same effects. However, the fact that Sanctorius included the quoted aphorism in the section on exercise and rest suggests that he counted tossing and turning in bed among exercises. Moreover, in an aphorism of the section on sleep and wake, Sanctorius compared the movement of the body in bed to a speedy run, which further implies that he considered movement in bed to be exercise. See: ibid.: 51r.

[88] This corresponds to the characterization of the early morning as the most propitious moment for purging put forward by the medieval rules of health (*regimina sanitatis*). At this particular time, the kidneys and bladder were thought to excrete superfluous material, which had been generated during the second stage of the digestive process. See: Sotres 1998: 311.

Forms of Exercise The variety of exercises to which Sanctorius referred in the *De statica medicina* was already mentioned above. But what was the most beneficial way of exercising? Which sports did Sanctorius recommend? With regard to the different ways of horseback riding, Sanctorius explained that trotting was the healthiest, while gallop was the least healthy. Walking was healthier than being transported in a palanquin or a boat, as it prepared the body better for the necessary perspiration. However, going by boat or by palanquin for a long time was very healthy, because only then did it dispose the body extraordinarily for the necessary perspiration. Going by carriage was the most violent movement of all, because it made the uncooked perspirable matter exhale and harmed the solid body parts, especially the kidneys, so Sanctorius. Discus exercise was good for perspiration and moderate dance without jumping was nearly as commendable as moderate walks, given that it expelled the cooked perspirable matter in moderation. In view of these statements, it seems that Sanctorius considered walking to be the best and healthiest exercise, also with regard to its effect on insensible perspiration, just like the physicians in Ancient Rome and the authors of the medieval rules of health (*regimina sanitatis*), to whom I refer in Sect. 4.1.2, (Sanctorius 1614: 66v–67v).

Exercises of the Mind Besides muscular activity, Sanctorius also wrote of exercising the mind, as was common in contemporary discussions of the topic. In doing so, he anticipated his treatment of the non-natural thing, affections of the mind.[89] The activity of the mind was important with regard to the excretion of insensible perspiration, because it especially evacuated insensible excrements, mainly from the heart and the brain, as Sanctorius explained. This meant, above all, that the smoky vapors, produced during the formation of the vital and animal spirits (Sect. 3.2.6) were excreted by the exercises of the mind. Among them, anger, great joy, fear, and sorrow made the spirits exhale the most. Along this line of thought, too much rest of the mind impeded perspiration more than too much rest of the body. And bodies that rested in bed, but were agitated by an intense emotion usually resolved more and lost more weight than bodies that were agitated by an intense physical activity, but with a calm mind. Here, too, moderation was the rule. Just like any violent exercise of the muscles, any violent activity of the mind made aging faster and dying sooner more likely (Sanctorius 1614: 64r–65r).

From the preceding paragraphs it becomes clear that Sanctorius followed traditional Galenic concepts with regard to exercise and rest. In accord with the non-naturals already considered, the novelty lies, here, too, in his focus on insensible perspiration and body weight. Given the fact that exercise and rest are the fourth non-natural pair scrutinized in this chapter, the complex interrelations between the different non-natural factors come more and more to the fore. Sanctorius's newly formulated rules, which were meant to guarantee a healthy insensible perspiration and, consequently, a healthy body weight and general well-being, must at times

---

[89] The physiological processes that were connected to the exercises of the mind will be explained below in the section on the affections of the mind, see Sect. 3.3.6.

have been difficult to reconcile with each other, in everyday life, as the proper timing of the different non-natural factors reveals. This is important when it comes (in Sect. 7.5) to consideration of the practical application of the *De statica medicina*, namely the questions of how the weighing procedures were conducted, and of the relation between theory and practice.

3.3.5 Coitus

In contrast to the common list of the six non-naturals (Sect. 3.1), the sixth section of the *De statica medicina* does not examine evacuation and repletion in general, but only with regard to the effects of sexual activity (Fig. 3.1).[90] In the *Commentary on Galen*, Sanctorius gave a possible explanation for this choice by stating that he reduced coitus not to excreted or retained matter, but to movement, that is, to those things that happened. The *Isagoge Johannitii*, which included coitus as a special category in the list of the non-naturals, may also have encouraged his separate treatment of sexual activity (Sect. 3.3.1). What is more, Sanctorius's general shift to the effect of the non-naturals on insensible perspiration and the potentially important influence of coitus on its excretion may likewise have contributed to this decision. It may also be seen as a manifestation of his social environment: sexual activity seems to have been an important aspect of the daily lives of Sanctorius's patients and readers, so it seems hardly surprising that he dealt with it in detail in his rules for a healthy life (Sanctorius 1612b: 90 f.; see also Sanctorius 1603: 166r).

The Role of Females Before turning to Sanctorius's physiological concept of coitus and its relation to *perspiratio insensibilis*, it is important to note that Sanctorius geared the *De statica medicina* to a male audience. Reflecting the tendencies of the dietary literature of the time, he made no mention of female arousal and did not specify women's needs in particular. Only one aphorism of the section on coitus refers to women, explaining that excessive coitus with the most coveted female does not make one feel exhausted. Hence, it is from a male perspective that Sanctorius reconsidered the rules for a healthy lifestyle; and certainly, his audience was predominantly male. In the *Commentary on Avicenna*, Sanctorius revealed a rather misogynist attitude to women, when he stated that female concupiscence was not directed toward sexual pleasure, but was merely a means of gaining tyrannical control over men (Sanctorius 1614: 69r; 1625: 384; Siraisi 1987: 303).

The Physiological Concept of Coitus In Renaissance physiology, generation was closely tied to nutrition. Sperm was produced during the final stage of the digestive process, and it was generated from an excess of blood remaining after the body had been nourished. This unused nutritive material that was equal to almost completely

---

[90] The topic of coitus was usually subsumed under the non-natural pair of evacuation and repletion. See: Sanctorius 1612b: 91, Sotres 1998: 312, Albala 2002: 143.

assimilated food was directly converted into sperm. This applied to both sexes. According to Galen, both males and females had a form of sperm. In the female body, however, abundant blood that was not converted into sperm was naturally evacuated in the form of monthly menses. During pregnancy, the menstrual blood fed the growing embryo and, after the delivery, it was transformed into milk and conveyed to the breast. Accordingly, sperm, blood, and milk resulted from the same basic substance, and all were the direct product of the food first ingested. Thus, the diet of both sexes directly influenced reproduction (Van't Land 2012: 363–74). When Sanctorius discussed generation, he followed this physiological thinking, which was largely based on ancient beliefs, especially on Aristotle. Unsurprisingly, in matters of dispute between Aristotelian and Galenic theories of generation, Sanctorius followed the latter, as for example with regard to whether females actively contributed to the formation of the fetus, which Aristotle had denied (Sanctorius 1612b: 98; 1625: 656 f.). Without diving into the vast topic of Renaissance embryology, the following passages will focus on Sanctorius's notions with regard to the importance of coitus for maintaining health and, most importantly, for the excretion of insensible perspiration.

Beneficial Effects of Coitus Sanctorius, referring to Galen, explained that coitus was healthy when it was done with sufficient pauses in between. Only superfluous semen should be expelled, in order that the body be relieved and its strength increased. But due to the variety of the individual complexions and the different foodstuffs consumed, it was difficult to generally determine the lengths of the intervals between sexual activities. Warm and moist bodies, for example, regenerated semen more quickly than warm and dry bodies, and people who ate oysters and cooked onions or capons were more quickly prepared for coitus than people who ate lettuce and cabbage. This was why Galen put forward two precepts from which everyone could derive the individual pause needed between one coitus and the next, per Sanctorius. First, if a person was lighter, more agile, and readier to fulfill all duties after coitus and, second, if inhalation was better and easier, one knew that there had been a suitable interval between sexual activities. The reasons for this were that a copious semen, if retained, choked heat and thereby diminished the animal, vital, and natural operations and, especially, slowed down respiration (Sanctorius 1612b: 90 f.).[91]

On the basis of his observations with the weighing chair, Sanctorius argued that a body did not only *feel* lighter and more agile after useful coitus, but that the actual bodyweight also always diminished after sexual activity. Interestingly, this weight loss had to be compensated during subsequent sleep, after which, as Sanctorius wrote in the *De statica medicina*, there should be no change in weight, if coitus was

---

[91] According to Galenic medicine, operations (*operationes*) were functions of particular organs. They were subdivided into animal, vital, and natural operations and included for example imagination, the five senses, movement, respiration, or digestion. These operations were associated with respective virtues and faculties (Sect. 3.3.3, fn. 77) (Sanctorius 1625: 91, Siraisi 1990: 107 ff.).

proper. However, Sanctorius was somewhat ambiguous on this point, because in another aphorism he explained that generally old people became heavier by a moderate use of coitus, while young people became lighter. Further salutary effects of coitus, which Hippocrates and Galen promised and which Sanctorius seems to have accepted, were that it made a person more daring and less irascible, that it procured sleep, prevented inflammations of the groin, and removed the heaviness of the head. It is remarkable that Sanctorius did not deal with the beneficial effects of coitus with regard to insensible perspiration per se in the *De statica medicina*. Instead, he focused on the dangers of excessive coitus, from which the salutary effects of sexual activity could only be deduced. This may reflect his experience as a physician, which probably taught him that many of his patients "used Venus" excessively (Sanctorius 1612b: 95, 99; 1614: 68v, 69v, 74r).

Harmful Effects of Coitus Right at the beginning of the section on sexual activity in the *De statica medicina*, Sanctorius wrote: "Both too much abstinence from coitus and the excessive use of it impede perspiration; but the excessive use, more so" (Sanctorius 1614: 68r). Hence, as with all non-natural factors, moderation was the key. A healthy body continuously produced an abundance of sperm, which, as soon as it built up an excess, required expulsion by means of sexual activity. Otherwise, the retained sperm would harm the body. Sexual desire signaled the build-up of sperm and was therefore a sign of useful coitus. The higher the libido, the healthier was the frequent use of coitus. However, if there was no excess of sperm, coitus was not required, and sexual activity was immoderate and harmful. Sanctorius explained that the afflictions which resulted from excessive coitus depended indirectly on impeded perspiration and directly, on a harmed digestive process. Immoderate sexual activity resolved spirits and heat, cooled the body, and led to the perspiration of crude matter. Innate heat was diminished, the stomach cooled and, therefore, the digestive power reduced. Consequently, less insensible perspiration was excreted. From this resulted tremor and flatulence. Besides the stomach, excessive coitus damaged mostly the eyes, per Sanctorius. It removed a large amount of spirits from the eyes, which rendered the tunics of the eyes very hard and rough, and the channels less penetrable. As a consequence of diminished perspiration, the fibers that formed the tunics of the eyes became opaquer. Therefore, vision occurred through very small passages, as if through a lattice. Glasses, which united the objects in a single point, were needed so that one might see distinctly through one space only. This explanation of the harmful effects of coitus on vision originated with Sanctorius. However, the fear of a weakening of the eyesight effectuated by an overdrying of the brain through immoderate sexual activity was common at the time (Sanctorius 1603: 123v; 1612b: 91; 1614: 68v, 69v–70v, 71v–72r).[92]

---

[92] For an account of early modern medical concepts of vision and of the general Galenic framework on which they are based, see: Boudon-Millot 2012, Vanagt 2012. Early modern medical perspectives on eyeglasses are dealt with in Vanagt 2010. For more information on Sanctorius's notion of optics, see Sect. 4.2.3.

On the other hand, abstinence from coitus entailed its own set of harms. Sanctorius alluded here again to Galen and described six effects of retained semen: heaviness of the head, disgust for food, risk of fever, diminished concoction and digestion, numbness, and fear. These afflictions could be traced back to the sympathy or consent that Sanctorius thought to exist between the different body parts (Sect. 3.2.10), as well as to acrimonious vapors, which were raised by the retained semen and perturbed the organs. No purging drugs nor other changing aids could help, as the only cure was coitus. But semen could not only harm with regard to its quantity, but also with regard to its quality. If semen was corrupted, poisonous vapors arose and, transmitted to the organs, corrupted them as well. This might generate very serious affections, such as strokes or catalepsy. Because good semen enhanced strength, its corruption resulted in the opposite, that is, in the worst afflictions, as Sanctorius explained. Healthy concoction and digestion as well as regular sexual activity were thus crucial for a suitable quantity and quality of semen (Sanctorius 1612b: 91, 93; 1625: 649).

According to the various constitutions of individuals, improper coitus could have numerous other harmful effects. It might diminish memory and strength, warm the liver and the kidneys, produce toothache, bad breath, or bloody spittle as well as nephritis or diseases of the bladder. In any complexion, however, excessive coitus ultimately cooled and dried, thereby accelerating the aging process, as the latter was also a matter of cooling and drying (Sanctorius 1612b: 99). In order to evade the harmful effects of sex and to make sure that it was healthy, proper timing was of course important.

Time for Coitus Following the teachings of Hippocrates, Galen, and Avicenna, Sanctorius defined the proper time for coitus as subsequent to sleep. The first and second stages of the digestive process had to be completed, while the third stage should be advanced but not finished. The reasons for this were that semen was made from the food ingested during lunch and supper, which was only concocted during sleep. After sleep, when the semen was concocted, the body was primed for reproduction. In fact, according to Sanctorius, this was also the time when coitus was most suitable for producing offspring, as the semen was not only well-cooked, but also stuck more tenaciously (Sanctorius 1612b: 93 f., 98–101; 1634: 62r).

Hence, in the section on coitus, Sanctorius again adhered to traditional Galenic conceptions, while shifting the focus to body weight and to the excretion of insensible perspiration. Contrary to the common list of the six non-natural things, Sanctorius identified coitus itself as a non-natural factor, which hints at its importance with regard to body weight and insensible perspiration. Moreover, his frequent warnings about excessive sexual activity probably reflect the sexual life of his distinguished Venetian clientele. The male perspective that Sanctorius adopted in this section reveals that he addressed the *De statica medicina* to a male audience and raises the question as to whether he included females in his weighing procedures—a question which will be considered in Sect. 7.5.4.

3.3.6 Affections of the Mind

In medieval and Renaissance medicine, mind and body were inextricably interwoven. There was thought to be a mutual influence: of the body upon the mind, and of the mind upon the body. Physical health and mental health could not be separated. On the one hand, the humors were directly linked with emotional states, character traits, and dispositions of the mind. Hence, the predominance of one humor in the body did not only determine whether some people were, for example, hotter and moister compared to others (Sect. 3.1.2), but also referred to psychological characteristics. According to their individual complexions, people could be described as sanguine, choleric, phlegmatic, or melancholic, with each of these attributes being connected to one of the humors: sanguine to blood, choleric to yellow bile, phlegmatic to phlegm, and melancholic to black bile (Fig. 3.2). On the other hand, changes in emotion altered the humors and digestion, which is why moderation was praised here, too, in order to keep the passions in balance. In fact, sudden emotions were thought to be especially dangerous and might even lead to death. To understand how emotions such as joy or sorrow could produce an alteration in the complexion, or actually pose a threat to life, a look at the associated physiological processes is required.

Physiological Concept of Emotions Sanctorius's treatment of exercises of the mind in the section on exercise and rest in the *De statica medicina* has already highlighted the connection between emotions and movements (Sect. 3.3.4). However, in this context movement does not refer to muscles, bones, or body parts, but to heat and the spirits. Following the common Galenic physiological understanding of emotions, Sanctorius thought that emotions could produce two different movements. Depending on the different mental affections, vital spirit and heat either moved from the heart to the extreme parts, or the other way around, from the exterior parts toward the center of the body. While the first movement was the natural movement of heat and therefore usually quite harmless for healthy people, the second movement was unnatural and rendered the body cool and dry. Excessive emotions suddenly moved the spirits and heat to such a degree that they harmed the body by corrupting or burning the spirits. Similarly, great joy could lead to death, as too much heat was moved to the exterior parts, whereby innate heat was extinguished (Sanctorius 1603: 116r; 1612b: 41 f., 89 f.; 1625: 66 f., 369; Ottosson 1984: 263).

The close relation between emotions and spirits may account for Sanctorius's conviction, mentioned above (Sect. 3.3.4), that exercises of the mind were important with regard to insensible perspiration, as they mainly excreted the smoky vapors produced during the formation of the vital and animal spirits. The fact that emotions were equally closely connected to the innate heat of the body, which determined the digestive power, explains how affections of the mind could disturb the digestive process and how a defective digestive process could produce harmful emotions. In his work *De remediorum inventione*, Sanctorius stated that passions of the mind rendered the stomach weak, sometimes because they scattered the heat flowing to the stomach, and sometimes because they corrupted the spirit, which together with innate heat, effected concoction (Sanctorius 1629b: 109). This direct causal relation between the digestive process and emotional wellbeing highlights the correlations between affections of the mind and insensible perspiration.

The Division of Emotions   In the *De statica medicina*, Sanctorius identified four basic emotions from which all the others could be inferred: anger (*ira*), great joy (*pericharia*), fear (*timor*), and sorrow (*maestitia*). These corresponded to the exercises of the mind that, according to Sanctorius, made the spirits exhale most (Sect. 3.3.4). He organized them into contrasting groups according to their effect on body weight. While anger and great joy rendered bodies lighter, fear and sorrow increased body weight. This was because in fear and sorrow only light matter was perspired, while heavier materials remained in the body. On the contrary, in joy and anger both, light and heavy matter was expelled. Along the same lines, an excess of retained heavy perspirable matter disposed a person to fear and sorrow, whereas an obstruction of lighter perspirable matter, to anger and joy. The *Pantegni*, one of the most influential general medical textbooks in the Middle Ages and the Renaissance, reduced the passions of the mind to six: joy (*gaudium*), distress (*tristitia*), fear (*timor*), anger (*ira*), anxiety (*angustia*), and shame (*verecundia*).[93] Of these, anxiety, joy, fear, and anger were the main four passions discussed by medieval and Renaissance Galenists. They were conceptualized as the "accidents of the soul" or "affections of the mind" and normally considered to be the sixth of the non-natural factors. Thus, Sanctorius's division of emotions was very much in line with traditional medical thought, even though he deviated slightly from the common list of the basic emotions, by referring to sorrow instead of anxiety.[94] However, the classification criterion put forward in traditional accounts of the affections of the mind was different from Sanctorius's: instead of body weight, the movement of the vital spirit was the decisive factor. Joy and anger were associated with the movement of the vital spirit from the heart to the extreme parts, and anxiety and fear, with the movement toward the heart. Sanctorius thus shifted the focus from the movement of the spirits to body weight and to the excretion of insensible perspiration, while still remaining in the traditional Galenic framework. It seems that the different movements of heat and spirit had, according to Sanctorius, a direct bearing on *perspiratio insensibilis* (Sanctorius 1614: 75r–76r).

---

[93] The *Pantegni* was a Latin rendition of ʿAlī ibn al-ʿAbbās al-Majūsī's (Lat. Haly Abbas, fl. tenth century) Arabic *Kitāb Kāmil aṣ-Ṣināʿa aṭ-Ṭibbiyya*, (The Complete Book of the Medical Art), written by Constantine of Africa (d. 1087). The work was largely based on Galen's writings and, together with Avicenna's *Canon* and the *Isagoge Johannitii*, numbered among the most influential general medical textbooks in the Middle Ages and the Renaissance (Siraisi 1990: 12, 14, 110).

[94] The emotion "anxiety" (*angustia*) was termed differently by different medieval and early modern medical authors and was for example often referred to as "distress" (*tristitia*). Thus, it can be assumed that Sanctorius had the same emotion in mind when using the term "sorrow" (*maestitia*). It has to be noted that modern English translations vary, too (e.g., *tristitia* is sometimes translated as sadness and *angustia* as distress). See: Knuuttila 2004: 215 f., Carrera 2013: 115–26.

Interestingly, Sanctorius related the division of the four basic emotions only in the *De statica medicina* to body weight and insensible perspiration, while he remained completely in the traditional scheme in his other works, including those published after 1614 (e.g., Sanctorius 1612b: 42; 1625: 66 f.). Instead of pursuing his characterization of the basic emotions according to their effect on body weight, he repeatedly explained that he followed Aristotle's twofold division, according to which all the affections of the mind could be reduced to pleasure and pain (Sanctorius 1612b: 89; 1625: 65).[95] This makes it difficult to understand which conception of emotions Sanctorius actually held and to what extent it was influenced by his observations with the weighing chair. The fact that he published in 1634 a second revised edition of the *De statica medicina*, in which he added one aphorism to the section on the affections of the mind, suggests that he still supported the views he expressed in the original work, even though he did not refer to them in his other books.

Healthy and Harmful Emotions   In correspondence to their effect on body weight and insensible perspiration, Sanctorius's two groups of emotions can be characterized as healthy (anger and great joy) and unhealthy emotions (fear and sorrow). This is analogous to the traditional classification of emotions according to the two movements of the vital spirits, which considered one movement to be natural (from the heart to the exterior parts) and hence positive, or healthful and the other to be unnatural (from the exterior parts toward the heart) and hence negative, or unhealthful. Accordingly, medieval and early modern medical authors commonly agreed on the harmful effects of fear and sorrow and praised joy as a healthy passion. The opinions with regard to anger were, however, varied. Usually conceived as a deleterious emotion that should be avoided, anger was also sometimes described as beneficial to health (Carrera 2013: 132–43). Yet, Sanctorius's positive view of anger as an emotion which was closely connected to joy, is exceptional. In the *De statica medicina*, he often contrasted anger and joy with fear and sorrow, explaining the healthy and harmful effects of these two groups of emotions. Angry or cheerful people, for example, did not feel fatigue when travelling, because their bodies easily excreted thick perspirable matter, contrary to people who were troubled by fear or sorrow. As the latter only excreted the lighter parts of insensible perspiration and the heavier parts remained in their bodies, they often suffered from obstructions, a hardening of the parts, and hypochondriac affections. While joy facilitated the diastole and the systole of the heart, sorrow and melancholy rendered these processes more difficult. This implies that joy promoted the formation of vital spirits that occurred during diastole as well as the excretion of the residual matter, smoky vapors, which took place throughout systole (Sect. 3.2.6). However, long-lasting joy could also be harmful, as it impeded sleep and took the forces away. In the same vein, any excess

---

[95] Notwithstanding that Sanctorius adopted Aristotle's division of emotions into pleasure and pain, he differentiated between the medical and the moral philosophical study of the mind. He was of the opinion that philosophers should consider the affections of the mind in order to acquire virtue, while it pertained to the physicians to deal with them in order to gain and to preserve health (Sanctorius 1612b: 89, Sanctorius 1625: 65).

of joy was unhealthy as it did not only evacuate the superfluous, but useful matter, too. The danger of sudden emotions was already mentioned above and so Sanctorius warned in the *De statica medicina* that unexpected joy provoked the exhalation not only of the excretions of the third stage of digestion, but also of vital spirits (Sanctorius 1614: 75v, 79v–80r; 1634: 68v). This shows that emotions, usually characterized as healthy, could have harmful effects, too, depending on the circumstances of their appearance. In fact, the last aphorism in the 1614 edition of the *De statica medicina* reads:

> Those who are sometimes cheerful, sometimes sad, sometimes angry, and sometimes afraid have a healthier perspiration than those who always enjoy one single affection, albeit a healthy one (Sanctorius 1614: 84r).[96]

Hence, a variety of different emotions from both groups was recommended. Anyway, it is hard to imagine that individuals' emotions do not change from time to time. But how can a well-balanced mind be acquired? By what means can imbalanced passions be corrected? How can one keep one's emotions in check?

As with the other non-natural things, it was contraries which effected a cure. Accordingly, anger and hope removed fear, while joy took away sorrow. However, due to the close relationship between emotions and insensible perspiration, fear and sorrow could also be taken away by an evacuation of thick perspirable matter. Contrariwise, anger and great joy were removed by the evacuation of thin perspirable matter. Sanctorius explained more generally that immoderate passions could be diminished or completely taken away by the evacuation of perspirable matter. Inversely, comfort of the mind made the body perspire most freely, as it opened the pores and produced abundant perspiration. Hence, according to Sanctorius, the monitoring and manipulation of insensible perspiration allowed the physician to draw direct conclusions on the emotional state of his patients. It seems then that by controlling insensible perspiration, emotions could be controlled, too. This goes of course hand in hand with the management of the other non-natural factors. Foodstuffs which opened the pores, for example, produced joy, while those which impeded perspiration provoked sorrow. And sleep was hindered by excessive joy which led to a removal of the forces (Sanctorius 1614: 76r, 77r, 78r–79r, 80r–81r, 83v–84r).

3.3.7   The Ars … de statica medicina *and Its Galenic Context*

By concluding my analysis of the final section of the *De statica medicina*, the whole section on Sanctorius's new interpretation of the six non-natural things likewise comes to an end. It has shown how Sanctorius conceptually integrated his novel finding, the high quantity of insensible perspiration, into this standard Galenic

---

[96] "Nunc hilares, nunc maesti, nunc iracundi, nunc timidi perspirationem magis salutarem habent, quam qui unico, licet bono, semper gaudeant affectu." See: Sanctorius 1614: 84r.

framework. But it has also shown the difficulties into which the historian plunges, when trying to reconstruct and understand the physiological foundations of static medicine. Inconsistencies, unsolved questions, and puzzling features came to the fore. It has become clear that the *De statica medicina* is about much more than a steelyard and quantitative values. It is a manifestation of the way in which contemporary dietetic guidelines coincide with new experiences and observations. In this conglomeration of traditional and innovative ideas it is anything but easy to disentangle the old from the new. And it might be even misleading. The point is this: However abstract and tiresome Sanctorius's intellectual background may seem from a modern perspective, it is inextricably linked to those of his activities which his followers and historiography have labelled innovations. In fact, the study not only of the famous *De statica medicina*, but also of Sanctorius's other books, especially those published after 1614, discloses a much more refined view of Sanctorius and his undertakings than is usually presented in the literature.

Given the central role which Sanctorius assigned to the monitoring of insensible perspiration for the preservation of health in the *De statica medicina*, one expects his general concept of medicine, or at least of dietetics, to be oriented to this physiological phenomenon and its weighing. However, perusal of the other publications reveals that Sanctorius did not always relate the six non-natural things to *perspiratio insensibilis* and its quantification. While these works add important additional information to the *De statica medicina* with regard to the physiological processes that characterized human involvement with the non-naturals, they often mention insensible perspiration, the weighing procedures, and the importance of quantification for physiology only marginally, if at all. Insensible perspiration and its quantitative investigation do not play a major role even in the *Commentary on Avicenna*, in which Sanctorius published all of his instruments, including the weighing chair. Therefore, static medicine cannot readily be identified as the overall framework of Sanctorius's written works. To further illustrate this point, his last publication, *De remediorum inventione*, makes no mention of either the observations or the findings with the weighing chair.

However, Sanctorius's written output is just one side of the coin. Static medicine was not only the product of intellectual activity, but had a practical dimension, too. Sanctorius's use of instruments, his interaction with patients, his observations and their interpretation, in short, the material dimensions of his medical research and practice, are crucial to a full appreciation of his endeavors. Yet, they cannot be isolated from their conceptual background. Starting from the medical context outlined in the preceding paragraphs, the next part of the work will analyze the practical side of Sanctorius's work and elucidate the correlations between the two realms, theoretical and practical. The close connection between the six non-natural things and insensible perspiration suggests that the former of these were more than just a structural element in the *De statica medicina*. In the following chapters, I will consider the role this doctrine has played in the preparation and conduct of Sanctorius's weighing experiments. Maybe it was reading contemporary dietetic handbooks that inspired Sanctorius to do research on insensible perspiration? Maybe the importance of moderation in quantities with regard to the use of the six non-natural things made him think of using a steelyard to define precisely what that meant in practice? From the perspective of Galenic dietetics, according to which balance and moderation were crucial to maintaining health, the step from the idea of balance to the use of a balance itself seems, at least in retrospect, quite natural.

References

Albala, Ken. 2002. *Eating Right in the Renaissance*. Oakland CA: University of California Press.

Alpini, Prospero. 1591. *De medicina aegyptiorum: libri IV*. Venice: Apud Franciscum de Franciscis Senensem.

Arrizabalaga, Jon. 1994. Facing the Black Death: Perceptions and Reactions of University Medical Practitioners. In *Practical Medicine from Salerno to the Black Death*, ed. Luis García-Ballester, Roger French, Jon Arrizabalaga, et al., 237–288. Cambridge: Cambridge University Press.

Arrizabalaga, Jon, Montserrat Cabré, Lluís Cifuentes, and Fernando Salmón, eds. 2002. *Galen and Galenism: Theory and Medical Practice from Antiquity to the European Renaissance*. Aldershot/Hampshire: Ashgate Variorum.

Baricellus, Iulius Caesar. 1614. *Doctoris Medici & Philosophi de Hydronosa Natura, sive sudore humani corporis*. Naples: Apud Lazarum Scoriggium.

Bigotti, Fabrizio. 2017. A Previously Unknown Path to Corpuscularism in the Seventeenth Century: Santorio's Marginalia to the *Commentaria in primam Fen primi libri Canonis Avicennae* (1625). *Ambix* 64: 1–14.

Boudon-Millot, Véronique. 1996. L'*Ars Medica* de Galien est-il un traité authentique? *Revue des Études Grecques* 109: 111–156.

———. 2012. Vision and Vision Disorders. Galen's Physiology of Sight. In *Blood, Sweat and Tears: The Changing Concepts of Physiology from Antiquity into Early Modern Europe*, ed. H.F.J. Horstmanshoff, Helen King, and Claus Zittel, 551–567. Leiden: Brill.

Bylebyl, Jerome J. 1971. Galen on "the Non-Natural Causes" of Variation in the Pulse. *Bulletin of the History of Medicine* 45: 482–485.

Carrera, Elena. 2013. Anger and the Mind-Body Connection in Medieval and Early Modern Medicine. In *Emotions and Health, 1200–1700*, ed. Elena Carrera, 95–146. Leiden: Brill.

Castiglioni, Arturo. 1931. The Life and Work of Santorio Santorio (1561–1636). *Medical Life* 135: 725–786.

Copenhaver, Brian P., and Charles B. Schmitt. 1992. *Renaissance Philosophy*. Oxford\New York: Oxford University Press.

Dalby, Andrew. 2000. *Dangerous Tastes: The Story of Spices*. Oakland CA: University of California Press.

Debru, Armelle. 1996. *Le corps respirant: la pensée physiologique chez Galien*. Leiden: Brill.

Farina, Paolo. 1975. Sulla Formazione Scientifica di Henricus Regius: Santorio Santorio e il "De Statica Medicina". *Rivista Critica di Storia della Filosofia* 30: 363–399.

Galen and Ian Johnston. 2016. *On the Constitution of the Art of Medicine; The Art of Medicine; A Method of Medicine to Glaucon*. Cambridge, MA\London: Harvard University Press.

———. 2018a. *Hygiene: Books 1–4*. Cambridge, MA\London: Harvard University Press.

———. 2018b. *Hygiene: Books 5–6; Thrasybulus; On Exercise with a Small Ball*. Cambridge, MA\London: Harvard University Press.

Galen. 1997a. De alimentorum facultatibus lib. II. In *Claudii Galeni Opera Omnia*. Vol. VI, ed. Carl Gottlob Kühn, 554–659. Hildesheim: Georg Olms Verlag.

———. 1997b. De pulsibus libellus ad tirones. In *Claudii Galeni Opera Omnia*. Vol. VIII, ed. Carl Gottlob Kühn, 453–492. Hildesheim: Georg Olms Verlag.

———. 1997c. De simplicium medicamentorum temperamentis ac facultatibus lib. VIII. In *Claudii Galeni Opera Omnia.* Vol. XII, ed. Carl Gottlob Kühn, 83–158. Hildesheim: Georg Olms Verlag.

Garber, Daniel. 2006. Physics and Foundations. In *The Cambridge History of Science*, ed. Peter J. Bowler, Lorraine Daston, David C. Lindberg, et al., 21–69. Cambridge: Cambridge University Press.

García-Ballester, Luis. 1992. Changes in the *Regimina sanitatis*: The Role of the Jewish Physicians. In *Health, Disease and Healing in Medieval Culture*, ed. Sheila Campbell, Bert Hall, and David Klausner, 119–131. Toronto: Centre for Medieval Studies.

———. 2002. On the Origin of the "Six Non-Natural Things" in Galen. In *Galen and Galenism: Theory and Medical Practice from Antiquity to the European Renaissance*, ed. Jon Arrizabalaga, Montserrat Cabré, Lluís Cifuentes, and Fernando Salmón, 105–115. Aldershot/Hampshire: Ashgate Variorum.

Gourevitch, Danielle. 1996. Wege der Erkenntnis: Medizin in der römischen Welt. In *Die Geschichte des medizinischen Denkens: Antike und Mittelalter*, ed. Mirko D. Grmek and Bernadino Fantini, 114–150. Munich: C.H. Beck.

Grendler, Paul F. 2002. *The Universities of the Italian Renaissance.* Baltimore: The Johns Hopkins University Press.

Hennepe, Mieneke te. 2012. Of the Fisherman's Net and Skin Pores. Reframing Conceptions of the Skin in Medicine 1572–1714. In *Blood, Sweat and Tears: The Changing Concepts of Physiology from Antiquity into Early Modern Europe*, ed. H.F.J. Horstmanshoff, Helen King, and Claus Zittel. Leiden: Brill.

Hippocrates, and William Henry Samuel Jones. 1923. *Hippocrates,* Vol. I. Cambridge MA/ London: Harvard University Press.

Hooykaas, Reijer. 1949. The Experimental Origin of Chemical Atomic and Molecular Theory before Boyle. *Chymia* 2: 65–80.

Jacquart, Danielle. 1984. De *crasis* a *complexio*: note sur le vocabulaire du tempérament en latin médiéval. In *Mémoires V: Textes Médicaux Latins Antiques*, ed. G. Sabbah, 71–76. Saint-Étienne: Université de Saint-Étienne.

Jarcho, Saul. 1970. Galen's Six Non-Naturals: A Bibliographic Note and Translation. *Bulletin of the History of Medicine* 44: 372–377.

Jouanna, Jacques. 1996. Die Entstehung der Heilkunst im Westen. In *Die Geschichte des medizinischen Denkens: Antike und Mittelalter*, ed. Mirko D. Grmek and Bernadino Fantini, 28–80. Munich: C.H. Beck.

———. 2012. The Legacy of the Hippocratic Treatise *The Nature of Man*: The Theory of the Four Humours. In *Greek Medicine from Hippocrates to Galen*, ed. Philip van der Eijk, 335–360. Leiden: Brill.

Knuuttila, Simo. 2004. *Emotions in Ancient and Medieval Philosophy.* Oxford: Clarendon Press.

Kollesch, Jutta. 1988. Anschauungen von den *apxai* in der *Ars medica* und die Seelenlehre Galens. In *Le Opere Psicologiche di Galeno. Atti del Terzo Colloquio Galenico Internazionale Pavia, 10–12 Settembre 1986*, ed. Paola Manuli and Mario Vegetti, 215–229. Naples: Bibliopolis.

Kuriyama, Shigehisa. 2008. The Forgotten Fear of Excrement. *Journal of Medieval and Early Modern Studies* 38: 414–442.

Lüthy, Christoph, John Murdoch, and William Newman, eds. 2001. *Late Medieval and Early Modern Corpuscular Matter Theories.* Leiden: Brill.

Morgagni, G. Battista. 1965. *Opera Postuma.* Vol. II: *Lezioni di Medicina Teorica, Commento a Galeno*, ed. Adalberto Pazzini. Rome: Istituto di Storia della Medicina dell'Università di Roma.

———. 1966. *Opera Postuma.* Vol. III: *Lezioni di Medicina Teorica, Commento a Galeno*, ed. Adalberto Pazzini. Rome: Istituto di Storia della Medicina dell'Università di Roma.

Niebyl, Peter H. 1971. The Non-Naturals. *Bulletin of the History of Medicine* 45: 486–492.

Nutton, Vivian. 2017. *Humours, Epidemics and a Connected Universe: Santorio's Plague Aphorisms in Context.* [Talk given at the conference "Humours, Mixtures, Corpuscles" in Pisa, 20 May 2017.]

Obizzi, Ippolito. 1615. *Staticomastix sive Staticae Medicinae demolitio*. Ferrara: Apud Victorium Baldinum.

Ongaro, Giuseppe. 2012. Morgagni, Giovanni Battista. In *Dizionario biografico degli italiani*. Vol 76. Rome: Istituto dell'Enciclopedia Italiana. http://www.treccani.it/enciclopedia/giovanni-battista-morgagni_(Dizionario-Biografico)/. Accessed 8 Feb 2019.

Orland, Barbara. 2012. White Blood and Red Milk. Analogical Reasoning in Medical Practice and Experimental Physiology (1560–1730). In *Blood, Sweat and Tears: The Changing Concepts of Physiology from Antiquity into Early Modern Europe*, ed. H.F.J. Horstmanshoff, Helen King, and Claus Zittel, 443–478. Leiden: Brill.

Ottosson, Per-Gunnar. 1984. *Scholastic Medicine and Philosophy: A Study of Commentaries on Galen's Tegni (ca. 1300–1450)*. Naples: Bibliopolis.

Preto, Paolo. 1984. La società veneta e le grandi epidemie di peste. In *Storia della Cultura Veneta: Il Seicento*, ed. Girolamo Arnaldi and Manlio Pastore Stocchi, 377–406. Vicenza: Neri Pozza Editore.

Rather, Leland. 1968. The "Six Things Non-Natural": A Note on the Origins and Fate of a Doctrine and a Phrase. *Clio Medica* 3: 337–347.

Renbourn, Edward Tobias. 1959. The History of Sweat and the Sweat Rash from Earliest Times to the End of the 18th Century. *Journal of the History of Medicine and Allied Sciences XIV*: 202–227.

———. 1960. The Natural History of Insensible Perspiration: A Forgotten Doctrine of Health and Disease. *Medical History* 4: 135–152.

Rocca, Julius. 2003. *Galen on the Brain: Anatomical Knowledge and Physiological Speculation in the Second Century A.D.* Leiden: Brill.

———. 2012. From Doubt to Certainty: Aspects of the Conceputalisation and Interpretation of Galen's Natural Pneuma. In *Blood, Sweat and Tears: The Changing Concepts of Physiology from Antiquity into Early Modern Europe*, ed. H.F.J. Horstmanshoff, Helen King, and Claus Zittel, 629–659. Leiden: Brill.

Rothschuh, Karl. 1978. *Konzepte der Medizin in Vergangenheit und Gegenwart*. Stuttgart: Hippokrates-Verlag.

Salmón, Fernando. 1997. The Many Galens of the Medieval Commentators on Vision. *Revue d'histoire des sciences* 50: 397–420.

Sanctorius, Sanctorius. 1603. *Methodi vitandorum errorum omnium, qui in arte medica contingunt, libri quindecim*. Venice: Apud Franciscum Barilettum.

———. 1612a. *Commentaria in Artem medicinalem Galeni*, Vol. I. Venice: Apud Franciscum Somascum.

———. 1612b. *Commentaria in Artem medicinalem Galeni*, Vol. II. Venice: Apud Franciscum Somascum.

———. 1614. *Ars Sanctorii Sanctorii Iustinopolitani de statica medicina, aphorismorum sectionibus septem comprehensa*. Venice: Apud Nicolaum Polum.

———. 1625. *Commentaria in primam Fen primi libri Canonis Avicennae*. Venice: Apud Iacobum Sarcinam.

———. 1629a. *Commentaria in primam sectionem Aphorismorum Hippocratis, & c. … De remediorum inventione*. Venice: Apud Marcum Antonium Brogiollum.

———. 1629b. *De remediorum inventione*. Venice: Apud Marcum Antonium Brogiollum.

———. 1634. *Ars Sanctorii Sanctorii de statica medicina et de responsione ad Staticomasticem*. Venice: Apud Marcum Antonium Brogiollum.

———. 1902 [1615]. Santorio Santorio a Galilei Galileo, 9 febbraio 1615. In *Le Opere di Galileo Galilei*, ed. Galileo Galilei, 140–142. Florence: Barbera. http://teca.bncf.firenze.sbn.it/ImageViewer/servlet/ImageViewer?idr=BNCF0003605126#page/1/mode/2up. Accessed 5 Nov 2015.

Sanctorius, Sanctorius, and Giuseppe Ongaro. 2001. *La medicina statica*. Florence: Giunti.

Secker, Thomas. 1721. *Disputatio Medica Inauguralis De Medicina Statica*. Gradus Doctoratus.

Siegel, Rudolph E. 1968. *Galen's System of Physiology and Medicine*. Basel\New York: S. Karger.

Singer, Peter. 2016. *Galen. The Stanford Encyclopedia of Philosophy*, ed. Edward N. Zalta. https://plato.stanford.edu/archives/win2016/entries/galen/. Accessed 10 Dec 2018.

Siraisi, Nancy. 1987. *Avicenna in Renaissance Italy: The Canon and Medical Teaching in Italian Universities After 1500*. Princeton: Princeton University Press.

———. 1990. *Medieval & Early Renaissance Medicine: An Introduction to Knowledge and Practice*. Chicago: University of Chicago Press.

———. 2012. Medicine, 1450–1620, and the History of Science. *Isis* 103: 491–514.

Smith, Wesley D. 1979. *The Hippocratic Tradition*. Ithaca\London: Cornell University Press.

Sotres, Pedro Gil. 1998. The Regimens of Health. In *Western Medical Thought from Antiquity to the Middle Ages*, ed. Mirko D. Grmek, Bernardino Fantini, and Antony Shugaar, 291–318. Cambridge, MA: Harvard University Press.

Stolberg, Michael. 2012. Sweat. Learned Concepts and Popular Perceptions, 1500–1800. In *Blood, Sweat and Tears: The Changing Concepts of Physiology from Antiquity into Early Modern Europe*, ed. H.F.J. Horstmanshoff, Helen King, and Claus Zittel, 503–522. Leiden: Brill.

Strohmaier, Gotthard. 1996. Die Rezeption und die Vermittlung: die Medizin in der byzantinischen und in der arabischen Welt. In *Die Geschichte des medizinischen Denkens: Antike und Mittelalter*, ed. Mirko D. Grmek and Bernardino Fantini, 151–181. Munich: C.H. Beck.

Temkin, Owsei. 1951. On Galen's Pneumatology. *Gesnerus – Swiss Journal of the History of Medicine and Sciences* 8: 180–189.

———. 1973. *Galenism: Rise and Decline of a Medical Philosophy*. Ithaca\London: Cornell University Press.

Touwaide, Alain. 1996. Heilkundliche Verfahren: die Arzneimittel. In *Die Geschichte des medizinischen Denkens: Antike und Mittelalter*, ed. Mirko D. Grmek and Bernadino Fantini, 278–292. Munich: C.H. Beck.

Treccani enciclopedia online. 2019. semplici, Lettura dei. http://www.treccani.it/enciclopedia/lettura-dei-semplici/. Accessed 5 Sep 2019.

Van't Land, Karine. 2012. Sperm and Blood, Form and Food. Late Medieval Medical Notions of Male and Female in the Embryology of *Membra*. In *Blood, Sweat and Tears: The Changing Concepts of Physiology from Antiquity into Early Modern Europe*, ed. H.F.J. Horstmanshoff, Helen King, and Claus Zittel, 363–391. Leiden: Brill.

Vanagt, Katrien. 2010. Suspicious Spectacles. Medical Perspectives on Eyeglasses, the Case of Hieronymus Mercurialis. In *The Origins of the Telescope*, ed. Albert Van Helden, Sven Dupré, Rob van Gent, et al., 115–127. Amsterdam: KNAW Press.

———. 2012. Early Modern Medical Thinking on Vision and the Camera Obscura. V.F. Plempius's *Ophthalmographia*. In *Blood, Sweat and Tears: The Changing Concepts of Physiology from Antiquity into Early Modern Europe*, ed. H.F.J. Horstmanshoff, Helen King, and Claus Zittel, 569–593. Leiden: Brill.

Vogt, Sabine. 2008. Drugs and Pharmacology. In *The Cambridge Companion to Galen*, ed. R.J. Hankinson, 304–322. Cambridge\New York: Cambridge University Press.

Wear, Andrew. 1981. Galen in the Renaissance. In *Galen: Problems and Prospects*, ed. Vivian Nutton, 229–262. London: The Wellcome Institute for the History of Medicine.

Webster, Colin. 2015. Heuristic Medicine: The Methodists and Metalepsis. *Isis* 106: 657–668.

Weyrich, Victor. 1862. *Die unmerkliche Wasserverdunstung der menschlichen Haut. Eine physiologische Untersuchung nach Selbstbeobachtungen*. Leipzig: Verlag von Wilhelm Engelmann.

Zitelli, Andreina, and Richard J. Palmer. 1979. Le teorie mediche sulla peste e il contesto veneziano. In *Venezia e la peste 1348/1797*, ed. Comune di Venezia: Assessorato alla Cultura e alle Belle Arti, 21–70. Venice: Marsilio.

Chapter 4
Sanctorius's Work in Its Practical Context

Abstract This chapter spotlights the practical context of the *De statica medicina* and explores Sanctorius's use of instrumentation. The investigation of the form and style of the *De statica medicina* and its relation to the literary genre of *Regimina sanitatis*—a medieval tradition of rules of health—allows important conclusions to be drawn on how Sanctorius shared his practical experiences, on his intended audience, and more generally, on the purpose of the publication. Complementary to established knowledge on Sanctorius, the analysis of his use of instrumentation focuses here not on the measuring instruments, but on the various other lesser-known devices that he developed, ranging from surgical devices to a special sick-bed. I examine the relation of these devices to Sanctorius's medical practice as well as to his teaching activities at the University of Padua. Even though—or exactly because—they were not part of the quantitative approach to physiology, their study helps to complement the picture of Sanctorius as a practicing physician. Moreover, it provides glimpses of the social context in which he developed and used his instruments and of how he used his head and hands in medicine. Finally, the results of this chapter allow the *De statica medicina* to be reviewed afresh within the broader practical context of Sanctorius's undertakings.

Keywords Early modern medical practice · Medical aphorisms · Medical instruments

The previous chapters spotlight the conceptual background of Sanctorius and analyze his work in relation to the medical tradition—Galenic medicine. Now, it is necessary to turn toward the practical and material resources of Sanctorius's endeavors in order to further investigate the processes that contributed to his innovative approach—the quantification of physiological phenomena. Like many of his colleagues, Sanctorius combined his activity as a university teacher of medicine with the practice of medicine. In doing so, he oscillated not only between these two occupations, but also between two important cities of Renaissance Italy: Padua and Venice. While the first was mainly known as a center of learning, with the University of Padua being one of the most famous universities in Europe at the time, the latter shined as the center of the mighty Republic of Venice and as a busy marketplace,

© The Author(s) 2023   97
T. Hollerbach, *Sanctorius Sanctorius and the Origins of Health Measurement*,
https://doi.org/10.1007/978-3-031-30118-6_4

where merchants from all over the world exchanged their commodities. Sanctorius's movement between these two worlds reflects in some ways the combination of theoretical and practical knowledge that shaped his works. On the one hand stands the professor of theoretical medicine, who wrote extensive commentaries on traditional university textbooks. On the other, the practicing physician, who devised an innovative weighing chair to observe the insensible perspiration of his patients.

However, as has become apparent, the categories of tradition and innovation cannot be clearly differentiated. Similarly, a simple dichotomy between theory and practice falls short of accounting for the complex interplay between the intellectual and the material, as well as their social dimensions. Instead of representing discrete and well-defined realms, these factors are one and the same phenomenon and should be analyzed as such (Valleriani 2017: vii). Therefore, it is the aim of the following chapters to deal with Sanctorius's introduction of quantitative research into physiology as something not distinct from, but complementary to the intellectual framework outlined in Chap. 3.

4.1   The *Ars … de statica medicina* and Its Practical Context

The starting point for the investigation is the analysis of the practical context of the *De statica medicina*. The published work of course does not offer a direct window onto Sanctorius's medical practice, and it has already been shown how strongly it was rooted in the medical tradition. Still, the choices Sanctorius made with regard to the presentation of his weighing procedures allow some important conclusions to be drawn on how he shared his practical experience. This sheds light on his intended audience and more generally, on the purpose of the publication. It gives a first insight into the way Sanctorius connected theory and practice.

4.1.1   The Aphoristic Form

Sanctorius wrote the *De statica medicina* in aphorisms. To modern eyes, these short and sententious sayings, which Sanctorius used in order to present the results of his weighing procedures, seem somewhat odd and foreign. In the preface of the *De statica medicina* he explained his choice of the aphoristic form with the following words:

> [it] seemed to me more reasonable to present [this art] in the form of aphorisms than in a descriptive form from beginning to end. [I did so], at first in imitation of our great Hippocrates, always priding myself on following in his footsteps; but then I was virtually driven by necessity to do so, since the same experiments, in which I was daily engaged for many years, through continual studies, virtually led me by the hand to this aphoristic form of the doctrine. Thus, I was able to arrange the aphorisms, which are interrelated to each other in this marvelous order, in exactly the same way as bees first pick at the honey of

diverse flowers and then, after having worked on it, arrange it in a marvelous order in their hives by means of the combs (Sanctorius 1614: Ad lectorem).[1]

Hence, on the one hand, Sanctorius saw himself in the tradition of Hippocrates and on the other, he stated that his weighing procedures led him naturally to the aphoristic form. This suggests that there was a close connection between Sanctorius's practice and its formal textual presentation. In the *Commentary on Hippocrates*, Sanctorius gave more insight into his understanding of aphorisms. Discussing the term "aphorism," he distinguished three levels of meaning—separation, definition, and selection—which he claimed corresponded to three conditions of aphorisms. First, that they were distinct sentences without a determined order. Second, that they were arranged, defined, and authoritative sentences, and certain explanations of things. Third, that they were selected sentences, which contained within themselves great power. Sanctorius further specified that aphorisms were phrases that were poor in terms of words, but rich in terms of sense. Their wording and content were carefully chosen and purified. In this context, he again put forward the analogy to the bees' production of honey. Just as bees collected the sweetest honey from the most excellent flowers, Hippocrates had chosen for his aphorisms the divine phrases from his other works. His intention had been, so Sanctorius, to select from the entirety of medicine those phrases most appropriate to the physician's use. But according to Sanctorius, aphorisms were useful not only for the physician, but for other skilled fields as well. There was, however, one prerequisite: the aphorisms must be selected from the respective field, hence, for example, political aphorisms for politicians. Only then would they unfold their great power (Sanctorius 1629a: 3 ff., 9 f.). This implies that Sanctorius intended the *De statica medicina* for physicians. Yet, whether this was really the case will be scrutinized later.

Even though aphorisms did not have a predetermined order, they still had to be ordered and the way this was done was important. Sanctorius differentiated here between two kinds of order. There was a universal order, which gave books their condition and form, and this was the resolutive (analytical) or the compositive (composed, synthetic) order.[2] But there was also a particular, or accidental order,

---

[1] "… quam consultius iudicavi doctrina Aphoristica quam diexodica describere, primò ad imitationem magni nostri Dictatoris, cuius vestigijs insistere gloriosum semper duxi: deinde id feci quasi necessitate impulsus, quandoquidem ipsa experimenta, quibus quotidie assiduis multorum annorum studijs incumbebam, ita me ad hanc doctrinae formam Aphoristicam manu quasi ducebant, ut Aphorismos optimè inter se connexos miro hoc ordine digesserim, eo plane modo quo apes primum mel ex varijs floribus delibant, & deinde in apiarijs per aedicularum suarum favos elaboratum miro ordine disponunt." See: ibid.: Ad lectorem.

[2] Sanctorius here tied in with Renaissance discussions of medical method, which were largely based on Galen's works and influenced by various conceptions, such as Aristotelian methodology and geometrical methods. Without delving deeper into this vast and complex topic, resolutive order was understood in this context as a form of teaching (*doctrina*) which begins with the idea of an aim and proceeds by way of resolution (in modern terminology "analysis"), while the compositive order was thought to proceed by way of composition, i.e., composition of the things discovered by resolution (in modern terminology "synthesis") (Edwards 1976: 285, Sanctorius 1612a: 25, 33). For further information on Renaissance discussions of medical method, see: Randall Jr. 1940,

which was neither resolutive nor compositive, but served for any occasion and for the memory. And this was the type of order that Hippocrates had used in his *Aphorisms*, and it constituted an alternative form of teaching—the *doctrina aphoristica*. The great significance that Sanctorius ascribed to Hippocrates's *Aphorisms* is apparent from his statement that this work embraced all the solid precepts of the medical art. In the *Methodi vitandorum errorum*, however, he was also critical of the *Aphorisms,* concurring with Galen that not all of them contained eternal truth (Sanctorius 1603: 25v–26r; 1629a: 9, 12). So, what conclusions can be drawn from these statements with regard to Sanctorius's use of the aphoristic form in the *De statica medicina*?

First of all, it must be noted that Hippocrates and the Hippocratic writings gradually gained importance in Western medical circles from the second half of the sixteenth century on. Galen's credibility as an interpreter and guide to scholars in Hippocratic studies began to decline and Hippocrates came more to the fore, slowly but steadily dissociated from Galen and the Galenic doctrine. This change of view occurred in a multifaceted process, which I shall not discuss here.[3] What is of interest in this context is that Hippocrates's *Aphorisms* were especially influential and the focus of medical academic attention. The preeminence of the work is illustrated by the fact that it held its place as one of the three set texts for the lectures on medical theory at the University of Padua until 1767 (Smith 1979: 13 f.; Nutton 1989: 422–31). In light of these circumstances, Sanctorius's ambition to follow in the footsteps of the great Hippocrates, as expressed in his preface to the *De statica medicina*, and, too, his choice of the aphoristic style, can be interpreted as a sign of the growing popularity of Hippocrates and the long-standing interest in the Hippocratic *Aphorisms*, which reputedly began well before the Renaissance.

However, it should not be forgotten that Sanctorius was a Galenist and still accepted the unity of the systems of Hippocrates and Galen. Perusal of his *Commentary on Hippocrates* shows that he strongly relied on Galen's interpretations of Hippocrates's teachings, as he frequently referred to the former's commentaries not only on the *Aphorisms*, but also on other Hippocratic works (Sanctorius 1629a: e.g., 7 f., 335 f., 409 f.). It is therefore in a Galenic spirit that Sanctorius praised Hippocrates as the author of essential precepts and as an excellent and sincere man of great talent and intelligence (Sect. 3.1.2) (Sanctorius 1629a: 7, 9). Hence, it might have been a combination of both, the growing contemporary interest in Hippocrates as well as Galen's veneration of the "Physician of Kos," which made Sanctorius wish to imitate the great master.

Randall Jr. 1961, Gilbert 1963, Wightman 1964, Randall Jr. 1976, Mugnai Carrara 1983. Most of the controversy over method in medicine turned upon the interpretation of the opening passage of Galen's *Ars medica* and this is also the place where Sanctorius discussed the issue, see: Sanctorius 1612a: 4 f., 10, 17–46.

[3] An analysis of the process that led to the change in opinion about Hippocrates and the Hippocratic writings and their emancipation from Galen and his doctrine can be found in: Smith 1979: 13–60, see also: Nutton 1989.

Moreover, Sanctorius's discussion of the meaning and conditions of aphorisms shows that he ascribed great power to this style of writing. According to him, aphorisms were especially rich in terms of sense and presented the most useful distillation of a skill, in this case the medical art, which was otherwise explained with many and, often, superfluous words. Aphorisms provided the reader with already *digested* content. In analogy to the nourishment of the body, Renaissance dieticians conceived of the nourishment of the mind—by reading and understanding a book, for example—as a matter of ingesting and incorporating knowledge, transforming it into bodily substance. This is still reflected in modern usage, when books or movies are referred to as *difficult to digest* (Sanctorius 1625: 24; Albala 2002: 141). Accordingly, by offering digested content, aphorisms conveyed knowledge that was easy for their readers to ingest and incorporate. This made them most appropriate to the use of the audience.

When Sanctorius discussed whether Hippocrates's *Aphorisms* had an order, he hinted at the functions of the aphoristic form. The accidental order which, so Sanctorius, had been used by Hippocrates for the arrangement of his aphorisms, helped the physician to memorize their content and made the aphorisms suitable for medical practice, as they could be applied to any situation. "Accidental" referred here not to the order itself, but to its purpose, namely to provide order for occasions and accidents. Sanctorius did not specify how this order actually looked, but gave the example of the head-to-toe arrangement of diseases which he said was used by practicing physicians. In fact, this classification was followed in many practically oriented manuals in the Middle Ages and the Renaissance, which were essentially based on experience.[4] Sanctorius explained that this ordering of illnesses was similar to the accidental order which Hippocrates had used in his aphorisms. In this context, it is interesting to note that Sanctorius chose for the *De statica medicina* an organizing principle whose occasional character he highlighted. As was mentioned earlier, Sanctorius described the involvement of the human body with the six non-natural things as purely fortuitous (Sect. 3.1.1). Thus, on the one hand Sanctorius seems to have considered the doctrine of the six non-natural things as an accidental order, suitable for structuring accidents and occasions such as a physician encountered in daily practice. On the other hand, he conceived of the six non-natural factors themselves as occasional causes of disease, which could be aptly described in aphorisms. In his view, it would seem, the doctrine of the six non-natural things and the aphoristic form informed each other (Sanctorius 1629a: 9).

Consequently, the conveyance of useful and compressed content, in connection with an easy intelligibility, memorability, and practical applicability, were central to Sanctorius's choice of the form and structure of the *De statica medicina*. This implies that he intended the work as a practical handbook for the daily use of

---

[4]The ordering of diseases from head to toe goes back to the *Kitāb al-Manṣūrī* (The Book of al-Mansūr, early tenth century) by the medical encyclopedist Rhazes (al-Rāzī, ca. 865–932), which was known to the West as the *Almansor*. Until the sixteenth century, it was frequently used as a university textbook in courses on practical medicine (Siraisi 1990: 12, 131, Grendler 2002: 324, Straface 2011: 7).

practicing physicians. But since he published the *De statica medicina* when a professor at the University of Padua, educational purposes might equally have been at
play. It seems then, that his usage of the well-known doctrine of the six non-natural
things as a means of ordering his aphorisms was intended to guarantee on the one
hand, that static medicine cover any occasion in dietetic practice, and on the other,
that it be easy to memorize, familiar as the scheme was.

The Tradition of Medical Aphorisms   Similar to the use of the six non-naturals as
a structural element in a dietetic treatise, the use of medical aphorisms was nothing
new or out of the ordinary. In fact, there was a tradition of medical aphoristic treatises that Sanctorius could tie in with. Given the persistent significance of
Hippocrates's *Aphorisms* and the popularity of the work, the application of such
terse statements by later physicians does not come much as a surprise. One of the
most famous representatives of the medical aphoristic writers is Moses Maimonides
(1138–1204). Probably around the end of the twelfth century, he wrote the *Aphorisms
of Moses* (*Aphorismi Rabi Moysi*, in Latin), which is the most voluminous of the ten
medical works he composed. It comprises approximately fifteen hundred aphorisms
based mainly on the writings of Galen, including the latter's commentaries on the
works of Hippocrates. Each of its twenty-five chapters deals with a different area of
medicine, ranging from anatomy to physiology, drugs, and medical curiosities
(Rosner 1998: 7–43; Maimonides and Bos 2004: xix–xxi). Interestingly, Maimonides
revealed his reasons for using the aphoristic form in the preface to the work. He
explained:

> People have often composed works in the form of aphorisms on [different] kinds of sci-
> ences. The science most in need of this is the science of medicine, because it has branches
> of knowledge that are difficult to conceptualize …, and [because] it has branches of knowl-
> edge that are difficult only with respect to remembering what has been written down about
> them …. As for the science of medicine, its conceptualization and the understanding of its
> concepts are not as difficult as in [the case] of the exact sciences. However, aspiring [to
> master] this science is difficult in most cases because it requires retaining a very large
> amount of memorized material, not merely of general principles but also [of] particu-
> lars, …. These works composed in the form of aphorisms are undoubtedly easy to retain;
> they help their reader to understand and retain their objectives. Therefore, the most eminent
> of the physicians, Hippocrates, has written his famous work in the form of aphorisms. Later
> on, many physicians followed his example and composed aphorisms, such as the *Aphorisms*
> of the famous al-Rāzī, the *Aphorisms* of al-Sūsī, the *Aphorisms* of Ibn Māsawayh, and oth-
> ers (Maimonides and Bos 2004: 1 f.).[5]

The citation shows that Maimonides considered the use of aphorisms especially
suitable for medicine, not only due to the field's complexity, but also and most
importantly, because the physician was required to know its contents by heart.
Standing at the bedside of a patient, there was hardly time to pore over lengthy
books. Thus, aphorisms should, so Maimonides, make it easier to grasp and memorize medical knowledge. The parallels to Sanctorius's argumentation are evident.

---

[5] The English translation is taken from the parallel Arabic-English edition of Moses Maimonides'
Medical Aphorisms, edited, translated, and annotated by Gerrit Bos (Maimonides and Bos 2004).

However, there seems to be a tension between the conciseness and brevity of aphorisms and their easy intelligibility. Maimonides wrote further below in the preface, "the intention of one who has composed aphorisms has not been to encompass everything that one needs in the field of that science ..." and "anyone who is like me or who is less knowledgeable than I am can benefit from them [the aphorisms]" (Maimonides and Bos 2004: 2, 4). But if aphorisms abridged or omitted content, the question comes to mind: How could they facilitate understanding for readers with no good grounding in Hippocratic–Galenic theory? Sanctorius did not seem to share this concern. Regarding the question of whether Hippocrates's *Aphorisms* served as an introduction to medicine, or were intended rather for advanced studies, he stated that they could be understood without the help of a teacher (Sanctorius 1629a: 11). This might be true for the *De statica medicina* as well. The clear practical orientation of the work may have pushed theoretical considerations into the background, contributing at the same time to Sanctorius's choice of the aphoristic form. The conciseness, memorability, and practicability of static medicine seem to have been more important to Sanctorius than its elaborate embedding in the theoretical context.[6] To follow his newly formulated rules of health, the information he provided in the aphorisms might well have been enough even for a less-educated audience, given that this audience too was most likely familiar with the aphoristic style. So, given the form of the *De statica medicina*, Sanctorius probably intended the work to be both: a handbook for experienced practicing physicians and a teaching tool or instruction manual for beginners.

Notwithstanding that Sanctorius did not refer to Maimonides in his works, it can be assumed that he was acquainted with the *Aphorisms of Moses*, as Latin editions of the work existed in his day. Originally written in Arabic, the work was translated into Latin in the thirteenth century and appeared as an incunabulum in Bologna in 1489, and in Venice in 1497, followed rapidly by numerous printed Latin editions.[7] The success and popularity of Maimonides' medical aphorisms in medieval western Europe may have drawn Sanctorius's attention to the work, albeit more than 300 years after the manuscript had first been published in Latin. In view of the fame and prestige of Hippocrates's *Aphorisms*, Sanctorius may have preferred to establish a direct connection between his static aphorisms and those of the great master, without bothering with other, more recent medical aphoristic writers. Maimonides, on the contrary, mentioned other followers of Hippocrates, who composed medical aphorisms: Rhazes (al-Rāzī, ca. 865–932), Abd Allāh ibn Muhammad al-Taqafī

---

[6] Ian Maclean has argued that the recommendation of the aphoristic form by medieval physicians foreshadowed some developments in the natural philosophy of the seventeenth century. Just as the presentation of medical precepts through the medium of aphorisms did not involve the elaboration of a complete system, in seventeenth century natural philosophy local explanations were suggested for phenomena, without any attempt to link these to a broader system of thought (Maclean 2002: 114). In his study on the *Aphorismi de gradibus* by Arnold of Villanova, McVaugh pointed out that late-thirteenth-century explanations and descriptions of medical practice only went into problems as far as was necessary to develop a solution, but did not try very seriously to incorporate these isolated cases into a general framework of medical thought (de Villanova et al. 1992: 89).

[7] For the bibliographical references, see: Dienstag 1983: 107 ff., Dienstag 1989: 455 f.

al-Sūsī (942–1012) and Mesue (Ibn Māsawayh, ca. 777–ca. 857). Interestingly, Sanctorius knew at least two of the three authors listed by the Jewish scholar, as he frequently referred to Rhazes and Mesue in his work (e.g., Sanctorius 1612a: 51, 468 f., 709; 1629a: 331, 500; 1629b: 120, 129). What is more, the name of Arnold of Villanova (ca. 1240–1311), another important proponent of medical aphorisms, appears in Sanctorius's commentaries, too. At the end of the thirteenth century, most probably in the 1290s, the renowned Catalan physician published the *Aphorismi de gradibus* (Aphorisms on measurement by degree), a treatise in which he set out a new theory of compound medicines (Sect. 5.2.2) (Sanctorius 1625: 410; 1629a: 389; de Villanova et al. 1992: 81 f.).

It is not my intention to dwell at any length on the different uses of the aphoristic form by these doctors, nor to compare them with Sanctorius's aphorisms. For the moment, it is enough to note that Sanctorius was certainly familiar with the tradition of medical aphorisms. Even though he did not discuss, and indeed rarely mentioned, the aphoristic works of anyone but Hippocrates, he must have been acquainted with them, at least to some degree. A systematic historical study on the use and function of medical aphorisms would help to contextualize the *De statica medicina* within this historical framework and possibly provide more insight into Sanctorius's adoption of the form. This, however, lies beyond the scope of this study. But the foregoing demonstrates that Sanctorius's use of aphorisms was closely related to the practical nature of the knowledge he conveyed in the *De statica medicina*, and that the two were interdependent. It is worth remarking here, that the appreciation of aphorisms not only as historical curiosities, but also as tools of medical education, is currently undergoing a revival.[8]

In his introduction to the 2001 edition of the *De statica medicina*, Giuseppe Ongaro opened up yet another aspect of Sanctorius's use of aphorisms, when he argued that it gives the work the character of a *Regimen sanitatis* (Sanctorius and Ongaro 2001: 40). In the next section, I will give an overview of these medieval hygienic writings and explore to what extent they are echoed in the *De statica medicina*.

---

[8] David Levine and Alan Bleakley have proposed a novel framework for aphorisms tailored to contemporary medical education and practice. In this context, aphorisms serve as rules of thumb in practice and as memory aids in medical education. The authors argue that aphorisms aid clinical judgement, reinforce professional behavior, and educate for narrative sensibility, which means to understand medicine not simply in technical-rational terms, but for example, to also listen carefully to patients' stories. Moreover, they identify aphorisms as a site of the clinician's identity construction and suggest that aphorisms be included in fictional accounts of medicine, such as television shows based on medical themes, to educate the public. See: Levine and Bleakley 2012.

4.1.2   The Medieval **Regimina sanitatis**

The medical literary genre of *Regimina sanitatis* is concerned with individual hygiene and served to give practical advice on diet and a healthy lifestyle. As innumerable and often very diverse texts are subsumed under its heading, the genre is somewhat complex and must be seen in the broader context of contemporary practical medical texts and dietary writings. It originated in the course of the second half of the thirteenth century and reached the peak of its popularity and diffusion at the end of the Middle Ages. However, dietary writings, more broadly conceived, continued to be in vogue well into the Renaissance, with output in the period from 1450 to 1650 proving the most prolific, numerically. This was a consequence not only of the invention of the printing press, but also of factors such as more widespread literacy or the medicalization of society.[9] Only in the later seventeenth century did the publications on diet decrease dramatically in number, the demand for dietaries apparently having been saturated by then.

Coming back to the medieval rules of health, a similar evolution of the genre can be detected. Initially directed to wealthy individuals, such as members of the civil or ecclesiastical nobility, or royalty, during the fourteenth century these writings came to be extended to the population in general, especially to the new urban social groups, such as merchants, craftsmen, or professionals. With the new consumers, a relatively large market for the genre began to grow and the regimens, originally mostly written in Latin, were increasingly translated into, or even written directly in the various vernacular languages. What is more, a growing number of them was composed in verse, which not only helped memorization, but also assisted the spread of the *Regimina*. Even though most of these texts were structured along the lines of the six non-natural things, the chapter on food and drink was particularly prominent and eventually became an independent medical genre in its own right. As mentioned above, these dietary writings gained particular importance and were popular until the end of the seventeenth century.

The authors of the *Regimina sanitatis* ranged from respected university physicians to anonymous writers, probably obscure doctors of no particular renown, whose names added nothing to the prestige of the work and were thus often overlooked and then forgotten. Contrary to this, the so-called *university* regimens were frequently linked to the teaching activity of their authors, which is why they addressed a larger audience from the start, and tended to consider all of the possibilities of human life, as for example the different ages, or complexions. During the plague of 1348, university physicians also composed so-called plague *regimina* to address laymen, reinforcing thereby the expansion of the medical literary genre (Sect. 3.3.1). Usually, the university regimens were not structured according to the six non-natural things, but contained scholastic elements, in particular *quaestiones*. Mostly published in the first half of the fourteenth century, these works combined

---

[9] For an overview of the broader cultural and social changes in the Western Middle Ages and early Renaissance as reflected in the history of medicine, see: Siraisi 1990.

profundity of content with simplicity of form. In contrast, the anonymous regimens appeared only later and were often characterized by the absence of an organizational scheme, especially in the fifteenth century (García-Ballester 1992: 119–22; Sotres 1998: 300–14; Albala 2002: 25–46).

The *De statica medicina* and Salerno's Regimen With the six non-natural things as organizational criterion, the *De statica medicina* followed the tradition of the medical school of Salerno. Situated in southern Italy, it was one of the first medical schools in Europe after the fall of Rome, famous for the expertise of its practitioners and key to the establishment of a standard education in medicine. In this context, Constantine of Africa (d. 1087) translated Arabic medical works into Latin, which profoundly influenced medieval hygiene and dealt with the topic in terms of the six non-natural things (Sect. 3.1.1). And in fact, in the *De statica medicina*'s section on food and drink Sanctorius referred to the *Regimen sanitatis salernitanum* (Salernitan Guide to Health). He explained that if unusual weight, gained from drinking the night before, would not be removed the day after, neither by the digestive power, nor by corruption, the following two verses were advised: "If drinking wine at night harms you, drink it again in the morning, and it will be medicine for you" (Sanctorius 1614: 47r).[10] This is one of the rare occasions, when Sanctorius gave insight into his literary sources in the *De statica medicina*, even citing directly from another work.[11] What is more, it hints at the connection between the *De statica medicina* and the genre of *Regimina sanitatis*.

The *Regimen sanitatis salernitanum* was a medieval medical poem and one of the most popular food and health guides up to and throughout the Renaissance. Its exact origin is, however, unknown. Probably, it was written by several anonymous authors associated with the school of Salerno, mostly in the late thirteenth century. Composed in catchy verse, it referred to the six non-naturals, but was not clearly structured along their lines. Rather, it was a miscellaneous collection of dietetic knowledge, uncomplicated and, often, witty, to which new verses were added progressively over the years (Wear 1993: 1288; Jacquart 1996: 224; Albala 2002: 24). Without overestimating its influence on the *De statica medicina*—the Salernitan poem was so famous and widespread that it was presumably known to most doctors—Sanctorius's citation of it shows that he was familiar with the genre of *Regimina sanitatis* and that he considered the work a reliable source, apt to complement his observations with the weighing chair. The orientation toward individual

---

[10] "Si nocturna tibi noceat potatio vini. Hoc tu manè bibas iterum, & fuerit medicina." See: Sanctorius 1614: 47r.

[11] Apart from the citation of the *Regimen sanitatis salernitanum*, Sanctorius directly mentioned the Roman encyclopedist Celsus (first century CE) and Hippocrates in the aphorisms of the *De statica medicina* (ibid.: 39v, 81r–81v). Furthermore, there are several indirect references to characters in the works of Hippocrates and Galen as well as one reference to "the philosopher," by which Sanctorius probably meant Aristotle (ibid.: 52v, 65r, Sanctorius 1634: 15r, 17r–17v, 40v). For the identification of Sanctorius's sources, see: Sanctorius and Ongaro 2001: 81, 85, 117, 129, 139, 157, 179.

hygiene and the use of verses to profit from the memory aids offered by rhythm are characteristics that the static aphorisms share with Salerno's regimen.

Similarities to University Regimens In other respects, however, Sanctorius's treatise more resembles the university regimens. First of all, his use of Latin suggests that he addressed the work to an audience within the realm of the university, to his students and colleagues. Outside of this context, the *De statica medicina* was reserved to learned physicians, scholars, or other well-educated, Latin-literate persons. Moreover, the aphorisms were designed for a broader public, not tailored to an individual's needs. Still, age and gender were seldom addressed by Sanctorius and it seems that his work was for the most part directed at middle-aged men.[12] Likewise, the medieval rules of health often overlooked childhood and old age and were mainly geared to a male audience. The activities Sanctorius mentioned in the section on exercise and rest imply that he envisaged a wealthy readership, who had time and money enough to play ball, dance, or travel in a palanquin. This is also supported by the fact that the *De statica medicina* did not refer to exercise performed by manual laborers. Interestingly, though, mental exercise, studying, and its relation to the affections of the mind was mentioned. Thus, the assumption that Sanctorius wrote the *De statica medicina* for a scholarly audience is further confirmed (Sanctorius 1614: 83r–84r; Sotres 1998: 314; Albala 2002: 151).

In this context it is important to bear in mind that Sanctorius was himself a university professor, when he published the *De statica medicina*. This urges the question: Was there a connection to Sanctorius's teaching activities? According to his own testimony, during his professorship at the University of Padua, he continually lectured on his instruments and static experiments in public as well as in private lessons. Furthermore, already 2 years before the publication of the *De statica medicina*, Sanctorius mentioned that one of his instruments, the thermoscope, could be admired by anyone who came to his house in Padua, and that he showed it to his disciples and taught them its use. The same instrument could also be detected in one of the static aphorisms. It seems then, that there was a connection between Sanctorius's lectures and the *De statica medicina*. But the latter being a published text, it is difficult to say to what extent it actually reflects Sanctorius's original university teaching. Were the procedures that he used in teaching only demonstrative, or did the students actively take part in his investigations? How was the balance between these innovative elements and more traditional features in his lectures on medical *theoria*? The relatively small role which insensible perspiration, the weighing procedures, and quantitative physiological reasoning more generally play in Sanctorius's voluminous commentaries imply that these aspects formed only a small part of his teaching overall. However, as the proceedings of the German Nation of Artists report, foreign students went to the University of Padua primarily for practical training and not for the formal lectures on the subject. In this light,

---

[12] The aphorisms of the *De statica medicina* that deal with age can be found here: Sanctorius 1614: 8v–9r, 19r, 31r, 42r, 74r, Sanctorius 1634: 12v–13r, 15r, 17v, 18v, 40v. Four aphorisms mention women; see: Sanctorius 1614: 15v, 69r, Sanctorius 1634: 13r–13v, 28v.

Sanctorius's statement that his lectures were crowded suggests that instrumentation and experimentation, in the sense of repeated and controlled observations, were an integral part of his teaching. But can we trust his words given the quarrels he had with the German students and the mysterious circumstances under which his professorship ended? Without being able to clarify these questions at this point, I will resume their discussion below, when dealing more closely with Sanctorius's quantitative approach to physiology and the practical and material dimensions of his work (Sanctorius 1612b: 62, 105; 1614: 21r; 1625: Ad lectorem; Bylebyl 1979: 351 f.).

Two further factors relate the *De statica medicina* to the category of university regimens. On the one hand, Sanctorius's later inclusion of the plague aphorisms reminds the medieval plague *regimina*, written by university physicians. On the other hand, the combination of profundity of content with simplicity of form, characteristic of university regimen, applies to Sanctorius's treatise as well. Thus, besides the parallels to Salerno's regimen, the *De statica medicina* also bears strong similarities to university regimens.

Health Handbooks and the Prominence of Food and Drink   The practical orientation of the *De statica medicina* and the lack of theoretical considerations associate the work not only with the medieval rules of health, but also with other contemporary practical medical texts, such as the *Tacuinum sanitatis*, or dietaries.[13] These works combine theoretical knowledge with knowledge gained from practice and observation. As a counterpart to scholarly tomes, their authors wanted to present medical knowledge in an abridged and concise way. Practical advice was the focus, not theoretical debate. This trend responded to a public eager for self-improvement. People became increasingly diet conscious and were interested in knowledge that would guide them to lead a healthy life. Accordingly, the *De statica medicina* centers around prevention rather than cure, even though Sanctorius occasionally referred to sick bodies, too (Sanctorius 1614: e.g., 11r, 19v, 55r–55v; 1634: e.g., 13v, 15r).

Following the prominence of food and drink in the *Regimen sanitatis* literature, this non-natural pair also takes up an important place in the static aphorisms. As the second largest section of the work, it is surpassed only by the first section, which deals with the weighing of insensible perspiration. It is striking that it was especially to those two sections that Sanctorius added aphorisms in later editions of the *De statica medicina*. Of the ninety-three additional aphorisms, the plague aphorisms excluded, forty-four belong to the first section and twenty-three to the third, that is, food and drink.[14] This may reflect Sanctorius's own research agenda and its results, an issue which will be scrutinized in later chapters. At the same time, however, it

---

[13] The *Tacuinum sanitatis* was a genre of richly illustrated guides to health that was popular in Western Europe in the late Middle Ages and addressed to a courtly audience. Like most of the *Regimina sanitatis*, the *Tacuinum sanitates*, too, was structured in line with the six non-natural things. For more information, see: Arano 1976, Bovey 2005.

[14] While Sanctorius highlighted most of the aphorisms "added by the author" to the 1634 edition of the *De statica medicina*, he overlooked to mark up two new aphorisms and one deletion, see: Sanctorius 1614: 34v–35r, 50r–50v, 84r, Sanctorius 1634: 31r, 43v, 68v.

may also have been a reaction to the needs of the readership. The great demand for health handbooks dealing with food may have prompted Sanctorius to not only republish the *De statica medicina*, but also to expand this topic, in order to increase sales of the work. This underlines the practical character of the work, as practical texts of the time were often revised for and by their users. Moreover, the many modifications of the *De statica medicina*—the added aphorisms, the section on plague, the response to the *Staticomastix* (Sect. 3.1, fn. 2)—illustrate the importance of the work for Sanctorius and imply that the work was discussed controversially at the time and that there was an audience thirsty for more. In contrast to the lengthy commentaries, the small book offered Sanctorius the opportunity to react quickly to external circumstances, be it the defense of his static doctrine, or the plague.

Concluding this chapter, it must be stressed that despite the many similarities between the medieval regimens of health and the *De statica medicina* there are also important differences: the focus on insensible perspiration, the related reinterpretation of the six non-natural things, and the quantitative approach to physiological processes. While one occasionally encounters the measurement of meals in the Renaissance dietary literature (Pontormo and Nigro 1988; Cornaro 1591; Lessius 1613), Sanctorius's observation of weight changes in human bodies by means of a steelyard was an absolute novelty. Just as the static aphorisms were a combination of old and new ideas, so, too, the form and style of the *De statica medicina* merged different characteristics of established genres of practical medical texts, such as the *Regimina sanitatis* and dietaries, and peppered them with a new element: the presentation of research results based on observation and quantification. It may have been exactly this mixture, which guaranteed the *De statica medicina*'s great fame and long-lasting popularity.

4.2   The Use of Instruments

The practical orientation of the *De statica medicina* is closely connected to Sanctorius's use of instruments, first and foremost the huge steelyard with which he intended to weigh insensible perspiration. But this was by no means the only instrument that Sanctorius proposed. The *Commentary on Avicenna*, the only work in which Sanctorius published the illustrations and explanations of his medical instruments, discloses a variety of devices that go far beyond the purpose of measuring *perspiratio insensibilis* (Fig. 4.1).[15] The book even has a special index of instruments which is so uncharacteristic of the genre as to differentiate Sanctorius's commentary from the entire previous tradition of commentary on Avicenna's *Canon*. In the preface to the work, Sanctorius explained that this was motivated by the fear of

---

[15] Sanctorius did refer to some of his instruments in other works, but neither explained them in detail nor illustrated them. Instead, he directed the reader to the *Commentary on Avicenna*, or to the *Book on Medical Instruments* (Liber de instrumentis medicis), which he probably never published (Sect. 2.6). See: Sanctorius 1603: 26v, 109r–109v, Sanctorius 1612b: 59, 62, 105, 136, 229, 374, Sanctorius 1614: 20v–21r, Sanctorius 1629a: 24, 137, 153, 164 f., 209, 326, 373 f., 378.

plagiarism. As was mentioned above (Sect. 2.3), he found out that some of his students were copying his instruments and claiming them as their own inventions. In response to this, he hurried to add illustrations of the instruments to his next publication, which were therefore, he apologized, in a rather rudimentary style. Originally, he had planned to publish elaborate illustrations and descriptions of all of his instruments in a separate book with the title *De instrumentis medicis*, but his teaching activity distracted him from finishing the work (Sect. 2.6). Sanctorius stated that he included in the *Commentary on Avicenna* solely those instruments pertaining to physiology—the subject matter of courses on theoretical medicine that were based on readings of Avicenna's *Canon* (Sanctorius 1625: Ad lectorem; Siraisi 1987: 181, 209).

The index of instruments contains thirty-four different items, which range from clysters and cupping glasses to a special sickbed, hygrometers, and thermoscopes (Sanctorius 1625: index instrumentorum). Most of them can be roughly summarized in the following categories: surgical instruments, measuring instruments, instruments for the improvement and alleviation of the sick, and instruments to demonstrate optical phenomena (Fig. 4.1). Thus, Sanctorius's development and use of instruments was not exclusively related to his doctrine of static medicine, but formed part of a larger effort—to improve therapeutics. The variety of the devices suggests a long and miscellaneous medical practice, during which Sanctorius gained experience in various medical fields, before striving to advance their practices with his instruments. In doing so, he may have followed the Galenic ideal of the medical man who provides theoretical and practical expertise in physic, surgery, and pharmacology (Nutton 1985: 80).

Existing studies on Sanctorius have tended to focus on the measuring instruments, among them the famous steelyard to weigh insensible perspiration.[16] Contrary to this, I will start my analysis of Sanctorius's use of instrumentation by exploring some of the lesser-known instruments and their relation to the physician's medical practice and teaching activities. Even though—or precisely because—these instruments were not part of the quantitative approach to physiology, studying them helps complement the picture of Sanctorius as a practicing physician. What is more, it allows the *De statica medicina* to be reviewed afresh within the broader practical context of Sanctorius's undertakings.

4.2.1   Surgical Instruments and Anatomy

In the *Commentary on Avicenna*, Sanctorius presented six instruments that I categorize as surgical instruments (Fig. 4.1). These are a syringe to extract bladder stones (Fig. 4.2), several trocars (Fig. 4.3), an uterus speculum (Fig. 4.4), a device for

---

[16] E.g., Mitchell 1892, Miessen 1940, Mulcahy 1997, Bigotti and Taylor 2017, Bigotti 2018, Hollerbach 2018. One exception is a monograph in Croatian by Mirko Grmek, in which the author also considered some of Sanctorius's surgical instruments and instruments for the improvement and alleviation of the sick. See: Grmek 1952: esp. 31–61.

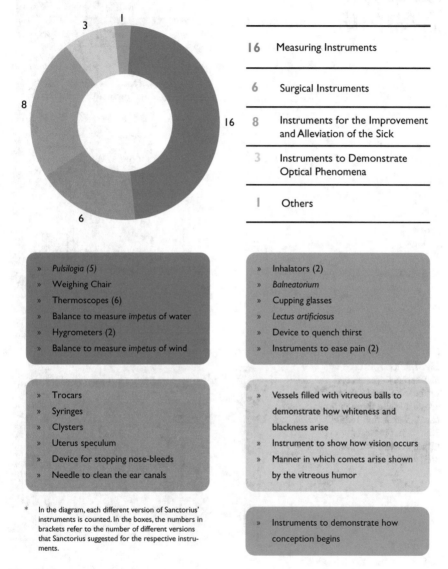

OVERVIEW OF SANCTORIUS'S INSTRUMENTS

16	Measuring Instruments
6	Surgical Instruments
8	Instruments for the Improvement and Alleviation of the Sick
3	Instruments to Demonstrate Optical Phenomena
1	Others

» Pulsilogia (5)
» Weighing Chair
» Thermoscopes (6)
» Balance to measure *impetus* of water
» Hygrometers (2)
» Balance to measure *impetus* of wind

» Inhalators (2)
» *Balneatorium*
» Cupping glasses
» *Lectus artificiosus*
» Device to quench thirst
» Instruments to ease pain (2)

» Trocars
» Syringes
» Clysters
» Uterus speculum
» Device for stopping nose-bleeds
» Needle to clean the ear canals

» Vessels filled with vitreous balls to demonstrate how whiteness and blackness arise
» Instrument to show how vision occurs
» Manner in which comets arise shown by the vitreous humor

*   In the diagram, each different version of Sanctorius' instruments is counted. In the boxes, the numbers in brackets refer to the number of different versions that Sanctorius suggested for the respective instruments.

» Instruments to demonstrate how conception begins

Fig. 4.1 Overview of Sanctorius's instruments

stopping nose-bleeds (Fig. 4.5), clysters (Fig. 4.6), and a special needle to remove cerumen from the ear canal (Fig. 4.7).[17] Of course, some of these instruments might

---

[17] The clyster on the right (Fig. 4.6) seems to be identical to the uterus speculum (Fig. 4.4) which is why I counted them as only one instrument in the circle diagram in Fig. 4.1.

Fig. 4.2 A three-pointed
syringe to extract bladder
stones (Sanctorius 1625:
302). (© British Library
Board 542.h.11, 302)

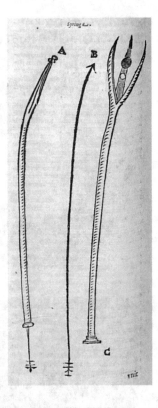

Fig. 4.3 Trocars used by
Sanctorius to prevent
suffocation and to draw off
dropsical fluid through the
navel (Sanctorius 1625:
363, 435). (© British
Library Board 542.h.11,
363, 435)

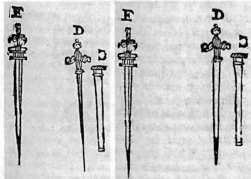

Fig. 4.4 Uterus speculum
to extract water from the
uterus and to cure internal
affections of the uterus
(Sanctorius 1625: 435).
(© British Library Board
542.h.11, 435)

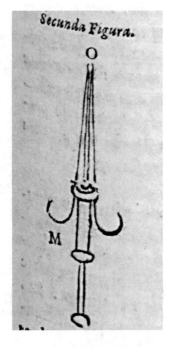

Fig. 4.5 A device for
stopping nosebleeds
(Sanctorius 1625: 596,
668). (© British Library
Board 542.h.11, 596)

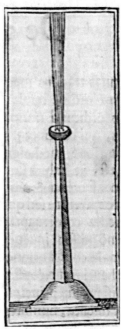

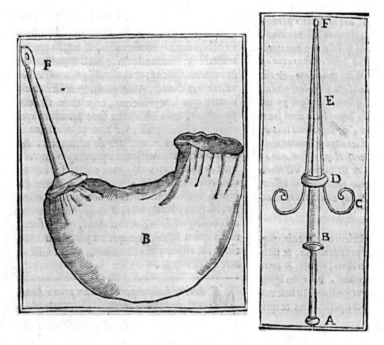

Fig. 4.6 Clysters (Sanctorius 1625: 596, 652). (© British Library Board 542.h.11, 596, 652)

have been used by physicians as well. However, with this classification I want to emphasize the role of "hands-on" medical practice and understand surgery as "the manual operations needed to restore health" (Grendler 2002: 322).

Thus, Sanctorius was clearly a medical man, who did not shy away from using his hands to perform operations himself. His expertise in surgery has already been mentioned (Sect. 2.6), and frequent references in his books to his surgical successes, along with the instruments, reinforce the impression that this field was an important part of his medical practice. In the description of one of his trocars, he wrote:

> But if there is no other remedy that helps infants and adults, who choke [because of excreta accumulated in the lungs], our perforation which is done below the larynx with instrument E revives the patient safely from immediate death to immediate health. If the choking matter is above the larynx, or above the perforation, … or in the lungs themselves, the perforation is useless. Instrument C is a perforated silver tube.[18] Instrument D is a pointed needle, which is inserted in instrument C … and this results in instrument E …. If we then want to perforate with this instrument, we have to ensure, first, that the patient leans the head backwards so that the trachea is distended. Then, after two, or three circles [probably fingers] beneath the larynx, we perforate the intermediate space of the circle. The following principle has to be followed, namely that as soon as the instrument begins to enter the cavity of the trachea, it needs to be immediately removed and the internal needle has to be separated from the tube, so that it does not sting the opposite part of the trachea. Done in this way, the tube is safely pushed inward. After that, inspiration and expiration freely occur

---

[18] The letter C is printed inversely (Fig. 4.3).

Fig. 4.7 Instrument to
clean the external ear
canals (Sanctorius 1625:
764). (© British Library
Board 542.h.11, 764)

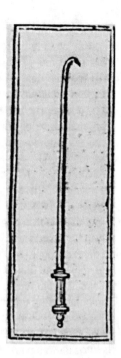

through the perforated tube from which the needle has been removed, and all suffocation is
prevented not only in angina, but in any similar affection (Sanctorius 1625: 363).[19]

Hence, Sanctorius gave his readers detailed information on how his trocar had to be
used in order to prevent suffocation, without forgetting to explain how the patient
should be positioned (Fig. 4.3). Interestingly, he did not describe how to hold down
the patient, who must have squirmed with pain during the operation. Given the fact
that the *Commentary on Avicenna* resulted from his teaching activity at the
University of Padua, this begs the question to what extent these explanations reflect
his courses on theoretical medicine. In Padua (as in other Italian universities), surgery was taught as a separate subject along with theoretical and practical medicine.

[19] "Sed pro infantibus & adultis, qui suffocantur, si nullum aliud remedium iuvet, nostra perforatio
facta infra laryngem cum instrumento E à subita morte ad subitam salutem tutò patientem revocat:
dummodo materia suffocans sit à larynge supra, vel supra perforationem, quia si infra, vel in ipso
pulmone existat vana redditur perforatio. Instrumentum C est fistula argentea perforata.
Instrumentum D est acus mucronata, quae intromittitur in instrumentum C quo tamen acus longior
est, & intromissa sit instrumentum E, quod cum illo sit ita unitum, ut tactui nulla occurrat asperitas:
imo instrumentum E unum continuum, & non duo esse videntur. Dum igitur volumus dicto instrumento perforare, prius curamus, ut patientes inclinent caput retrorsum, hoc fine, ut aspera arteria
distendatur: deinde sub larynge post duos, vel tres circulos, circuli intermedium perforamus: hac
lege servata quod dum incipit instrumentum ingredi cavitatem tracheae statim retrahatur, & aufera-
tur ab ipsa fistula acus interna, ne pungat partem oppositam tracheae: quo peracto fistula tutò inti-
mius impellitur: inde per fistulam perforatam acu ablata, libera sit inspiratio, & expiratio,
omninoque prohibetur suffocatio non solum in angina suffocante, sed in quocumque simili
affectu: …". See: Sanctorius 1625: 363.

However, surgery had been the least important medical university chair and grew in importance only in the sixteenth century, when it was combined with anatomy. But from this time on, anatomy began to oust surgery and grew increasingly distinct, until the two chairs were separated in the second half of the seventeenth century. One can imagine the surprise of Sanctorius's students, who expected to hear lectures on *theoria* but were instead confronted with explanations and maybe also demonstrations of surgical operations—a medical field which was neither prestigious nor popular within the university context (Facciolati 1757: 385–98; Grendler 2002: 322–34).

Still, there is also another side to this issue. Medical historians have argued that the contrast between surgeons and physicians was not as great as has often been suggested. Especially in Italy, where academic training was available for surgeons, a graduate with a degree in surgery would have much in common with a learned physician. He might have had the same lecturers and studied similar texts as the medical student proper. Moreover, students of surgery also took courses in other branches of medicine, and it is thus possible that among Sanctorius's audience were prospective surgeons, probably attracted by his novel method of teaching *theoria*. The outstanding importance that anatomy enjoyed at the time paired with an interest in practical training on the part of the students, might also explain why students were intrigued to learn more about the use of trocars, or clysters. The fact that Sanctorius was named as *promotore* by a doctoral candidate for his examination in surgery suggests that Sanctorius's expertise in this field was known to students, either through their own attendance at his lectures, through his reputation as a practicing physician, or through his *Commentary on Avicenna*, which was already published by then (Sect. 2.6). One of Sanctorius's students, Johan van Beverwijck, recalled in 1633 that he continually followed "his most famous doctor Sanctorius" on visits to the sick in Padua (Beverovicius 1638: 216); this, in the context of a discussion on the causes of kidney stones that is included in Beverwijck's work *De calculo renum & vesicae* (On kidney and bladder stones), which also contains a *consilium* (piece of advice) from Sanctorius on removing bladder stones.[20] It therefore seems highly likely that Sanctorius took his students outside the classroom and let them attend his medical practice, including surgical procedures, such as lithotomy (Sect. 2.6).

Teaching anatomy, rather than surgery, became increasingly important, as it was mostly anatomical research, which enabled professors to make new discoveries. However, as Vivian Nutton has pointed out, in the sixteenth century there was "humanist surgery," too, that is, surgery based on classical texts, which could lead to practical as well as intellectual benefits. Thus, also surgeons laid claim to successful innovation, from time to time, especially with regard to the invention of surgical procedures and instruments, control of pain, and wound management. Yet, it is difficult to say whether these were merely variations on techniques and

---

[20] "Quae mihi in memoriam revocarunt clarissimi Doctoris mei Sanctorii, quem Patavii olim ad aegros sectatus sum, …". See: Beverovicius 1638: 216.

instruments described in textbooks, let alone whether they substantially improved surgical procedures and their outcome (Nutton 1985: 75–87). With regard to Sanctorius, it is not my intention to investigate his claims to novelty and success in surgery, nor to discuss his surgical instrumentation and its practical application at any length.[21] But it is important to note that by highlighting the originality of his surgical instruments and techniques, Sanctorius might have aroused the interest of medical students who were not aiming for a degree in surgery. At a time when anatomical studies were flourishing, one can easily imagine that new approaches to dangerous operations such as the tracheotomy in children attracted attention. These, together with his other instruments, may well have lent his lectures an aura of novelty.

Whatever the reactions of the students may have been, Sanctorius's inclusion of surgical instrumentation in his *Commentary on Avicenna* shows that he considered surgical training important for aspiring physicians and wanted to share his experience in this field. To what extent this reflects his own interests, or educational purposes is difficult to say. In any case, the descriptions and illustrations of his instruments are closely connected to the passages of text and commentary in which they appear. The implication—given that the first part of the first book of Avicenna's *Canon* that Sanctorius commented on was used as a medical physiology textbook on courses in theoretical medicine—is that Sanctorius believed that surgical therapeutics might well be integrated into physiological theory (Siraisi 1987: 10, 210).

Leaving Sanctorius's teaching activities aside, for a moment, and focusing on him as a practicing physician, it was not uncommon in Italy for university-trained doctors to practice surgery. In fact, a physician skilled in surgery was often held in high regard. According to Richard Palmer, physicians and surgeons cooperated freely in Venice, and certain doctors were members, at separate times, of both medical colleges: the College of Surgeons of Venice and the College of Physicians of Venice.[22] However, while physicians were allowed to practice surgery, surgeons could certainly not practice medicine. This was prohibited by the statutes of the colleges and by civic law. Moreover, the salaries of surgeons were generally lower than those of physicians. In contrast, doctors in charge of the plague were usually surgeons, since the treatment of the disease was regarded as a primarily surgical operation. Still, despite these differences, there was a close relationship between the two branches of the profession in daily medical practice in Venice (Palmer 1979: 451–60).

Sanctorius's works show that he regarded this with a certain ambivalence. On the one hand, he frequently denounced errors committed by surgeons, or mentioned their inexperience. He contrasted this with his own surgical experience and the presentation of his instruments. Furthermore, his involvement in the treatment of the plague suggests that he was known and trusted for his surgical experience. Despite

---

[21] Pietro Castagna and Mirko Grmek ascribed the invention of the trocar to Sanctorius. See: Castagna 1951, Grmek 1952: 53–6.

[22] I refer here to collegiate surgeons, who were distinct from the guild of barber-surgeons and formed a professional elite in Venice (Palmer 1979: 456).

his previous denial of the existence of the disease in Venice (Sect. 2.7), he was assigned the care of a Venetian district—the *Sestiero di Cannaregio*. On the other hand, Sanctorius never explicitly referred to himself as a surgeon and was not a member of the *College of Surgeons of Venice*. In the *Commentary on Galen,* he explained how he had examined a corpse, while leaving the cutting to a surgeon. Thus, it is somewhat unclear to what extent Sanctorius was willing to get his hands bloodied and how often he asked for the assistance of a surgeon. In my opinion, Sanctorius saw himself as a physician, who fulfilled the Galenic ideal of the unity of medical knowledge (in physic, surgery, and pharmacology) according to which a medical man used his head and his hands alike, to carry out complicated and diffi-cult tasks in all medical fields. Certainly, the higher social standing of physicians compared to surgeons may have made Sanctorius hesitate to refer to himself as a surgeon. From his publications, it is clear that Sanctorius had surgical experience and conducted surgical operations. But it remains unclear how much he was assisted by other surgeons, even though his frequent complaints about incompetent surgeons imply that he was constantly in touch with members of this profession (Sanctorius 1603: 37v–38r; 1612b: 220, 237, 335; 1625: 12 f., 36; ASVe-a (n.d.): 60r–61v).

All in all, Sanctorius's interest and experience in surgery was not uncommon at the time, for a university-trained physician in Italy. But his development of surgical instruments and their presentation in a university textbook on theoretical medicine was a new departure.

Anatomy The close connection between surgery and anatomy was already men-tioned above. In his publications, Sanctorius stated that surgeons needed to have anatomical knowledge to properly carry out their work. Anatomists, on their part, needed to have skills in surgery.[23] In this context, Sanctorius mentioned an error made by famous anatomists regarding the surgical operation of lithotomy. He wrote:

> … Colombo slipped [into error] in the fifth book of his anatomy in Chapter 26, when deal-ing with the position of the muscles of the neck of the bladder. [In this book] he holds that a lithotomist, who has no knowledge of the bladder, sometimes cuts the muscle of the neck transversely, whereupon a new disease is introduced; for (as he says), the muscle having been cut, the urine can no longer be contained. Because this opinion was held by the most educated anatomists, I am ashamed to refute it. But if Colombo had ever observed a lithoto-mist's incision, he would have changed his opinion. There is no risk of the bladder neck being cut, in that case, as it lies at a distance of around half the span of a hand from the lithotomist's incision. Moreover, this latter wound [the incision] arrives only at the tube through which the urine is released. And it cannot penetrate any deeper since it is impeded by a syringe inserted beforehand by the lithotomist. Since Colombo never observed this, it is no surprise that he made an inexcusable error (Sanctorius 1612a: 662 f.).[24]

---

[23] Contrary to Sanctorius's emphasis on the importance of anatomical studies for surgeons, there were also attempts by Venetian surgeons in the late sixteenth century to distinguish learned surgery from anatomy. As Cynthia Klestinec has shown, anatomy had become a conflicted resource by then (Klestinec 2016).

[24] "… sicuti lapsus est Columbus lib. 5. suae anatomiae cap. 26. agens de musculi colli vesicae situ, ubi habet haec aliquando a Lithocomo, qui scitum vesicae ignorat, musculum cervicis transversim incidi, & inde novum morbum induci, quoniam dicit secto hoc musculo urinam non amplius posse contineri, quam sententiam licet fuerit Anatomici eruditissimi, pudet me refellere: quia si semel

Fig. 4.8 Illustration of the bladder neck, indicating the point at which the lithotomist made an incision in order to extract a bladder stone (Sanctorius 1603: 65v). (Bayerische Staatsbibliothek München, 2 Med.g.149, p. 65v, urn: nbn:de:bvb:12-bsb10942689–8)

Hence, according to Sanctorius, the famous anatomist and surgeon Realdo Colombo (ca. 1516–1559), who succeeded Andreas Vesalius (1514–1564) as professor of surgery at the University of Padua in 1543, had never observed the work of a lithotomist, nor performed a lithotomy himself. Without discussing whether the criticism was legitimate, the citation shows that Sanctorius was of the opinion that even the most learned anatomists could learn something from lithotomists, medical practitioners who were often itinerant and probably had never entered a university. According to him, thus, surgical techniques were closely related to anatomical knowledge and only a combined expertise in both fields enabled the doctor to fulfill his duties, even though this meant leaving the universities' anatomical theatres and lecterns to become acquainted with the daily work of medical practitioners. To his first work *Methodi vitandorum errorum*, Sanctorius even added a figure (Fig. 4.8) that shows the bladder neck (E) and the point at which the lithotomist made his incision (o) (Sanctorius 1603: 65v; 1625: 12 f.; Colombero 1982).

Sanctorius's strong interest and expertise in lithotomy has already become apparent, in view of the fact that he composed a *consilium* on this surgical operation and presented a special syringe to extract bladder stones in the *Commentary on Avicenna* (Fig. 4.2, Sect. 2.6). This might well be connected to his medical practice, as in the Renaissance, many people suffered from stone in the bladder. How often Sanctorius actually performed a lithotomy, which involved a high level of risk, and how often he made use of his syringe is not known. In any case, he was familiar with the work of lithotomists and did not shy away from learning their skills. His inclusion of a figure that illustrates the sections of the bladder in relation to a male genital organ

---

Columbus inspexisset Lithotomum seccantem, mutasset sententiam: quia collum vesicae incidi non potest, cum distet à vulnere Lithotomi dimidij palmi interstitio circiter. Praeterea illud vulnur vulnus solum pervenit ad fistulam, qua defertur lotium: neque ultra penetrare potest: quia impeditur à syringa prius à Lithotomo immissa: quod cum non viderit Columbus non est mirum, si in errorem inexcusabilem inciderit." See: Sanctorius 1612a: 662 f.

Fig. 4.9 Illustration of the stomach (Sanctorius 1603: 64v)

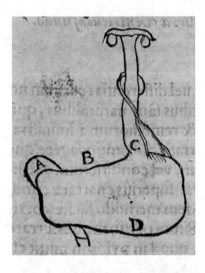

Fig. 4.10 Illustration of the veins of the arms, indicating their relation to the liver (Sanctorius 1603: 76r). (Bayerische Staatsbibliothek München, 2 Med.g.149, pp. 64v, 76r, urn: nbn:de:bvb:12-bsb10942689–8)

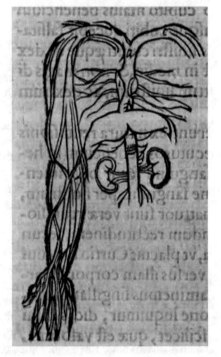

highlights his firm belief that anatomy and surgery were closely intertwined. Even though it is only a very rough illustration, it hints at Sanctorius's own anatomical experience—together with two other figures in the *Methodi vitandorum errorum*, one of the stomach and the other of the veins of the arms and their relation to the liver (Figs. 4.9 and 4.10) (Sanctorius 1603: 64v, 65v, 76r).

The illustration and discussion of the stomach (Fig. 4.9) is particularly interest-ing, because Sanctorius mentioned here an anatomical error made by Galen: the Greek physician had maintained that the pylorus was at the bottom of the stomach. But experience contradicted this, so Sanctorius, since one could observe in human cadavers that the pylorus was on the right-hand side of the stomach, nearly the span of a hand away from the bottom of the stomach.[25] And if, as Galen had thought, the pylorus was at the bottom of the stomach, its contents would easily spill out, Sanctorius continued. He concluded:

> We show in the figure of the stomach [Fig. 4.9] that things behave this way, as was observed by Vesalius and by other anatomists of our time. Is it not obvious to the senses that where A is, is the beginning of the pylorus? And is the position of D, which is the bottom part of the stomach, not distant from A? Thus, Galen saw in apes and dogs that the beginning of the pylorus was at the bottom [of the stomach] and believed that in the same was true of humans, too (Sanctorius 1603: 64v).[26]

The citation makes clear that Sanctorius accepted the achievement of Renaissance anatomists and agreed that they had successfully contradicted Galen. Perusal of Sanctorius's works reveals that this is not the only passage, in which he refuted Galen on anatomical matters. On the question as to whether the cerebellum or the cerebrum was harder, Sanctorius again followed "proper inspection" and the opin-ion of "Vesalius, Colombo, and many others" and held that the cerebellum was not harder than the cerebrum, which was contrary to Galen's teaching. According to Sanctorius, this error had resulted in further mistakes, such as the assumption that the nerves originated in the cerebellum, whereas they actually originate in the cere-brum. Other instances of Sanctorius contradicting Galen concern the position of the stomach and the kidneys, to name but two examples (Sanctorius 1603: 70v; 1612a: 279; 1612b: 144, 231; 1625: 331 f.).[27]

But, as was common among Galenic physicians and anatomists at the time, Sanctorius excused Galen's mistakes by pointing out his limited access to human cadavers. According to him, Galen had observed only two corpses in his lifetime, and imperfect (i.e., damaged) ones at that. What is more, in contradicting Galen, Sanctorius actually believed himself to be following one of the Greek physician's own important precepts, namely: "rather, those are imprudent, who put more faith

---

[25] In the illustration, the pylorus is marked on the left part of the stomach (instead of on the right) with the letter A. I think this is an error due to the inversion of the illustration for printing, even though the letters are illustrated correctly. In fact, this suggests that the letters were not part of the woodcut, but were printed separately. Another possible reason for the error might have been negli-gence on the part of the woodcutter, whose identity is not known to me and might therefore even have been Sanctorius himself.

[26] "Quod igitur ita se res habeat, ventriculi figuram observatam à Vesalio, & a caeteris anatomicis nostri temporis proponamus, an ne sensu patet ubi est a, ibi esse pylori exordium? an ne locus ubi est d, distat ab a, quod est ima ventriculi pars? videns igitur Galenus in simijs, & canibus esse pylori initium in fundo, credidit in eodem situ, esse quoque in hominibus." See: Sanctorius 1603: 64v.

[27] For another important instance, in which Sanctorius accepted recent anatomical findings and refuted Galen, see below, Sect. 4.2.3.

in authorities than in experience and reason" (Sanctorius 1603: 74v).[28] Thus, according to Sanctorius, the anatomists and he himself were following the teachings of Galen in trusting more to observation than to Galen's authority. It was in the Galenic spirit that they refuted Galen. In his *Methodi vitandorum errorum*, Sanctorius even complained that scholars at many European universities currently believed more in Aristotle, Galen, and Hippocrates than in what their own senses told them. Interestingly, Sanctorius still greatly relied, nonetheless, on established authorities in anatomical matters, as a look at his commentaries shows. Here, for example, his defense of the Galenic notion that the veins and, consequently, the heat of the blood originated in the liver and not in the heart paints an entirely different picture. Sanctorius referred to the problem in the traditional form, as a dispute between the philosophers and the physicians.[29] To support his view, he drew not on anatomical observations, but on the established authority of Hippocrates and Galen (Sanctorius 1603: 74r–74v; 1612a: 429; 1612b: 144, 231).

Hence, Sanctorius's criticism of trusting more in authority than in one's own senses can be turned against himself. Apparently, his willingness to accept very recent anatomical findings was limited to topics that did not touch upon the major traditional matters of controversy between Aristotelians and Galenists. As Nancy Siraisi has pointed out, Sanctorius, when arguing in his *Commentary on Avicenna* against the Aristotelian view of the primacy of the heart, relied almost exclusively on texts by Galen. Even with regard to experience (*experimenta*), he referred to Galen, explaining, for example, that, since a tortoise whose heart has been removed is still able to walk, movement and the senses must originate in the brain and not in the heart. A similar passage can be found also in Sanctorius's *Commentary on Galen* (Sanctorius 1612a: 251–5; 1625: 627–33; Siraisi 1987: 323 f.).[30]

The topic of *spiritus* is likewise connected to the issue of the anatomy and function of the heart and the brain. As was mentioned above (Sect. 3.2.6), the heart and the brain were the organs in which the spirits were generated, according to Galenic physiology, and the spirits were responsible for sensation, voluntary motion, and the heating of the body. In the sixteenth century, however, the anatomist Giacomo Berengario da Carpi (ca. 1460–ca. 1530) and shortly after him, Vesalius denied the existence of the retiform plexus (*rete mirabile*) in the brain, where the animal spirits were thought to be prepared. Even though both anatomists argued that it could not be observed in the human brain, Sanctorius held that the retiform plexus is

---

[28] "… illos potius, qui magis credunt auctoribus, quam experientiae, & rationibus, esse temerarios: …". See: Sanctorius 1603: 74v.

[29] Galen had differed from Aristotle on some basic issues, such as the seat and division of the soul, the relative functions of the brain, heart, and liver, as well as male and female seed. Here, Galen's authority conflicted with Aristotle's authority, which fueled a dispute between their respective followers. Generally, philosophers inclined toward their authority, Aristotle, and physicians toward theirs, Galen, even though there were many philosopher-physicians, who tried to find a compromise between the two (Temkin 1973: 73).

[30] On the role of authoritative argument as a form of knowledge in Renaissance medicine, see: Maclean 2002: 191 ff.

conspicuous. How he came to this conclusion, iscompletely unclear, however. With regard to the vital spirits, Sanctorius also followed the traditional theory, according to which blood passed from the right ventricle of the heart through pores in the septum to the left ventricle, where it, when mixed with inhaled air, served to generate vital spirits. Even though anatomists like Vesalius and Colombo denied the existence of the pores in the interventricular septum, Sanctorius claimed that he had personally been able to observe these pores in dissection, although he assumed that they must be more open in living bodies than in corpses (Sanctorius 1612a: 260; 1625: 746; 1629a: 363; Siegel 1968: 68, 70) (Sect. 3.2.6, fn. 34).[31]

Thus, Sanctorius endorsed recent anatomical findings only as long as they could be accommodated within Galenic theory. Any new material that did not fit within a Galenic framework was not integrated into his own work. When he observed human cadavers himself, he seems to have worn Galenic lenses, so to speak, through which he could detect the retiform plexus and the pores in the septum. Or, he might have actually relied on others' observations, of instead of his own, without mentioning it. In fact, Sanctorius's ambiguous attitude toward firsthand observation and his ongoing reliance on authority with regard to medical theory was typical for university-trained physicians and anatomists at the time. Original findings from observation were often not carried over into theory and Galenic medicine was still regarded as a reliable framework into which any novel observations should be integrated (Wear 1981: 233–53).

Regarding Sanctorius's own anatomical experience, it can be assumed that he observed and anatomized human and animal cadavers himself. In the *De remediorum inventione*, Sanctorius wrote for example that he had opened the body of several persons killed by malignant fever and thereby discovered in their liver a small, entirely blackish gangrene. In the *Commentary on Avicenna*, he described how he had observed the brain of a lamb, which was tepid, not cold, to the touch. Moreover, Sanctorius mentioned that he reproduced anatomical procedures previously conducted by others. In the *Commentary on Galen*, he referred to a dissection around 1611 by a certain Aloysius Regocia (life dates unknown), who removed the bowels from a cadaver to show that a clyster cannot pass through the upper parts of the intestine because of the valves between the colon and the cecum. In a later passage of his commentary, Sanctorius claimed to have performed the same demonstration himself, several times. Without going into the details of the procedure, it is interesting to note that it bears similarities to an anatomical demonstration of the ileocecal valve by Caspar Bauhin (1560–1624). The physician from Basel was the first to describe this valve in detail, and published his findings in the treatise *De corporis humani partibus externis* (On the External Human Body Parts, 1588). Even though Sanctorius did not mention Bauhin in his works, it can be assumed that by the time Aloysius Regocia performed his dissection, word of the discovery of the ileocecal valve had already been spread in Venetian–Paduan medical circles, just like Bauhin's

---

[31] Consequently, Sanctorius also refuted Realdo Colombo's hypothesis that blood passed from one side of the heart to the other via the lungs. For more instances of Sanctorius's defense of Galenic ideas against the work of sixteenth-century anatomists, see: Siraisi 1987: 309–44.

anatomical demonstration of it. Thus, Sanctorius was actively engaged in dissections that referenced recent developments in the field of anatomy (Sanctorius 1612b: 196 f., 293; 1625: 318; 1629b: 70; Stolberg 2010: 9 f.).

Sanctorius's mention of the anatomical demonstration by Aloysius Regocia is interesting also for another reason. Apparently, Sir Henry Wotton (1568–1639), the English ambassador to Venice, was present. Sanctorius wrote that with him there were also "other very distinguished barons, to whose pleasure I held a broad discourse on anatomy's hidden secrets" (Sanctorius 1612b: 197).[32] Hence, Sanctorius attended public anatomical dissections, which were conducted within the most illustrious circles of the Republic of Venice. In fact, Sir Henry Wotton was closely associated with Paolo Sarpi and supported him in defending the Republic of Venice in its diplomatic quarrel with the Papal Curia at the beginning of the seventeenth century.[33] As was mentioned earlier, Sanctorius was a close friend of Sarpi and also involved in a dispute with the Church regarding his presidency of the *Collegio Veneto*—the first institution to confer doctorates without ecclesiastical intervention (Sect. 2.4). Accordingly, Sanctorius discussed recent anatomical findings not only in the university context, with students and colleagues, but also in the highest political and diplomatic circles of the Venetian Republic, whose members were critical of the Church and the Pope (Wootton 1983: 93 f.).

Besides the anatomical procedures that Sanctorius claimed to have conducted, he also attended the annual public anatomies in Venice. In the *De remediorum inventione*, he praised Ioannes Baptista Doleonius (life dates unknown), the physician who was elected by the Venetian College of Physicians to lecture on anatomy in 1629, for his accuracy and conciseness. It should be recalled here that Sanctorius, too, had been proposed as a lecturer in public anatomy, in 1613, but had refused (Sect. 2.6) (Sanctorius 1629b: 35).

The preceding paragraphs have shown that Sanctorius considered anatomy as very important for the physician. In the opening section of the *Commentary on Avicenna*, he even included a defense of the place of anatomical studies in medicine and asked: "Should anatomy pertain to the physician?" (Sanctorius 1625: 101 ff.).[34] At a time when anatomy was an integral part of the medical university curriculum, this question seems somewhat obsolete. Still, Sanctorius obviously found it necessary to stress that anatomy was not based on the senses alone, but involved reasoning (*ratiocinium*), too, which did not mean, however, that anatomy properly belonged only to natural philosophy. According to Sanctorius, the physician had to use his hands and his head in anatomical studies. He did not go so far as to call anatomy a science on the grounds that the anatomist performed mental activities, but he claimed in the *Commentary on Galen* that the *medicus anatomicus* often

---

[32] "… & alij percelebres Barones in cuius gratiam ego fusa oratione de anatomiae arcanis sermocinabar: …". See: Sanctorius 1612b: 197.

[33] For more information on Paolo Sarpi's role in the Venetian Interdict, see: Cozzi and Cozzi 1984: 47–52.

[34] "Quaest. XV. An anatomia pertineat ad Medicum." See: Sanctorius 1625: 101 ff.

obtained mathematical certainty in his inquiry into diseases and causes.[35] It will be shown later, how this understanding of anatomy fit into Sanctorius's general concept of medical knowledge and his answer to the traditional question, as to whether medicine is an art or a science (Sect. 6.2) (Sanctorius 1612a: 736).[36]

While the presentation of surgical instruments and techniques on university courses in theoretical medicine was highly unusual, the inclusion of anatomy was not. In fact, physiology, an important subject matter of *theoria*, went hand in hand with anatomy, as physiological theory usually took its sensory information from anatomy. Textbooks like Avicenna's *Canon* dealt with topics for which anatomical considerations were highly relevant, such as the parts of the body, humors, or spirits. Accordingly, professors of theoretical medicine and their students could not remain indifferent to anatomical work. Still, it was of course not the responsibility of professors of theoretical medicine to give detailed lectures on anatomy or to conduct anatomical demonstrations. Thus, their willingness to integrate anatomical considerations into their teaching was determined by individual interests and competences, which in turn depended on past careers and the anatomical training they had received in their own student days. Moreover, teaching by means of commentary on classical textbooks meant that the appeals to anatomical experience were often limited to those aspects perceived to have some bearing on the standard topics of physiological debate (Siraisi 1987: 324–33; Cunningham 2003: 52).

Striking, in this context, is Sanctorius's emphasis on the importance of anatomy for medicine and his broad knowledge of ancient as well as contemporary anatomical work. Besides Vesalius and Colombo, he referred to numerous other sixteenth-century anatomists, among them André du Laurens (1558–1609), Gabriele Falloppia (1523–1562), Laurent Joubert (1529–1582), Leonardo Botallo (1530–1587) and Bartolomeo Eustachi (ca. 1500–1574).[37] Hence, Sanctorius's students could learn about the views of modern anatomical and physiological writers on a variety of topics, even if they also learnt that most of these authors were often wrong. But according to Sanctorius, reading books was not enough. In the *Commentary on Galen*, he

---

[35] Based on the authority of Aristotle, who presented mathematics as the demonstrative science par excellence in his work *Analytica posteriora* (Posterior Analytics, ca. 350 BCE), and of Averroes—as well as of all their Greek and Latin interpreters—mathematics was generally considered a certain science. However, in the second half of the sixteenth century, a dispute arose over the question of the causes and foundation of this certainty and the way in which it was interpreted (De Pace 1993: 9). For more information on this debate, see: ibid.

[36] Nancy Siraisi has written that Sanctorius defended in this *quaestio* "the standing of anatomy as a science" (Siraisi 1987: 327). I think this is misleading, as Sanctorius did not use the term *scientia* here, but rather referred to *ratiocinium* (reasoning), when explaining why anatomy pertained to the physician. Moreover, as will be shown below (Sect. 6.2), Sanctorius conceived of medicine not as a science (*scientia*), but as an art (*ars*), albeit one that could approximate, if not attain, certainty. This implies that anatomy, according to Sanctorius, likewise ranked among the *artes* (ibid.: 236 ff.).

[37] For references to these anatomists in Sanctorius's published works, see e.g., Sanctorius 1612a: 204, 260, 281, 286, 465, 528, 556, 565, 706, Sanctorius 1612b: 148, 200, 237–40, 302, Sanctorius 1625: 102, 615, 672, 746, 764, 799, Sanctorius 1629a: 342, Sanctorius 1629b: 158.

stated that it did not suffice to know the works of Galen and of modern anatomists; practice in the dissection of bodies was necessary, too. "One learns more in one day from this exercise, than from studying anatomy for years, without direct observation" (Sanctorius 1612b: 366), so Sanctorius.[38] Thus, he encouraged his students not only to attend lectures, but also to conduct dissections themselves, in line with his belief that anatomy is based on reasoned argument as well as on the senses.

We can therefore conclude that Sanctorius was very interested in physiological-anatomical problems, but usually defended the traditional Galenic assertions on anatomical points. As Nancy Siraisi has pointed out, Sanctorius "was indeed capable of writing on occasion as if 'Galen and anatomy' were almost interchangeable terms" (Siraisi 1987: 336). Even though he strongly emphasized the importance of the evidence of the senses, he frequently returned to authoritative positions, especially on matters of theory. Anatomical considerations appeared alongside other kinds of argument, and were not necessarily the most convincing form of evidence, per Sanctorius. Yet, anatomy was a medical field, in which he explicitly contradicted Galen, without, however, contesting the underlying Galenic framework. While this treatment of anatomy is fairly characteristic of the ways in which contemporary learned physicians responded to the new developments in this medical field, what stands out is that Sanctorius decidedly brought anatomy into the context of medical practice. In his lectures on theoretical medicine, he repeatedly appealed to his own anatomical experience and prompted his students to conduct dissections themselves. What is more, he demanded that anatomists and surgeons should directly pool their experience, given the strong correlations between anatomical knowledge and practical surgical skills. It remains unclear whether Sanctorius's reproductions of anatomical demonstrations found their way into the classroom, but alone the fact that a teacher on medical *theoria* referred to his personal experience of dissection can be considered unusual.

4.2.2   Instruments for the Improvement and Alleviation of the Sick

The devices, which I categorize as instruments for the improvement and alleviation of the sick, are two inhalators (Fig. 4.11), a mobile bath (Fig. 4.12), a perforated ball to quench the thirst of fever patients (Fig. 4.13), cupping glasses (Fig. 4.14), a hanging bed (Fig. 4.15) and two instruments to ease pain (Figs. 4.1 and 4.16). As this list suggests, the devices served multiple purposes, but were all used for dietetic-therapeutic measures. Some of them helped Sanctorius manage his patients' involvement with the six non-natural things. The inhalators, for instance, produced vapors that warmed, cooled, moistened, or dried the air, excited sleep, or were filled with remedies against certain diseases such as phthisis. They allowed Sanctorius to

---

[38] "… magis nam unica die in hoc exercitio addiscet, quam si anni spatio sine inspectione studeret anatomiae." See: Sanctorius 1612b: 366.

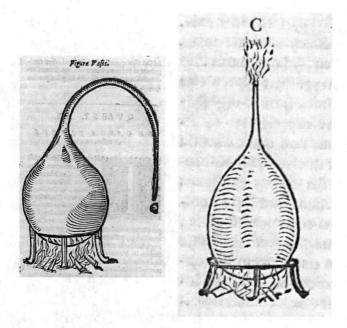

Fig. 4.11 Two inhalators to change the air (Sanctorius 1625: 129, 406). (© British Library Board 542.h.11, 129, 406)

artificially change the qualities of (i.e., condition) the air in a room, according to his patients' needs. The *balneatorium*, a mobile bath, served as a corrective for people with an overly dry complexion, such as occurred in hectical fevers. If different substances were added to the water, it could be used also to treat other afflictions. The hanging bed was supposed to rock the patient to sleep. Remarkably, despite their strong relation to the six non-natural things, Sanctorius did not consider the influence of these instruments on insensible perspiration, nor did he refer to them in the *De statica medicina*. This is especially interesting with regard to the hanging bed, as Sanctorius was not the first physician to suggest such a device (Sanctorius 1612b: 59; 1625: 129, 405 f., 439, 636, 674).

The physician Asclepiades of Bithynia, who was already mentioned in the context of the medical school of the Methodists (Sect. 3.2.11), is said to have recommended hanging beds as a form of passive exercise. According to Pliny the Elder (23–79 CE), Asclepiades devised suspended beds that could be rocked and thus served to mitigate disease or promote sleep. They helped render the pores and passages of the body more open and restore an interrupted flow of corpuscles, which was, according to Asclepiades, the immediate cause of most disease. In the sixteenth century, Girolamo Mercuriale described Asclepiades' hanging bed (*lectus pensilis*) on the basis of ancient reports, but asserted that few, or no physicians have dealt with the shape of the instrument, or its usage. According to him, the hanging beds were mostly unknown. However, with their description in his famous work *De arte gymnastica* (On the Art of Gymnastics, 1569), Mercuriale must have helped to popularize them (Mercuriale 1569: 176 ff.; Kamenetz 1977: 18; Wazer 2018: 83).

Fig. 4.12 The instrument *balneatorium*—a mobile bath (Sanctorius 1625: 405, 439). (© British Library Board 542.h.11, 405)

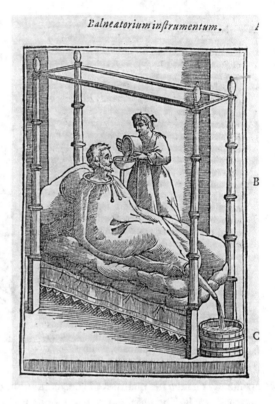

Thus, it can be assumed that Sanctorius was acquainted with Asclepiades's device, either through the work of Mercuriale (who had been Sanctorius's teacher at the University of Padua), or through the accounts of ancient authors like Plinius, or Aulus Cornelius Celsus (first century CE). But given that the descriptions of Asclepiades' hanging bed are rather imprecise and no illustrations of the device existed at the time it might have served, at the most, as inspiration for Sanctorius's *lectus artificiosus*. According to Mercuriale, Plinius explained that ropes were attached to the four corners of the roof of the bed in such a way was as to raise it a little from the ground, so it seemed to hang in the air. Looking at the illustration of Sanctorius's hanging bed, it is clear that his device was much more sophisticated than this. The bed was not simply elevated by ropes. Instead, there was a crank mechanism above the bedroom ceiling to lift and lower the bed (Fig. 4.15). Moreover, as will be shown below, swinging was by no means the device's only function. What is striking, however, is that, contrary to Asclepiades, Sanctorius did not consider lying in a swinging bed as a passive movement, or as a form of transportation. He did not associate it with *perspiratio insensibilis*, or consider any impact it might have on the pores and passages of the body, even though he praised, like Asclepiades, its soporific effect. In view of the importance of sleep for the

Fig. 4.13 Device to
quench the thirst of fever
patients (Sanctorius 1625:
499). (© British Library
Board 542.h.11, 499)

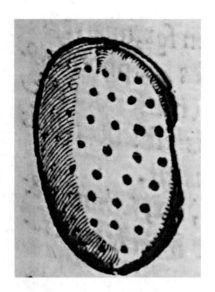

excretion of insensible perspiration, one can but wonder why Sanctorius did not
include the hanging bed in his program of static medicine (Mercuriale 1569: 177).[39]

   In addition to therapeutics, the instruments in this category had a clear practical
orientation. The *balneatorium* was mobile and could be used in the sickbed itself to
prevent, as Sanctorius explained, the patients' strength being weakened by the
movement required to carry them to the bath (Fig. 4.12). For this, he claimed, was
why the sick had more inconvenience than relief from ordinary baths and, as a consequence, bathing had been abolished for the treatment of diseases such as hectical
fevers. By contrast, the use of Sanctorius's instrument allegedly had neither harmful
nor inconvenient effects. Sanctorius claimed that it enabled him even to heal
patients' hectical fevers already declared incurable by other physicians. The *lectus
artificiosus* bed had a mechanism that turned it into a chair, so allowing the patient
to sit up during the day, without having to leave his bed (D).[40] This was important,
so Sanctorius, because if sick persons lie all the time stretched out in bed, they
become faint and their natural and animal faculties (Sect. 3.3.5, fn. 91) diminish
dangerously. Moreover, the patient could also eat and go to the toilet in bed, as a
table and a lavatory were installed in the device (C).[41] Attached to the ropes, which
connected the bed to the crank above the ceiling, there were small spheres (*globuli
aerei*) (B), which produced a sound, when the bed swung in the air, and helped

---

[39] Warm baths had been an important form of therapy also for Asclepiades, who used them to open
the pores and provoke sweat. However, Sanctorius did not refer to similar effects with regard to his
balneatorium and did not consider it in the context of *perspiratio insensibilis* (Sanctorius 1625:
405, 439, Benedum 1967: 95 f.).

[40] The following letters in brackets refer to the illustration of the *lectus artificiosus* in Sanctorius's
Commentary on Avicenna, see Fig. 4.15.

[41] The table is not indicated by a letter, but is located under the arm of the chair in the front.

Fig. 4.14 Cupping glasses
(Sanctorius 1625: 512,
680). (© British Library
Board 542.h.11, 512)

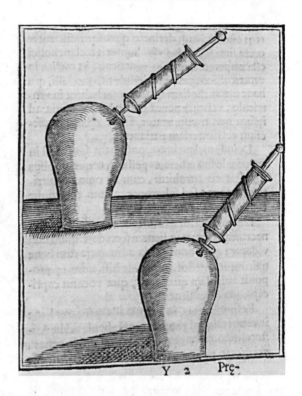

induce sleep. Finally, the bed could be attached by means of bolts to another similar bed, so enabling the patient to easily be rolled from one into the other, for a fresh change of bedclothes. Sanctorius stated that he successfully used the bed for a lot of patients, among them paralyzed people, or people with podagra (Sanctorius 1625: 405, 636 f., 673).

I refer to these two instruments in some detail, because they reveal a further dimension of Sanctorius's instrumentation: the improvement of patient care. The mobile bath and the hanging bed were useful especially for critically ill patients with limited mobility, but they eased strain for the caregivers, too. No longer did they need to carry the weight of the patient, be it at bathing time or when changing the bedclothes. This and the attention Sanctorius paid to the many difficulties that bedridden patients faced in order to satisfy their basic needs, imply that he spent many hours at the bedside of the sick. By including the two devices in his lectures on theoretical medicine, he drew his pupils' attention to very practical aspects of daily medical life. It is, of course, doubtful whether Sanctorius really built and used these instruments. In the late sixteenth century, Mercuriale described hanging beds as a curiosity. Despite them being popularized through his work *De arte gymnastica*, it is very likely that they retained this status until Sanctorius's publication of the *Commentary on Avicenna* in 1625, and beyond. To my knowledge, there is no evidence of similar devices being used in Renaissance Italy. In any case, if Sanctorius used his hanging bed and the *balneatorium*, which was probably seen as being just

Fig. 4.15 The instrument
lectus artificiosus—a
hanging bed (Sanctorius
1625: 636, 674). (© British
Library
Board 542.h.11, 636)

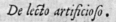

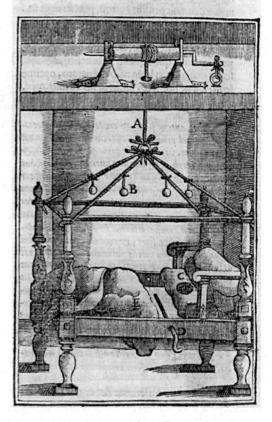

as peculiar as the bed, it must have been within the walls of one of the *palazzi*, in
which his wealthy Venetian clients lived (Mercuriale 1569: 176).

The two instruments that Sanctorius designed in order to ease pain were probably connected to his surgical activities, as control of pain was particularly important
here (Fig. 4.16). Without describing the devices in detail, it is pertinent to mention
a statement that Sanctorius made in the *Commentary on Hippocrates* with regard to
one of them (Fig. 4.16, right). He explained that he had found a procedure to remove
pain, which impressed anyone who saw it. He would take a cow's bladder filled with
a lot of snow or ice and wrap it in a handkerchief, so that neither the sick person nor
the attendants would notice it. Then, Sanctorius would suddenly apply the covered
bladder to the aching part and the pain would immediately cease. He personally
would have preferred not to have to cover up the bladder, but, as Sanctorius
explained, this was necessary in view of the fact that people easily spurn the things
they know. Hence, Sanctorius felt that he needed to impress his audience with tricks,
because otherwise they would not approve of his healing methods. It is clear that the
secrecy shrouding his use of the cooling cow bladder had nothing to do with

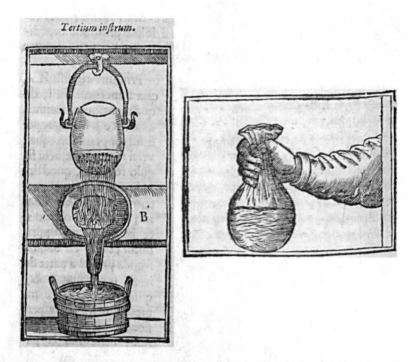

Fig. 4.16 Two instruments to ease pain (Sanctorius 1625: 668 f., 726). (© British Library Board 542.h.11, 668 f., 726)

protecting technical secrets, as he had already published the device and an explanation of it 4 years earlier, in the *Commentary on Avicenna*. Thus, it was rather the mockery of non-professionals that Sanctorius was concerned about. I will resume this discussion later, as it helps explain how Sanctorius's instruments were received, to whom they were addressed, and in which contexts they were used. For the moment, it is enough to note that the mechanisms driving the hanging bed or the famous weighing chair are hidden behind a ceiling (Fig. 4.15 and Fig. 7.23) (Sanctorius 1629a: 373 f.).

All in all, the therapeutic measures connected to Sanctorius's instruments for the improvement and alleviation of the sick are not original. In a later annotation to the *Commentary on Avicenna*, Sanctorius mentioned Galen as the source of his inhalator. He wrote:

> Such a vessel has been related by Galen in his *Method* (bk. 9, chap. 14) where he says that *in a hot and dry disease the air must be cold and humid*. He places it such that a cold breeze from the Euripus blows on it, calling it Euripus because of the tight channel through which the air comes out (Sanctorius 1625: 406).[42]

---

[42] "Simile vas p[ro]ponit[ur] a G[alen]o 9 meth[od]i 14 ubi dicit in morbo cal[id]o et sicco aer debet esse frig[idu]s et hum[idu]s. Subdit e[tiam] ut ex Euripo aura fr[igid]a inspiret—vocat Euripum ob viam angustam p[er] quam egred[itu]r aer." See: Sanctorius 1625: 406. The transcription and English translation are taken from: Bigotti 2017: 4.

In fact, in Sanctorius's time, instruments such as perfume-burners, or lanterns containing domestically produced scented substances were commonly used in households to sweeten the air by. The therapeutic use of baths, the pain-relieving effect of cold applications, and the soporific effects of sounds and rocking motions were well known (Cavallo 2011: 194). However, Sanctorius tried to refine these measures by means of his instruments and to facilitate their use in daily medical practice. It is remarkable that he integrated these instruments into courses on theoretical medicine, sensitizing his students to the very practical needs of patients as well as to the challenges their caregivers faced. It is likewise remarkable that, even though most of the instruments strongly relate to the six non-natural things, Sanctorius did not consider their effects on insensible perspiration. Notwithstanding that they were all used for dietetic-therapeutic measures and had a clear practical orientation, Sanctorius made no connection between his instruments for the improvement and alleviation of the sick and the *De statica medicina*.

4.2.3 Instruments to Demonstrate Optical Phenomena

The subject matter of Renaissance optics included a wide variety of topics, such as theories of vision, the nature of light and colors, or the anatomy and physiology of the eye. Perusal of Sanctorius's works shows that he was interested in the subject and in the *Commentary on Avicenna*, he presented three instruments to demonstrate optical phenomena (Figs. 4.1, 4.17, 4.18, and 4.19).

The first instrument (Fig. 4.17), consisting of two vessels filled with small vitreous balls, served Sanctorius to demonstrate how the colors black and white were generated. He explained that if the balls were filled with water (A), light could only penetrate at one point and the color black occurred. The reason for this was, so Sanctorius, that the light was pointed to only a small part, whereas in the other parts there was darkness out of which blackness emerged. If, however, the balls were only filled with air (B), light did not escape and no darkness occurred. In the *Commentary*

Fig. 4.17 Two vessels filled with small vitreous balls to demonstrate how whiteness and blackness arise (Sanctorius 1625: 460). (© British Library Board 542.h.11, 460)

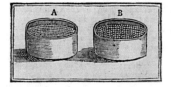

Fig. 4.18 Instrument
aimed at showing that
vision occurs through
crosswise divided rays
(Sanctorius 1625: 760).
(© British Library Board
542.h.11, 760)

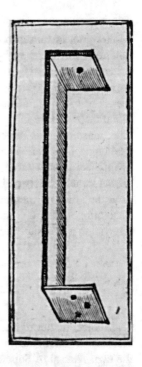

Fig. 4.19 Illustration to
demonstrate through the
examination of a comet's
tail (*cometae caudati
inspiciantur*) that the
vitreous humor acts like a
dark room and diminishes
transparency (Sanctorius
1625: 762). (© British
Library Board
542.h.11, 762)

on Galen, Sanctorius mentioned a similar experience and admitted that the generation of colors was very difficult to explain. But he followed, as he wrote, the *per-
spectivi*, who held that blackness emerged from the refraction of light from countless
surfaces. Whiteness, on the contrary, occurred if light was refracted from only a few,
but clean and polished surfaces. The generation of the colors black and white was
particularly important to Sanctorius, as all other colors were derived from them.
According to him, colors emerged not from the four primary qualities, but from a
mixture of darkness and transparency (Sanctorius 1612a: 320 f.; 1625: 460 f.).

Sanctorius was referring here to the medieval optical (*perspectiva*) tradition,
which was based on Aristotelian theories of light and colors. In this context, colors
were usually identified as modifications of white light produced by refraction or
reflection by other bodies. They were seen as innate properties of white light.
Interestingly, Sanctorius not only refuted the idea of a link between colors and the
four primary qualities, but also the view that colors came from a mixture of dense
and rare. In doing so, he disagreed with one of his famous teachers at the University
of Padua: Giacomo Zabarella. It is not my intention here to investigate Sanctorius's
concept of colors and their generation at any length.[43] Rather, I want to call attention
to the fact that Sanctorius was responsive to philosophical debates on the origin of
colors, of which he most probably learned during his studies in Padua. It was an
important topic to him, which he dealt with already early in his career and repeatedly in his published works. It was a field which he tried to support by means of
experimenta (Sanctorius 1612a: 318 f., 322 f.; Mancosu 2006: 597–628; Baker
2015: 162 f.).

As early as the 1580s, Sanctorius held a public lecture in the Istrian Accademia
Palladia with the title "What Every Color Really Means" (*Che cosa veramente sig-
nifichi ciascun colore*). This text is largely unknown as it was not published under
Sanctorius's name, but as part of a collection of public discourses held at the academy. Contrary to his other treatments of colors, the focus here is on the metaphorical meaning of colors and on the opinions of poets such as Vergil, Ovid, or Horaz on
the subject (Sect. 2.2). In the *Methodi vitandorum errorum*, Sanctorius's first published book, the discussion of colors and their generation is dealt with in the framework of semiology. Sanctorius investigated to what extent colors indicate the nature
and the course of a disease. He argued that, while a complexion could not be inferred
from colors alone colors did have a certain importance in diagnostics. In the
Commentary on Galen, he stated that hair color could lead to a knowledge of the
complexion of the brain, insofar as it was part of a syndrome of signs that, taken all
together, indicated the brain's complexion. Thus, Sanctorius's interest in colors was
very broad, ranging from their generation, to their meaning as metaphors, to their
value in diagnostics. The various experiences to which he referred follow those
made by painters and dyers and connect to the works of the medieval perspectivists
such as Alhazen (Ibn al-Haiṭam, d. 1041) and Witelo (ca. 1220/30–after 1277). A
detailed analysis of these *experimenta* is, however, beyond the scope of this work

---

[43] For a more detailed account of Sanctorius's thoughts on colors, see: Del Gaizo 1891: esp. 24–7.

(Sanctorius 1603: 110v–113r; 1612a: 317, 322; Vida 1621: 76r–86v; Del Gaizo 1891: 25 ff.).

While, with regard to his concept of colors and their generation, Sanctorius remained in a traditional framework, there was another field of optics in which he departed radically from tradition. In the *Commentary on Avicenna*, he argued that the retina and not the crystalline humor (now called the crystalline lens) was the principal organ of sight. In doing so, he contradicted Galen's teachings as well as the opinions of many subsequent medical writers and the medieval perspectivists. In the discussion, he gave seventeen reasons for his opinion. In contrast to his usual treatment of questions (*quaestiones*), Sanctorius hardly referred to other authors, but presented most of the arguments as his own. What is more, he completely omitted the traditional position, leaving his audience simply with his stance on the topic. Only in the last paragraph did he give a hint at his sources, when he referred to Christoph Scheiner (1573–1650) and his "most accurate experiments and demonstrations" (Sanctorius 1625: 763).[44] Aristotle, Galen and Avicenna, on the contrary, had all been wrong, so Sanctorius, as they identified the crystalline humor as the principal organ of sight; and because Scheiner had tried to reconcile his own findings with those of Galen by pointing to a work in which the latter had ascribed some role in vision to the retina, he had evidently neither read nor understood Galen. Notwithstanding that Sanctorius defended Galen in most other matters, with regard to this question, he was sure that Galen had erred and was ready to openly dismiss his opinion. Again, it was in the context of a recent anatomical finding that Sanctorius refuted Galen (Sect. 4.2.1) (Sanctorius 1625: 758–63).

In fact, Modestino del Gaizo and Nancy Siraisi have argued that Sanctorius drew heavily on Scheiner's work *Oculus* (The Eye, 1619) in his *quaestio* on the subject of vision. "Sanctorius not only takes the new doctrine of vision from Scheiner, but also the words," concluded Del Gaizo, and Siraisi stated that Sanctorius's "main arguments are highly simplified, nonmathematical, and abbreviated versions of propositions put forward in Scheiner's technical treatise" (Del Gaizo 1891: 38; Siraisi 1987: 343).[45] Christoph Scheiner was a Jesuit mathematician active in Rome, who described in the aforementioned work how he had verified in anatomical dissections that the retina is the visually sensitive part of the eye while the crystalline humor functions as a lens. By scraping the rear surface of an eyeball, leaving only a thin layer, he could directly observe the inverted image on the retina. Already before Scheiner, the Swiss physician Felix Platter (1536–1614) had maintained that the retina was the principal organ of vision and Francesco Maurolico (1494–1575), a Sicilian mathematician, had treated the crystalline of the eye as a convex lens. Moreover, the Neapolitan polymath Giambattista della Porta (1535–1615) popularized in his optical works certain new optical subjects, such as the analysis of

---

[44] "… Scheiner … in suis experimentis, & demonstrationibus exactissimus, …". See: Sanctorius 1625: 763.

[45] "Santorio prese da Scheiner non pure la dottrina nuova della visione, ma persino le parole; …". See: Del Gaizo 1891: 38.

radiation through lenses and the camera obscura.[46] And there was, of course, Johannes Kepler (1571–1630), who proposed a new theory of vision based on an understanding of the eye as an optical instrument. It is therefore possible that Sanctorius was familiar with their works, too, even though Scheiner is the only contemporary author whom Sanctorius cited by name. However, it can be assumed that he was not acquainted with Kepler's work on the retinal image, published in 1604, or in any case did not grasp it, as Sanctorius held that the vitreous humor was responsible for the righting of the image, and not, as Kepler maintained, processes of reflection and refraction (Sanctorius 1625: 761 ff.; Del Gaizo 1891: 24–38; Siraisi 1987: 343 f.).

In view of Sanctorius's rudimentary treatment of the issue of vision and his strong dependence on the work of Christoph Scheiner, it is difficult to assert whether the optical experiences to which he referred in his argumentation in the *Commentary on Avicenna* were his own. Only in one instance did Sanctorius explicitly claim to be drawing on his personal experience, but it is quite possible that he was referring merely to a mental process. In order to demonstrate that the vitreous humor had the property to right an image, he used a vitreous lens, placing it in front of an opening in his house, which was situated next to a river, and observing underwater objects through it. These, he reported, appeared upright, and not reversed. With regard to the two further instruments with which Sanctorius proposed to demonstrate optical phenomena, he was less clear about his personal use of them (Figs. 4.18 and 4.19). The first served him to explain the refraction of visual rays in the eye according to the varying transparency of the media they traversed, i.e., the aqueous humor, or the crystalline lens. The second was rather an observation than an instrument, as according to Sanctorius, the examination of comets' tails illustrated that the vitreous humor acted like a dark room and diminished transparency.[47] However, it is very difficult to interpret this observation, since Sanctorius wrote of a comet (*cometa*), but the image shows only a sun (Fig. 4.19). Nevertheless, even without a detailed analysis and some ambiguities concerning the demonstrations cited, one can assume that Sanctorius made at least some optical observations himself.[48] In view of the flourishing glass industry in Venice, lenses must have been easily accessible to him. In addition, Sanctorius moved in social circles that included other scholars with an interest in optics. Paolo Sarpi, Galileo Galilei, and the Venetian patrician Agostino da Mula (1561–1621) were all very much involved in optical studies at the time and

---

[46] A camera obscura is an instrument, such as a darkened room, with a tiny hole in one of the walls, through which external light passes and projects an image, upside down, on the opposite wall (Mancosu 2006: 613).

[47] Sanctorius wrote: "Eadem ratione in cubiculo obscuro Cometa caudatus ostenditur, si radij Solis ingrediantur per foramen fenestrae, in quo sit fumus ex palea accensa, ibi Cometa caudatus pulcherrimus apparet, ut in icone." See: Sanctorius 1625: 762 f.

[48] For more information on the arguments that Sanctorius presented in favor of the retina as the principal organ of vision, see: Del Gaizo 1891: 27 ff., Siraisi 1987: 342 f. Among the optical observations and demonstrations that Sanctorius described is also one made with a camera obscura. See: Sanctorius 1625: 761.

frequented the *Ridotto Morosini*. Even though they did not always share the same opinions and Sarpi, for example, still considered the crystalline lens as the principal organ of vision, discussions with these scholars might have aroused Sanctorus' interest in optics. Furthermore, the home of the Morosini might have been a place, where he participated in optical observations (Sanctorius 1625: 760–3; Cozzi 1986; Sarpi and Cozzi 1996: e.g., xxxvii, XLI–XLII).[49]

To conclude, although Sanctorius did not make original contributions to optics and did not refer to all recent developments in the field, he still was among the few who accepted that the retina was the main organ of sight and thereby clearly and unambiguously contradicted Galen. By including this notion in his *Commentary on Avicenna*, he provided his audience with a rather advanced theory of vision, compared to those available at other universities at the time, as, for example, in Basel. In his last work, *De remediorum inventione*, Sanctorius highlighted the importance of optical knowledge for the practicing physician, as according to him, only those versed in optics were able to correctly recognize the meaning of colors in diagnosis. In this context, it is interesting to note that Sanctorius bequeathed in his testament a "copy of one hundred optical problems not communicated to others" to his friend Hieronymus Thebaldus.[50] The fate of this document is unknown, as are its contents, but Sanctorius's mention of it illustrates, once again, his pronounced interest in the subject of optics (Sanctorius 1629b: 121 f.).

4.3   A New Approach to *theoria*—Head and Hand?

In the dedication and preface to the *Commentary on Avicenna*, Sanctorius confidently proclaimed that he was offering a new approach to the teaching of theoretical medicine. Contrary to his predecessors, he did not base his explanations of the subject solely on reason and on the authority of Hippocrates and Galen, but confirmed theory by practice—by experience (*experimenta*), instruments, and static art. According to him, theory was meaningless and useless, if it was not confirmed, a posteriori, by practice. Correspondingly, practice could not be understood if it was not corroborated, a priori, by theory. The preceding paragraphs have shown that Sanctorius took this seriously. Practical experiences, observations, and instrumentation repeatedly enter the otherwise very theoretical physiological discussions regarding Avicenna's *Canon*. In this context, it has to be noted that physiology, as taught at the time at universities, was a highly theoretical discipline. Therefore, Sanctorius's inclusion of instruments in his teaching on physiology was a bold new

---

[49] Sanctorius does not seem to have been involved in the development and use of the telescope. In the *Commentary on Avicenna*, he mentioned the "newly invented spyglass," only then to dismiss its ability to make visible changes in the moon, as he was convinced of the division of the sublunary and the celestial spheres. See: Sanctorius 1625: 141, 154, Siraisi 1987: 274 f.

[50] "Allo Eccellentissimo Tebaldi le sia dato copia da mio nepote de cento problemi de optica non comunicati ad altri." See: ASVe-g (n.d.). The transcription is taken from: Ettari and Procopio 1968: 140.

departure, which set his *Commentary on Avicenna* apart from all the other commentaries on this traditional textbook (Sanctorius 1625: dedication, Ad lectorem).

Sanctorius's attempt to reform the teaching of *theoria* might be connected to his earlier career. In 1611, when he was appointed first ordinary professor of theoretical medicine at Padua, he was a man of fifty who had spent most of his adult life in medical practice. According to his own testimony, he had by then already been busy for years with his weighing procedures, endeavoring to measure insensible perspiration (Sect. 2.2). Moreover, 8 years earlier, in his first published book, *Methodi vitandorum errorum*, Sanctorius not only referred to some of his instruments, but also mentioned the work *De instrumentis medicis* that he aimed to publish next. Indeed, while working as a practicing physician, Sanctorius devised and used various instruments and already began conducting his famous quantitative studies of physiological phenomena. The fact, that he had planned to publish a book on his instruments at least since 1603, implies that this was an area of great interest to him. The professorship at Padua interfered with this aim, as the duties the position involved prevented Sanctorius from finishing the illustrations of his instruments. His novel way of teaching *theoria* was a means to combine his own interests, experiences, and physiological ideas with his obligation to lecture on Avicenna's *Canon,* which was a set text in his students' curriculum (Sanctorius 1603: 26v, 109v; 1625: Ad lectorem).

But there was also yet another dimension to this. Sanctorius held that the division of medicine into theory and practice in the university curricula was itself improper. In keeping with Aristotle and Galen, he argued that the purpose of theory was truth (*veritas*), while the purpose of medicine was operation (*opus*), that is, to preserve and to restore health. Moreover, so Sanctorius, the term *theorica* meant speculation and not operation, nor the act of making or doing (*factio*). Practice (*praxis*), on the contrary, meant action and therefore also differed from medicine, which was not active, but operative (*factivus*) and restoring (*resarcitivus*). However, Sanctorius admitted that medicine could be "somewhat" (*aliquo modo*) divided into theory and practice, as the physician first explored the truth and then directed it to action, that is, to the preservation or restoration of health. Still, medicine proper was by its nature not theory. Hence, Sanctorius's new approach to teaching theoretical medicine was also motivated by his rejection of the disciplinary division of medicine into theory and practice. Bound by the university statutes to teach the traditional set of texts laid down for courses on *theoria*, Sanctorius tried to challenge the disciplinary boundaries from within, by linking his lectures on Avicenna's *Canon* to practical applications and by using evidence drawn from *practica* to confirm theory. The result is a seemingly peculiar mixture of highly traditional theoretical discussions with completely new elements relating to medical practice (Sanctorius 1625: 37 ff.).

Sanctorius's effort to confirm physiological theory by means of practical evidence has to be seen against the larger backdrop of a revaluation of practical medicine that had begun in the fifteenth century. As was mentioned earlier, practical training gained considerably in importance in the sixteenth century and, indeed, became foreign students' main motivation for studying at the University of Padua (Sect. 4.1.2). Therefore, Sanctorius was certainly not alone in highlighting the importance of the practice of medicine. However, his attempt to combine teaching by commentary with teaching by practical demonstration is exceptional. Hence, in Sanctorius's works the relation between theory and practice is ambiguous and cannot be described as a simple dichotomy. Notwithstanding that medical knowledge contained both, contemplation and action, he argued, it also differed from both, due to its operative and restorative character. Moreover, the previous sections have shown that empirical evidence must not necessarily result from personal firsthand experience, but might well be based on reports by others. This serves to further blur the lines in medical practice, between textual or theoretical knowledge, acquired using the head (as in: the mind), and experiential knowledge, acquired through hands-on practice. It is important to keep this in mind when addressing Sanctorius's quantitative approach to physiology.

References

Albala, Ken. 2002. *Eating Right in the Renaissance*. Oakland CA: University of California Press.

Arano, Luisa Cogliati. 1976. *The Medieval Health Handbook: Tacuinum Sanitatis*. New York: George Braziller.

ASVe-a: Provveditori e Sopraprovveditori alla Sanità, Busta 562, Giudizi sulla peste.

ASVe-g: Sezione Notarile Testamenti (Testamenti Crivelli), Busta 289, n. 537.

Baker, Tawrin. 2015. Colour in Three Seventeenth-Century Scholastic Textbooks. In *Colour Histories: Science, Art and Technology in the 17th and 18th Centuries*, ed. Magdalena Bushart and Friedrich Steinle, 161–177. Berlin: De Gruyter.

Benedum, J. 1967. Die "balnea pensilia" des Asklepiades von Prusa. *Gesnerus – Swiss Journal of the History of Medicine and Sciences* 24: 93–107.

Beverovicius, Iohannes. 1638. *De Calculo Renum & Vesicae liber singularis. Cum epistolis & consultationibus magnorum virorum*. Lugdunum Batavorum: Ex Officina Elseviriorum.

Bigotti, Fabrizio. 2017. A Previously Unknown Path to Corpuscularism in the Seventeenth Century: Santorio's Marginalia to the *Commentaria in primam Fen primi libri Canonis Avicennae* (1625). *Ambix* 64: 1–14.

———. 2018. The Weight of the Air: Santorio's Thermometers and the Early History of Medical Quantification Reconsidered. *Journal of Early Modern Studies* 7: 73–103.

Bigotti, Fabrizio, and David Taylor. 2017. The Pulsilogium of Santorio: New Light on Technology and Measurement in Early Modern Medicine. *Society and Politics* 11: 55–114.

Bovey, Alixe. 2005. *Tacuinum Sanitatis: An Early Renaissance Guide to Health*. London: Sam Fogg.

Bylebyl, Jerome J. 1979. The School of Padua: Humanistic Medicine in the Sixteenth Century. In *Health, Medicine and Mortality in the Sixteenth Century*, ed. Charles Webster, 335–370. Cambridge/London: Cambridge University Press.

Carrara, Daniela Mugnai. 1983. Una polemica umanistico-scolastica circa l'interpretazione delle tre dottrine ordinate di Galeno. *Annali dell'Istituto e Museo di Storia della Scienza di Firenze* 8: 31–57.

Castagna, Pietro. 1951. La storia del trequarti e Santorio Santorio. In *Pagine di Storia della Scienza e della Tecnica, allegato agli Annali di Medicina Navale e Tropicale* n/a, 49–59.

Cavallo, Sandra. 2011. Secrets to Healthy Living: The Revival of the Preventive Paradigm in Late Renaissance Italy. In *Secrets and Knowledge in Medicine and Science, 1500–1800*, ed. Elaine Leong and Alisha Rankin, 191–212. Farnham/Burlington: Ashgate.

Colombero, Carlo. 1982. Colombo, Realdo. In *Dizionario biografico degli italiani*. Vol. 27. Rome: Istituto dell'Enciclopedia Italiana. http://www.treccani.it/enciclopedia/realdo-colombo_%28Dizionario-Biografico%29/. Accessed 5 Sep 2019.

Cornaro, Luigi. 1591. *Discorsi della vita sobria*. Padua: Appresso Paolo Miglietti.

Cozzi, Gaetano. 1986. Da Mula, Agostino. In *Dizionario biografico degli italiani*. Vol. 32. Rome: Istituto dell'Enciclopedia Italiana. http://www.treccani.it/enciclopedia/agostino-da-mula_ (Dizionario_Biografico)/. Accessed 5 Sep 2019.

Cozzi, Gaetano, and Luisa Cozzi. 1984. Paolo Sarpi. In *Storia della Cultura Veneta: Il Seicento*, ed. Girolamo Arnaldi and Manlio Pastore Stocchi, 1–36. Vicenza: Neri Pozza Editore.

Cunningham, Andrew. 2003. The Pen and the Sword: Recovering the Disciplinary Identity of Physiology and Anatomy Before 1800 II: Old Anatomy–the Sword. *Studies in History and Philosophy of Biological and Biomedical Sciences* 34: 51–76.

De Pace, Anna. 1993. *Le matematiche e il mondo: Ricerche su un dibattito in Italia nella seconda metà del Cinquecento*. Milan: Franco Angeli.

de Villanova, Arnaldus, Michael McVaugh, and Luis García-Ballester. 1992. *Aphorismi de gradibus*. Barcelona: Publ. de la Univ. de Barcelona.

Del Gaizo, Modestino. 1891. Alcune Conoscenze di Santorio Santorio intorno ai fenomeni della visione ed il testamento di lui. In *Atti della Accademia Pontaniana*, 21–48. Naples: Tipografia della Regia Università.

Dienstag, Jacob I. 1983. Translators and Editors of Maimonides' Medical Works: A Bio-Bibliographical Survey. In *Memorial Volume in Honor of Prof. S. Muntner*, ed. Joshua O. Leibowitz, 95–135. Jerusalem: Israel Institute of the History of Medicine.

———. 1989. Bibliography of the Medical Aphorisms of Maimonides. In *The Medical Aphorisms of Moses Maimonides*, ed. Fred Rosner, 455–471. Haifa: The Maimonides Research Institute.

Edwards, William F. 1976. Niccolò Leoniceno and the Origins of Humanist Discussion of Method. In *Philosophy and Humanism: Renaissance Essays in Honor of Paul Oskar Kristeller*, ed. Edward Mahoney, 283–305. Leiden: Brill.

Ettari, Lieta Stella, and Mario Procopio. 1968. *Santorio Santorio: la vita e le opere*. Rome: Istituto nazionale della nutrizione.

Facciolati, Iacopo. 1757. *Fasti Gymnasii Patavini*. Padua: Apud Joannem Manfrè.

García-Ballester, Luis. 1992. Changes in the *Regimina sanitatis*: The Role of the Jewish Physicians. In *Health, Disease and Healing in Medieval Culture*, ed. Sheila Campbell, Bert Hall, and David Klausner, 119–131. Toronto: Centre for Medieval Studies.

Gilbert, Neal Ward. 1963. *Renaissance Concepts of Method*. New York: Columbia University Press.

Grendler, Paul F. 2002. *The Universities of the Italian Renaissance*. Baltimore: The Johns Hopkins University Press.

Grmek, Mirko D. 1952. *Santorio Santorio i njegovi aparati i instrumenti*. Zagreb: Jugoslav. akad. znanosti i umjetnosti.

Hollerbach, Teresa. 2018. The Weighing Chair of Sanctorius Sanctorius: A Replica. *NTM Zeitschrift für Geschichte der Wissenschaften, Technik und Medizin* 26: 121–149.

Jacquart, Danielle. 1996. Die scholastische Medizin. In *Die Geschichte des medizinischen Denkens: Antike und Mittelalter*, ed. Mirko D. Grmek and Bernadino Fantini, 216–259. Munich: C.H. Beck.

Kamenetz, Herman L. 1977. History of Exercises for the Elderly. In *A Guide to Fitness after Fifty*, ed. Lawrence J. Frankel and Raymond Harris, 13–33. New York/London: Plenum Press.

Klestinec, Cynthia. 2016. Renaissance Surgeons: Anatomy, Manual Skill and the Visual Arts. In *Early Modern Medicine and Natural Philosophy*, ed. Peter Distelzweig, Benjamin Goldberg, and Evan R. Ragland, 43–58. Dordrecht: Springer.

Lessius, Leonardus. 1613. *Hygiasticon*. Antwerp: Ex officina Plantiniana, apud Viduam & filios Io. Moreti.

Levine, David, and Alan Bleakley. 2012. Maximising Medicine through Aphorisms. *Medical Education* 46: 153–162.

Maclean, Ian. 2002. *Logic, Signs, and Nature in the Renaissance: The Case of Learned Medicine*. Cambridge/New York: Cambridge University Press.

Maimonides, Moses, and Gerrit Bos. 2004. *Medical Aphorisms: Treatises 1–5*. Provo: Brigham Young University Press.

Mancosu, Paolo. 2006. Acoustics and Optics. In *The Cambridge History of Science*. Vol. 3: *Early Modern Science*, ed. Katharine Park and Lorraine Daston, 596–631. Cambridge/New York: Cambridge University Press.

Mercuriale, Girolamo. 1569. *De arte gymnastica libri sex*. Venice: Apud Iuntas.

Miessen, Hermann. 1940. Die Verdienste Sanctorii Sanctorii um die Einführung physikalischer Methoden in die Heilkunde. *Düsseldorfer Arbeiten zur Geschichte der Medizin* 20: 1–40.

Mitchell, S. Weir. 1892. *The Early History of Instrumental Precision in Medicine. An Address before the Second Congress of American Physicians and Surgeons, September 23rd, 1891*. New Haven: Tuttle, Morehouse & Taylor.

Mulcahy, Robert. 1997. *Medical Technology: Inventing the Instruments*. Minneapolis: The Oliver Press.

Nutton, Vivian. 1985. Humanist Surgery. In *The Medical Renaissance of the Sixteenth Century*, ed. A. Wear, R.K. French, and Iain M. Lonie, 75–99. Cambridge/New York: Cambridge University Press.

———. 1989. Hippocrates in the Renaissance. In *Die Hippokratischen Epidemien. Theorie–Praxis–Tradition*, ed. Gerhard Baader and Rolf Winau, 420–439. Stuttgart: Franz Steiner Verlag.

Palmer, Richard. 1979. Physicians and Surgeons in Sixteenth-Century Venice. *Medical History* 23: 451–460.

Pontormo, Jacopo, and Salvatore Silvano Nigro. 1988. *Il libro mio: Aufzeichnungen 1554–1556*. Munich: Schirmer/Mosel.

Randall, John Herman, Jr. 1940. The Development of Scientific Method in the School of Padua. *Journal of the History of Ideas* 1: 177–206.

———. 1961. *The School of Padua and the Emergence of Modern Science*. Padua: Editrice Antenore.

———. 1976. Paduan Aristotelianism Reconsidered. In *Philosophy and Humanism: Renaissance Essays in Honor of Paul Oskar Kristeller*, ed. E.P. Mahoney, 275–282. Leiden: Brill.

Rosner, Fred. 1998. *The Medical Legacy of Moses Maimonides*. Hoboken, NJ: KTAV Publishing House.

Sanctorius, Sanctorius. 1603. *Methodi vitandorum errorum omnium, qui in arte medica contingunt, libri quindecim*. Venice: Apud Franciscum Barilettum.

———. 1612a. *Commentaria in Artem medicinalem Galeni*. Vol. I. Venice: Apud Franciscum Somascum.

———. 1612b. *Commentaria in Artem medicinalem Galeni*. Vol. II. Venice: Apud Franciscum Somascum.

———. 1614. *Ars Sanctorii Sanctorii Iustinopolitani de statica medicina, aphorismorum sectionibus septem comprehensa*. Venice: Apud Nicolaum Polum.

———. 1625. *Commentaria in primam Fen primi libri Canonis Avicennae*. Venice: Apud Iacobum Sarcinam.

———. 1629a. *Commentaria in primam sectionem Aphorismorum Hippocratis, &c. ... De remediorum inventione*. Venice: Apud Marcum Antonium Brogiollum.

———. 1629b. *De remediorum inventione*. Venice: Apud Marcum Antonium Brogiollum.

———. 1634. *Ars Sanctorii Sanctorii de statica medicina et de responsione ad Staticomasticem*. Venice: Apud Marcum Antonium Brogiollum.

Sanctorius, Sanctorius, and Giuseppe Ongaro. 2001. *La medicina statica*. Florence: Giunti.

Sarpi, Paolo, and Luisa Cozzi. 1996. *Pensieri naturali, metafisici e matematici*. Milan/Naples: Riccardo Ricciardi Editore.

Siegel, Rudolph E. 1968. *Galen's System of Physiology and Medicine*. Basel/New York: S. Karger.

Siraisi, Nancy. 1987. *Avicenna in Renaissance Italy: The Canon and Medical Teaching in Italian Universities After 1500*. Princeton: Princeton University Press.

———. 1990. *Medieval & Early Renaissance Medicine: An Introduction to Knowledge and Practice*. Chicago: University of Chicago Press.

Smith, Wesley D. 1979. *The Hippocratic Tradition*. Ithaca/London: Cornell University Press.

Sotres, Pedro Gil. 1998. The Regimens of Health. In *Western Medical Thought from Antiquity to the Middle Ages*, ed. Mirko D. Grmek, Bernardino Fantini, and Antony Shugaar, 291–318. Cambridge, MA: Harvard University Press.

Stolberg, Michael. 2010. Die Basler Universitätsanatomie in der Frühen Neuzeit. In *Universität Basel 1460–2010*, ed. Historisches Seminar Basel, 1–16. Basel: Historisches Seminar Basel. https://unigeschichte.unibas.ch/cms/upload/Aufbrueche_Stagnationen/Downloads/Stolberg_Anatomie.pdf. Accessed 24 Aug 2019.

Straface, Antonella. 2011. Abū Bakr al-Rāzī, Muḥammad ibn Zakarīyā' (Rhazes). In *Encyclopedia of Medieval Philosophy*, ed. Henrik Lagerlund, 6–10. Dordrecht: Springer.

Temkin, Owsei. 1973. *Galenism: Rise and Decline of a Medical Philosophy*. Ithaca/London: Cornell University Press.

Valleriani, Matteo, ed. 2017. *The Structures of Practical Knowledge*. Cham: Springer International Publishing.

Vida, Hieronimo. 1621. *De' cento dubbi amorosi*. Padua: Gasparo Crivellari.

Wazer, Caroline. 2018. Pleasure and the Medicus in Roman Literature. In *Pain and Pleasure in Classical Times*, ed. William V. Harris, 83–92. Leiden: Brill.

Wear, Andrew. 1981. Galen in the Renaissance. In *Galen: Problems and Prospects*, ed. Vivian Nutton, 229–262. London: The Wellcome Institute for the History of Medicine.

———. 1993. The History of Personal Hygiene. In *Companion Encyclopedia of the History of Medicine*, Vol. II, ed. William Bynum and Roy Porter, 1283–1308. London/New York: Routledge.

Wightman, William P.D. 1964. Quid sit Methodus? "Method" in Sixteenth Century Medical Teaching and "Discovery". *Journal of the History of Medicine* 19: 360–376.

Wootton, David. 1983. *Paolo Sarpi: Between Renaissance and Enlightenment*. Cambridge: Cambridge University Press.

Chapter 5
Quantification in Galenic Medicine

Abstract The central theme of this chapter is to identify and explore various forms of quantification in the medical tradition, on which Sanctorius might possibly have drawn for his quantitative approach to physiology. First, I address theories and practices connected to dietetics and pharmacology as well as the Galenic concept of a latitude of health, which assumed a certain graduation of the state of health. Second, I reconsider the relation of Sanctorius's work to two earlier authors who are commonly associated with Sanctorius and his static medicine: the Alexandrian physician Erasistratus (third century BCE) and the German cardinal and scholar Nicolaus Cusanus (1401–1464). Both were early proponents of quantitative approaches to medical problems. Third, I outline instances of quantitative physiological reasoning in Galen's work as well as in the works of Renaissance scholars, and I analyze their possible connection to Sanctorius.

Keywords Nicolaus Cusanus · Pharmacology · Quantification

As demonstrated in the previous passages, Sanctorius had a wide-ranging interest in various medical fields, was extremely well read in both ancient and contemporary literature, and promoted the use of various instruments, mostly to improve therapeutics, but also for demonstration purposes, as in the case of optics. He was a practicing physician, who unexpectedly and, it is alleged, only reluctantly took up the first chair in theoretical medicine at the University of Padua. While certainly not dissatisfied with the prestige and the money that went with the professorship, his true interest seems to have always been in medical practice and instrumentation. Yet, an integral aspect of Sanctorius's undertakings still remains to be considered: his quantitative approach to physiology based on the use of a series of measuring instruments.

In Chap. 3, the fusion of quality and quantity in Galenic medicine was clarified. Humoral theory, according to which balance was crucial to maintain health, necessarily involved a quantitative element, namely the proportions between the different variables involved, such as humors, qualities, ingestion, and excretion. Quantity was important also with regard to the six non-naturals. Given that these factors could change the primary qualities and thus influence the state of humoral balance, they had to be used in due quantity and quality. In order to fully appreciate

© The Author(s) 2023
T. Hollerbach, *Sanctorius Sanctorius and the Origins of Health Measurement*,
https://doi.org/10.1007/978-3-031-30118-6_5

Sanctorius's quantitative approach to physiology, it is necessary to scrutinize these forms of quantification in more detail. Some of them are dealt with in existing studies on Sanctorius, but a systematic analysis, bringing together the different threads of quantitative ideas and practices prevailing before, and contemporary to Sanctorius, is still lacking.[1] In the following, I provide such an analysis, and hope to thereby cover many different facets of quantification in Galenic medicine, although an exhaustive account is beyond my present means. The questions that guide this chapter are: Was there a mathematical tradition in Galenic medicine on which Sanctorius could draw? In what way was quantification part of medical theory and practice? In which medical fields was quantification used, and by whom?

5.1   The Quantification of Food and Drink

Out of the six non-natural things, discussions of food and drink, in particular, involved quantitative statements. As Melitta Adamson's study of medieval dietetics has shown, the quantity and proportion of food and drink was a standard topic dealt with in the *Regimen sanitatis* literature. While Adamson gave no example of the nature of these quantitative statements, their level of specification, or the measures used, Sanctorius referred in his *Commentary on Hippocrates* to a famous dietary discourse with the title *Trattato della vita sobria* (Treatise on the Sober Life, 1558), in which the author, Luigi Cornaro (ca. 1484–1566), had exactly specified the healthy quantities of food and drink to be consumed per day.[2] Cornaro determined a daily ration of twelve ounces of food and fourteen ounces of wine in order to conserve his health and lead a long life.[3] Sanctorius did not comment on the quantities given by Cornaro, but highlighted the importance of a steady routine. He thereby drew on Cornaro's report, that his friends had made him increase his daily food intake, arguing that he needed to adapt his eating habits to his old age. But this change in habits, so Cornaro, had a harmful rather than healthful effect, and provoked illness. In 1613, the Flemish Jesuit Leonardus Lessius (1554–1623) translated into Latin Cornaro's treatise, originally composed in the vernacular, and included it in his own treatise on hygiene, *Hygiasticon*. Echoing the themes mentioned by Cornaro, Lessius also included quantitative statements. Even while

---

[1] For studies dealing with some quantitative aspects related to Sanctorius, see e.g., Castiglioni 1931: 748, Bylebyl 1977, Grmek 1990: esp. 1–43, 71–89, Sanctorius and Ongaro 2001: 21 f., 42, Bigotti 2016a: 242–52.

[2] The *Trattato della vita sobria* was later published as part of the *Discorsi della vita sobria* (Discourses on the Sober Life, 1591) by Cornaro's grandson. In addition to the *Trattato,* this work contained three other essays written by Luigi Cornaro: a *Compendio della vita sobria* (Compendium on the Sober Life), a *Lettera al Sig. Barbaro* (Letter to Signor Barbaro) and an *Amorevole essortatione* (A Loving Exhortation) (Walker 1954: 529 f., Milani 2014: 3).

[3] The unit of measurement and of weight varied from one Italian state to another. The most common was the Roman *libra* (pound), which was equal to 327 g and divided into twelve *unciae* (ounces). See: Cardarelli 2003: 74.

acknowledging the difficulties in determining universally healthy amounts of nutrition, given the vast variations in human bodies, Lessius identified twelve to fourteen ounces of food and drink per day as the proper quantity, especially for the elderly and for those with a weak complexion (Cornaro 1591: 5v; Lessius 1613: 15; Sanctorius 1629a: 122; Gruman 1961: 225 f.; Adamson 1995).[4]

Moreover, the personal notes of the Italian artist Jacopo Pontormo (1494–1557) illustrate that measuring meals was not uncommon in the Renaissance. During the last years of his life, Pontormo systematically recorded his food intake along with other aspects of his daily routine, such as the weather and his medical condition. It is not entirely clear whether he used a balance to weigh the food before he consumed it, as he often referred to imprecise quantities. It is striking that he specified quantities especially for bread, as the following example shows: "Monday for dinner fourteen ounces of bread, pork loin, grapes and cheese and endive" (Pontormo and Nigro 1988: 43). Interestingly, Pontormo not only quantified bread in terms of weight, i.e., in ounces and pounds, but also in terms of price: "Friday evening salad, pea soup and one egg fish and bread for five kreutzer (quattrino)" (Pontormo and Nigro 1988: 37).[5] The inconsistent use of diverse measurements imply that Pontormo neither had a uniform measuring method for his food intake, nor considered this important. Yet, his notes reveal that he thought about nutrition in quantitative terms.

These three examples illustrate that the quantification of food and drink was part of the Renaissance dietary literature, and occasionally quite specific. And indeed, nutrition was quantified already much earlier, long before the first *Regimina sanitatis* were written and the doctrine of the six non-naturals was systematized. As Sanctorius mentioned in the *Commentary on Hippocrates*, the author of the appendix to the Hippocratic treatise *De victus ratione in morbis acutis* (On Regimen in Acute Diseases) specified that patients should be given twelve cotyles of ass's milk.[6] Thus, even though Sanctorius certainly did not know of Pontormo's still unpublished notes and made no reference to Lessius's *Hygiasticon,* he was well acquainted with ancient as well as contemporary attempts to quantify and, as in the case of Cornaro, also to stabilize food intake (Sanctorius 1629a: 421; Hippocrates and Potter 1988: 235).

With respect to the three Renaissance authors mentioned, Cornaro, Lessius, and Pontormo, it is important to note that none of them was a physician. This testifies to the popularity of hygiene at the time, which extended well beyond medical circles. Sanctorius's use of the *Trattato della vita sobria* to support his argumentation reveals that Cornaro, a Venetian nobleman, was to him, the learned physician, a

---

[4] Leonardus Lessius published his work *Hygiasticon* in Antwerp, so was very probably referring to Belgian units of weight. One Belgian *livre* (pound) was equal to 489.5 g and divided into sixteen *once* (ounces). However, Lessius compared his quantities to those mentioned by Cornaro without commenting on the possible discrepancies between the units of weight they used. See: ibid.: 84.

[5] A kreutzer (*quattrino*) was a small copper coin, the sixtieth part of one Tuscan lira. See: Pontormo and Nigro 1988: 36, fn. 16.

[6] Cotyle (*cotyla*, from the Greek for *cup*) was a measure used by the ancient Romans and Greeks, equivalent to nearly half an English pint, or ca. 250 ml. See: Smith 1848b.

trusted source. Thus, while management of the six non-natural factors was one of the most important tasks of the physician, it was also an area in which laypeople could gain a certain degree of authority. And so Cornaro wrote "a man can have no better doctor than himself, and no better medicine than the temperate life" (Cornaro 1591: 6v).[7] In this spirit, he offered simple rules to the general public without bothering with complex theoretical considerations of the body, or the properties of different food types. Instead, he placed his trust exclusively in his own common sense and self-knowledge, gained by means of trial and error. Accordingly, what he conveyed in his work was a kind of "self-help" approach to health suitable for everyone. This is in line with the public eager for self-improvement, mentioned earlier (Sect. 4.1.2). Yet, Cornaro's work was also subject to debate and criticism and Tessa Storey has cast doubt on Cornaro's popularity among the Italian public, by noting that his approach was not emulated in Italian vernacular health advice. But however popular Cornaro's work might have been, the fact that Pontormo, a contemporary of Cornaro, meticulously recorded his food intake implies that there was a general awareness, at the time, of the importance of regulating food intake not only in qualitative, but also in quantitative terms (Mikkeli 1999: 89–96; Storey 2017: 221–4).

What is more, the efforts of non-physicians to regulate their private hygiene raised the question of whether or not hygiene should be solely in the hands of the medical profession. Cornaro's answer to this was clear: physicians were necessary only for those who did not lead a sober life. Sanctorius, a university professor of medicine, writing a dietary treatise in Latin, the *De statica medicina*, could hardly have agreed. Can the *De statica medicina* therefore be seen as an attempt to reclaim authority in a medical field which was becoming more and more popularized? Was it influenced by his fear of losing patients, given that people were being encouraged to heal themselves? Or was the contrary the case: Did Sanctorius's instruments enable people to be their own physicians? I will resume these questions later, when more has been said on the material dimensions of Sanctorius's static medicine.

5.2   Degrees, Computation, and Proportions

In Galenic medicine one often encounters the term "degree" (*gradus*). It was used to express the range (*latitudo*) between health and disease (Sect. 3.1.2) as well as the range (*latitudo*) of qualities related to the properties of drugs. Thus, the Paduan professor of medicine Giambattista da Monte (1498–1551) wrote in his commentary on Galen's *Ars medica*: "Medicine is knowledge of all things in their latitude, from the first to the last degree" (Da Monte 1556: 151).[8] The idea of a latitude which permits of degrees was hotly debated by physicians in the Renaissance and had

---

[7] "Non havendo adunque l'huomo miglior Medico di se stesso, nè miglior medicina della vita ordinata, …." See: Cornaro 1591: 6v The English translation is taken from: Mikkeli 1999: 92.

[8] "est enim medicina scientia omnium in latitudine, & à primo gradu usque ad ultimum: …." See: Da Monte 1556: 151. For the English translation, see: Maclean 2002: 256.

given way in the fourteenth century to attempts to quantify qualitative changes by
the so-called Oxford calculators. It is not my intention here to describe these complex discussions at any length, but rather to briefly point out certain aspects that I
feel may better elucidate Sanctorius's quantitative approach to physiology.[9]

5.2.1   The Latitude of Health

In the Galenic corpus, the most important formulation of a gradual intension or
remission of health, the so-called *latitude of health*, appears in the *Ars medica*.
Thus, in his *Commentary on Galen*, Sanctorius referred to this idea and explained,
for example, that the primary and optimal degree of health was the body *simpliciter
salubre semper* (healthy in a general sense always), which served as the norm for all
other complexions. The second degree of optimal health was to be found in bodies
that were *salubri ut multum* (healthy for the most part). This terminology relates to
Galen's categorization of the latitude of health, according to which a body could be
healthy, morbid, or neutral, either *simpliciter* (in a general sense) or *ut nunc* (with
application to the present). Moreover, the body could be in these states either *semper* (always) or *ut multum* (for the most part). But it is not fully clear how Galen
understood these terms, or how they were received, since his *Ars medica*, gave rise
to different interpretations. Without going into the latter in detail, it is of interest
here to note that the latitude of health pertained to gradual differences between *types
of body*, which were introduced in the *Ars medica* but were not described in numerical values. Hence, the degrees of health, of disease, and of a neutral state were not
defined quantitatively—they were labeled rather than measured (Sanctorius 1603:
4r; 1612a: 102, 116).

However, the graduation of health and disease involved certain forms of computation and diagrammatic representation, as illustrated for example by Sanctorius's
count of ninety-six degrees of contra-natural bad complexions (Fig. 5.1) (Sanctorius
1603: 4r; 1612a: 121, 133).

In fact, similar combinatory calculations were used to determine the many variables involved in medicine. Thus, Girolamo Cardano (1501–1576) calculated that
considering astronomical, environmental, and physiological variables in diagnosis
and prognosis would amount to taking 2936 (later recalculated as 3194) equiprobable outcomes into account and, even then, there would be exceptions. Along similar lines, Sanctorius tried to compute the number of combinations of two, three, and
four corrupted humors in animals, of which there were, according to him, 165 in all.
He put forward the figure of 80,084 possible equiprobable mixtures of up to four

---

[9] For more information on the concept of the "latitude of health" and especially on the problem of
the neutral state of health, see Joutsivuo 1999, who also analyzed Sanctorius's view on the issue as
expressed in his *Commentary on Galen* and his *Commentary on Avicenna*. See also Ottosson 1984:
178–94, Maclean 2002: 139 f., 177–81, 256–9. On the Oxford calculators see, for example: Sylla
1973, Trzeciok 2016.

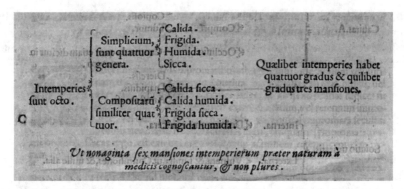

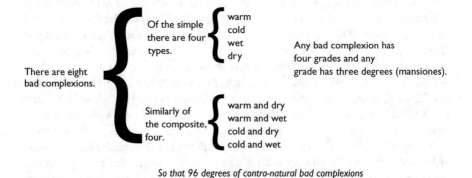

So that 96 degrees of contra-natural bad complexions
are recognized by physicians and not more.

Fig. 5.1 Table illustrating Sanctorius's count of ninety-six degrees of contra-natural bad complex-ions (Sanctorius 1603: 4r). The English translation of the table is taken from: (Wear 1973: 352)

(out of 165) corrupted humors.[10] The enormous numbers show that such forms of quantification were purely intellectual and had no practical application (Sanctorius 1603: 149r–151r; Maclean 2002: 175 f.).

5.2.2   Pharmacology and the Latitude of Qualities

Another area in which degrees, computation, and proportions were important is pharmacology. It was connected to the concept of a latitude of qualities, according to which there existed four degrees of strength, or intensity of the primary qualities. These degrees were introduced by Galen in order to understand the interactions

---

[10] Ian Maclean has shown that Sanctorius employed a wrong method for his calculation and that the real figure is even much higher. See: Maclean 2002: 176, fn. 120.

between body and drug, as well as to classify the powers and effects of drugs. As was mentioned earlier (Sect. 3.1.3), drugs were complexionate, meaning that they were characterized by the four primary qualities of hot, cold, moist, and dry. In addition, according to the Galenic doctrine, they possessed so-called *derivative* qualities, which are the effects a substance can be observed to have on the body: heating and cooling, drying and moistening, but also purging, burning, or the like. These effects cannot be determined in themselves, but only in relation to a body. Pepper, for example, is cold to the touch but has a hot taste and a heating effect on the body. So, just like the complexion of human bodies (Sect. 3.1.2), the complexion of medicines was relative. Therefore, the effect of a drug could change from body to body, too, and a drug could be hotter in relation to one patient than it was in relation to another. Its effect depended on the complexion of the body it acted on and was thus determinable solely in relation to this body. According to Galen's pharmacology, it was therefore necessary that the physician not only detect the properties of a drug (e.g., hot or cold), but also ascertain the individual strength of the substance.[11] These were defined in terms of four degrees of intensity: (1) weak, (2) obvious, (3) strong, (4) massive. Hence, in choosing the healing drug, it was not enough for the physician to find a substance matching that of the patient's state of complexional imbalance; in order to guarantee a healing effect, he also had to ensure that the degree of intensity between the two of them was inversely equal. So if a drug was characterized as hot in the first degree and dry in the third, a physician would have known that it helps against a cold and moist disease, if the moisture strongly predominated. Attention to the intensities was also important with regard to preserving health. While cure was effected by contraries, similars were thought to preserve health and it was necessary, therefore, that their degree of intensity be equal to that of the patient's normal complexion. Accordingly, Sanctorius wrote in the *Commentary on Hippocrates*: "If something is warm in the second degree it will certainly not be preserved in the same state by something that is warm in the first degree, but cooled down" (Sanctorius 1629a: 407 f.).[12]

However, given the variability and relativity of complexions, how could the physician determine the complexional balance, or imbalance of a patient and the degree of intensity of a drug? For Galen, the yardstick was the normal temperate complexion of a human body. As the most temperate part of the body was, according to Galen, the skin which covers the hands, the only means to decide complexion a patient had was touch. Likewise, the hand provided the standard by which medicinal complexions could be evaluated. As the physician would normally take his own hand as the reference point for assessing a remedy, or the complexion of a patient,

---

[11] In using the term "Galen's pharmacology," I follow Sabine Vogt, who has argued that even though the word "pharmacology" in today's sense was unknown to Galen and his contemporaries, it is correct to speak of Galen's "pharmacology," in view of his theoretical approach to drug-lore. See: Vogt 2008: 305.

[12] "calidum ut duo certè per calidum ut unum non conservari in eodem statu, sed refrigerari: ...." See: Sanctorius 1629a: 407 f. Sanctorius discussed the Galenic degrees of intensity also in his *Commentary on Galen*, see e.g., Sanctorius 1612b: 224–7.

he had to take into account his own remoteness from the temperate condition. This he could learn, so Galen, only through a long experience of touching different bodies. There remained, however, the problem of finding some sort of objective reference, by which to judge whether a complexion was temperate. Regarding the primary qualities of hot and cold, Galen proposed measuring the temperate complexion as the midpoint between the extremes found in reality—ice and boiling water or fire. With regard to dry and moist, the temperate was that which appeared neither hard nor soft to the touch. Hence, determining the intensities of qualities, be they in a drug or in a human body, was a difficult calculation involving several rather vaguely determined factors. This shows that Galen tried to quantify drug action and recognized the need for a point of reference when comparing complexions, but also that he did not go so far as to develop methods to measure the underlying qualities, such as heat, by constructing measuring instruments, for example, the thermometer (Ottosson 1984: 134–210; Vogt 2008: 308–10).

As Galen's pharmacological theory was restricted to the so-called *simple remedies*, consisting in a single substance, attempts were made in the Middle Ages to devise mathematical rules to determine the complexion of a compound medicine. The two central problems were (1) to account for the degree of intensity and (2) to determine the effect of varying the weight of each of the ingredients. At the end of the thirteenth century, Arnold of Villanova developed a theoretical system which enabled him to calculate both variables, the degree of intensity of, and the quantity of the ingredients in a compound drug.[13] However, Villanova himself seems to have thought his rules too complicated for any practical application. The nature of qualitative change and the problem of expressing the intensive effect of a qualitative force in quantitative terms were discussed by also philosophers of nature. In fact, Michael McVaugh has argued that Villanova's system had a strong natural-philosophical orientation and that the medical tradition on which it was based might have anticipated developments in natural philosophy. According to McVaugh, Villanova's system either directly, or indirectly influenced the so-called Oxford calculators, a group of thinkers at Oxford University in the mid-fourteenth century, most but not all of whom were associated with Merton College, and hence were earlier called the Merton School. They developed different concepts of a "latitude of forms," understood as an abstract range within which a given form, *complexio*, quality, or quantity can vary. These theories have been ably described elsewhere and need no further analysis here.[14] What must be stressed, however, is that the late medieval debates on the latitude of forms were intellectual endeavors, which did not consider the possibilities of practical application (Sylla 1973: 228–76; Temkin 1973: 111–4; Siraisi 1990: 146; Villanova et al. 1992; Sylla 2011: 903).

To sum up, the concept of a latitude of qualities put forward by Galen in relation to the properties of drugs led in the Middle Ages to attempts to quantify qualitative

---

[13] For a detailed analysis of Arnold of Villanova's system, published in the *Aphorismi de gradibus*, and for the development of medieval pharmacological theory, see: Villanova et al. 1992. For the relation of Villanova's theory to medical practice, see: McVaugh 1969.

[14] On the Oxford calculators and medieval theories of the latitude of forms, see: Sylla 1973, 2011.

changes. But these attempts were purely theoretical: the latitude of a quality or form remained a conceptual or abstract construct. Still, they may well have influenced Sanctorius in his quantitative approach to physiology, even though I was unable to find in his work a direct reference to any of the Oxford calculators.[15] I did find a reference to Arnold of Villanova, however. It is interesting to note that the Catalan physician published his pharmacological theory in aphorisms (*Aphorismi de gradibus*, Sect. 4.1.1)—the very form used by Sanctorius in the *De statica medicina*. Yet, it is difficult to detect any direct influence of Villanova's aphorisms on Sanctorius's static medicine, as Sanctorius neither referred to the *Aphorismi de gradibus* nor dealt specifically with the pharmacological theory it contained. Instead, as will be shown below, Sanctorius connected his quantification efforts with Galen and the concept of the latitude of health (Sanctorius 1625: 410; 1629a: 389).

5.2.3 Pharmacological Practice

In pharmacological practice, the quantification of medicinal substances by weight was an integral part of the daily work not only of pharmacists, but also of physicians; indeed, both professions were often practiced by one and the same person. In sixteenth century Venice, pharmacies were at the heart of medical practice. It was herethat doctors and surgeons met and, in all likelihood, also received their patients. Thus, physicians and pharmacists were in day-to-day contact and doubtless provided a mutual stimulus. Moreover, merchants and traders in *materia medica* from all over the world met in Venice, making it a hub of botanical and pharmacological exchange. Padua, too, was an important center for pharmacological research and practice, thanks to its botanical garden and the foundation there, in 1533, of the first ever chair of simples in Europe. In this light, it is no surprise that Sanctorius had a sound knowledge of medicines and presented them in his works. In the *Commentary on Galen* he wrote, for example:

> Apart from paracentesis, I use three ounces of juice of irises with two ounces of manna, dissolved in water, and flowers of citrus, to evacuate dropsical fluid. In this use, the root of jalapa which has been recently brought from India miraculously effectuates evacuation by a drachm.[16] These [things] evacuate water more safely than paracentesis. For the same use, medicated wine is prepared from jalapa, which is the most pleasing for the removal of dropsical fluid. For its preparation I use the work and diligence of the pharmacopoeia at the "sign of the ostrich" [pharmacy] of Albertus Stechinus, who in the preparation of this and

---

[15] Fabrizio Bigotti has argued that Sanctorius based his practice and conclusions with regard to the *pulsilogium* and the thermoscope on developments of the scholastic theory of the latitude of forms. See: Bigotti and Taylor 2017: 60, 65 f., 74, Bigotti 2018: 94 f.

[16] The *Mirabilis jalapa* plant, also known as the four o'clock flower, is named after Xalapa, the capital city of the Mexican state of Veracruz. It was already used by the Aztecs for medicinal purposes and was grown commercially in India. See: Neumann 1752: 149–63, Anagnostou 2008: 125. One Roman *drachma* (drachm) was equal to 3.39 g and the eighth part of one ounce (Robens et al. 2014: 57).

other medicines is so learned and diligent that he deserves the highest praise ever bestowed in this most distinguished city (Sanctorius 1612b: 315).[17]

The citation illustrates how specific quantities were carefully weighed out when preparing remedies for which the proportion and the quantity of the different ingredients were integral. A pair of scales (or: a balance) was a vital piece of equipment in any apothecary and certain physicians—not the least, Sanctorius—probably had one of their own. In preparing his remedies, as explained above, he relied on the pharmacopeia of Alberto Stecchini (life dates unknown). Stecchini worked at one of the most celebrated pharmacies of the later sixteenth century in Venice, namely, the aforementioned *ad Signum Strutij* or Struzzo pharmacy. The book to which Sanctorius referred, *Avvertimenti nelle compositioni de' medicamenti per uso della spetiaria* (Advice on the composition of medicines for use in pharmacy) was originally published in 1575 by the Struzzo founder Georg Melich (life dates unknown). Alberto Stecchini published revised editions of it in 1605 and 1627, adding new recipes and his own opening discourse on the art of the apothecary. Given Sanctorius's great admiration for Stecchini, it seems likely that he consulted not only his published works but also the man himself, frequenting the *Struzzo* pharmacy to this end. According to Richard Palmer, physicians in sixteenth century Venice often attached themselves to particular pharmacies and it seems that Sanctorius chose to buy the ingredients for his remedies from the Struzzo. Maybe this was also the place where he first learned about the exotic root of jalapa and its purgative effects, which he mentioned in the above citation. In any case, it is remarkable that Sanctorius praised Stecchini's pharmacopeia to the skies, in all three of his commentaries; and, too, it is indicative of the, often, close relationships between physicians and pharmacists, and the accordingly intense exchange of knowledge and experience between the two professions. In this context, optimal dosage was ascertained not by means of mathematical theories, but rather by hands-on testing paired with text-based knowledge. This included common knowledge acquired during the pharmacist's apprenticeship as well as knowledge gained from reading the work of predecessors, both ancient and medieval (Sanctorius 1625: 748 f.; 1629a: 489 f.; Palmer 1985: 101–16; Parrish 2015: 7; Leong and Rankin 2017: 157; Pugliano 2017: 249).

---

[17] "Ego omissa paracentesi pro aqua hydropicorum evacuanda utor uncijs tribus succi ireos cum duabus uncijs mannae dissolutae in aqua è floribus citri: in hunc usum radix salapae nuper ex India delatae ad drachmam mirificam efficit evacuationem: tutiusq; his evacuari poterit aqua, quam paracentesi. Ad eundem usum paratur vinum medicatum ex salapa iucundissimum in auferenda hydropicorum aqua; in eo conficiendo utor opera, & diligentia Alberti Stechini pharmacopeia ad Signum Strutij, qui in hac, & in alijs praeparandis medicinis adeo eruditus, & diligens est, ut hactenus summas laudes in amplissima hac Civitate meritus sit." See: Sanctorius 1612b: 315.

5.2.4   Pharmacology and Dietetics

As explained in Sect. 3.1.3, there was a close connection between drugs and food-stuffs. The latter were used not only to preserve health, but also to heal diseases. Accordingly, Galen applied the degrees of intensity for the potency of drugs also to food and drink. Parsley, for example, was thought to be hot in the third degree and dry in the third degree and could therefore strongly alter the body. Similarly to com-pound drugs, different foods were combined in a single meal, either to guarantee that it was temperate and would therefore not change the complexion of a body, or to counteract an imbalanced complexion of a body. Thus, while hot and dry pepper could be used to treat a phlegmatic person, it could also be used to render cold and moist fish more temperate. Ken Albala has argued that, even though the mathemati-cal theories developed in the Middle Ages in order to quantify qualitative changes (Sect. 5.2.2) were not used in dietetics, Renaissance dieticians did comprehend the basic idea of varying amounts having varying effects and applied it informally in their work. Without using mathematics, they knew, for example, that pepper, which was hot in the fourth degree and cinnamon, hot in the first degree, together formed a condiment somewhat less intense than the same amount of pepper alone. Similarly, they were aware that in combinations of opposite qualities, the food in the greater quantity would remain predominant (Albala 2002: 84–91).

The preceding paragraphs have shown that Galenic medicine as practiced in the early modern period involved certain ideas of quantification related to degrees, computation, and proportions. In the Middle Ages and the Renaissance, there was an increasing desire to quantify data and, especially with regard to therapeutics, the concept of a latitude of qualities led to the elaboration of a quantified medical the-ory, which was, however, not applied in practice. Still, the daily work of physicians, pharmacists, and dieticians was shaped by the management of quantities, be it in the composition of a remedy or the compilation of a balanced diet. The close connec-tion between drugs and foodstuffs and the applicability of pharmacological theory to food and drink probably inspired non-physicians like Pontormo and Cornaro to quantify their food intake. What is more, these aspects might well have given Sanctorius the idea of using a balance not only to weigh the ingredients of drugs, but also to measure the effects of the six non-natural things on insensible perspiration. Yet this assumption is thrown into doubt by the fact that Sanctorius referred neither to quantification nor to his measuring instruments in his last publication dealing with the invention of remedies (*De remediorum inventione*). Instead, he made recourse to Aristotelian syllogistic logic and directed the reader to his first published work *Methodi vitandorum errorum*, in which he described his method for finding the specific differences in diseases (*affectus*), which were, he claimed, the only indi-cation for remedies.[18] Whether he actually used this method rather than quantitative examinations in his medical practice is, of course, a different question entirely, and will be addressed later (Sanctorius 1629b: 1–12).

---

[18] On Sanctorius's first work *Methodi vitandorum errorum* and his medical logic, see: Maclean 2002.

5.3   Erasistratus and Nicolaus Cusanus—Two Early Quantitative Approaches

In addition to the quantitative tendencies just outlined, there are two names which are more closely associated with Sanctorius and his static medicine: Erasistratus (third century BCE) and Nicolaus Cusanus (1401–1464). The former was a major exponent of the ancient medical school of Alexandria and was already mentioned above, in connection with early ideas on *perspiratio insensibilis* (Sect. 3.2.1). The latter was one of the most important German thinkers of the fifteenth century, whose activities ranged from theology, law, and philosophy to mathematics and astronomy. Due to the fact that both men were early proponents of quantitative approaches to medical problems, the secondary literature has often related their undertakings to Sanctorius and his use of quantification. In the following, I reconsider such possible links, in chronological order.

5.3.1   Erasistratus

Around the third century BCE, the physician and anatomist Erasistratus demonstrated that animals give off invisible emanations:

> If one were to take a creature, such as a bird or something of the sort, and were to place it in a pot for some time without giving it any food, and then were to weigh it with the excrement that visibly has been passed, he will find that there has been a great loss of weight, plainly because, perceptible only to the reason, a copious emanation has taken place (Anonymus and Jones 1968: 127).

The similarity to Sanctorius's description of his own weighing procedures is instantly striking. Although separated by nearly two thousand years, the two physicians were interested in the same physiological phenomenon and used a balance to examine it. At first glance, it seems obvious: Sanctorius built his static medicine on the findings of Erasistratus. At second glance, however, the picture changes. As all of Erasistratus' works have been lost, they are known solely thanks to the references made by his successors, primarily Galen. But even though Galen referred to the problem that Erasistratus had tackled he made no mention of Erasistratus's quantitative observation. In his work *De naturalibus facultatibus* (On the Natural Faculties), Galen wrote:

> Now, the amount of urine passed every day shows clearly that it is the whole of the fluid drunk which becomes urine, except for that which comes away with the dejections or passes off as sweat or insensible perspiration. This is most easily recognized in winter in those who are doing no work but are carousing, especially if the wine be thin and diffusible; these people rapidly pass almost the same quantity as they drink. And that even Erasistratus was aware of this is known to those who have read the first book of his *General Principles* (Galen and Brock 1916: 109 ff.).

Hence, the lack of a reference on Galen's part to Erasistratus's demonstration of insensible perspiration provides strong evidence that Sanctorius was ignorant of it.[19] In fact, it was only at the end of the nineteenth century that a Greek papyrus from the second century CE, in which an anonymous author wrote of Erasistratus and his examination of weight changes in fowls, was rediscovered in the British Museum. The papyrus contains nothing but the brief paragraph cited above: the observation made by Erasistratus. From the evidence at hand, it seems that the Alexandrian physician'sconcern was to prove the existence of insensible perspiration rather than to systematically measure it. This would explain the absence of numerical values in the account of his observation. To summarize, long before Sanctorius, Erasistratus had put forward the idea of measuring insensible perspiration by means of weighing procedures. Yet, Sanctorius was most probably unaware of Erasistratus' quantitative procedure and it can therefore be assumed that their undertakings were not related. Moreover, the two physicians followed different approaches. Contrary to Erasistratus, Sanctorius was not out to prove the existence of *perspiratio insensibilis*, which, for him, was beyond any doubt. However, he did try to establish its quantity by means of systematic observation (Grmek 1990: 36 ff.; Bigotti 2016b: 5, fn. 14).

5.3.2 Nicolaus Cusanus

A more contemporary author, who might have been a reference point for Sanctorius, is Nicolaus Cusanus. His name often appears in historical accounts of the work of Sanctorius, but the question of a possible relation between Cusanus and Sanctorius is usually dealt with only briefly, in a few sentences.[20] However, in my opinion, the issue is by no means trivial and deserves a more in-depth look. As the following analysis will show, there are striking similarities in the work of the two authors, which hence good reason to assume that their quantitative approaches were closely related. Before addressing this point, I briefly examine Cusanus and his work, with a focus on the aspects I consider relevant with regard to Sanctorius. Furthermore, I compare their quantitative methods and, on this basis, review the likelihood of a connection between their efforts.

The Quantitative Approaches of Cusanus and Sanctorius Compared The son of a prosperous German merchant, Nicolaus Cusanus first studied at the University of Heidelberg before moving to the University of Padua in 1417, from where he graduated six years later as a doctor in canon law (*decretorum doctor*). The

---

[19] As will be shown below, Sanctorius knew Galen's work *De naturalibus facultatibus* and was familiar with Galen's statement, quoted here, that the entirety of any drink consumed became urine, except for those parts of it excreted as feces, sweat, or insensible perspiration (Sect. 5.4.1).

[20] See e.g., Del Gaizo 1889: 21, Castiglioni 1931: 748, Ettari and Procopio 1968: 27, Sanctorius and Ongaro 2001: 21 f., 41 f.

nomination as cardinal in the late 1440s marked the climax of his career and was soon followed by the publication of his famous series of papers under the title *Idiotae libri quatuor* (The Idiot in Four Books, 1450), which was written in the form of a dialogue between a layman (*Idiota*) and a Roman orator (*Orator*). Interestingly, the work included a paper on static experiments (*Idiota de staticis experimentis*).[21] This title alone, *de staticis experimentis*, sounds suspiciously as if it may have inspired the title of Sanctorius's *De statica medicina*. Both titles feature the New Latin word *staticus*, which can be translated as "relating to weighing," and derives directly from the Greek term *statikós*. This implies that weighing and, thus, the use of a balance was fundamental not only to Sanctorius's treatise, but played an important part in Cusanus's work, too. In fact, perusal of the latter's text shows that the similarities between the two works go well beyond their titles. At the beginning of the dialogue, the idiot, who is understood to be not a foolish person, but simply a layman, explained:

> It seems to me that by reference to differences of weight we can more truly attain unto the hidden aspects of things and can know many things by means of more plausible surmises (Cusanus 1983: 222).[22]

And a few lines later, the idiot continued:

> For identical sizes, of whatsoever different things, are not at all of the same weight. Accordingly, since the weight of blood or the weight of urine is different for a healthy man and for a sick man or for a youthful man and an elderly man or for a German and an African, wouldn't it be especially useful to a physician to have all these differences recorded? (Cusanus 1983: 222).[23]

The orator strongly agreed with the idiot who, in addition to the weighing of blood and urine, also proposed to record the weight of herbs. According to him, comparing the weights of the herbs administered with the weight of the patient's blood or urine would enable the physician to determine the correct dosage of a drug. Whether he meant absolute or relative weights is unclear, however. Thus, through the voice of the idiot, Cusanus had already postulated the importance of quantitative studies in medicine and suggested weighing procedures to realize them. These involved not only the fluids of the human body, but also medicaments. Notwithstanding that neither the idiot nor the orator explicitly referred to pharmacological theories, the close connection to contemporary discussions on the latitude of qualities outlined above is obvious (Sect. 5.2.2). Moreover, historiographical studies have shown that Cusanus was influenced by the works of the Oxford calculators, which gives grounds

---

[21] The following account of Cusanus's paper *Idiota de staticis experimentis* is based on the English translation of the work by Jasper Hopkins, see: Cusanus and Hopkins 2001: 602–30.

[22] "Per ponderum differentiam arbitror ad rerum secreta verius pertingi et multa sciri posse verisimiliori coniectura." See: Cusanus 1983: 222. The English translation is taken from: Cusanus and Hopkins 2001: 606.

[23] "Nam nequaquam est eiusdem ponderis identitas magnitudinis quorumcumque diversorum. Unde cum aliud sit pondus sanguinis et urinae hominis sani et infirmi, iuvenis et senis, Alemanni et Afri, nonne maxime conferret medico habere has omnes differentias annotatas?" See: Cusanus 1983: 222. The English translation is taken from: Cusanus and Hopkins 2001: 607.

to assume that his quantitative ideas were informed by discussions on the latitude of forms. In any case, in the *De staticis experimentis*, weighing became the crucial method for the physician to find the proper remedy. By adding "experiments done with weight-scales" to common methods of diagnosis based on the examination of, for example, color or taste, the physician was able to achieve greater precision in his judgements (Lohr 1988: 556–94; Vescovini 2002: 93; Miller 2017; Dictionary. com 2020).

Cusanus also considered the possibility of weighing a whole man in order to compare his weight with the weight of other animals. To this purpose, man and animal were to be placed successively on a balance-scale. Then, in a second round of measurements, both animal and man, were to be immersed in water and the differences in weight noted. Regardless of the difficulties entailed by such a procedure, it is interesting to note that Cusanus applied here to living bodies the by then well-known Archimedean principle of specific gravity. As specific gravity is the ratio of the weight of a body to its volume, Cusanus's suggestion of weighing bodies in water implies that he intended to compare not only the absolute weights of animals and men, but also their densities, i.e., their composition. This is further indicated by his proposal of an alternative way to assess the weights of men and animals. After having measured their bodyweights outside of water by means of a balance, the man and the animal were to be submerged in a tub of water and the water thereby displaced and caused to overflow was to be collected and weighed. Here again, Cusanus drew on a widely known Archimedean principle to compare the composition of human and animal bodies. In fact, in a later passage of the *De staticis experimentis*, the German thinker even suggested a method of measuring the elements contained in an object by means of a balance. It seems that he thought that this method, too, could be applied to men and animals, although he did not explicitly state this. Without going into the details of this method, which would be too great a digression from the present topic, it is pertinent to mention that Cusanus regarded the weighing procedures as a means to elaborate the average weight of a temperate man respectively of various species of animal, and did not foresee any diagnostic use of them, such as determining complexional imbalances.

Although Sanctorius made no reference in the *De statica medicina* to weighing human bodies (*viventia corpora*) in water, he did so in his *Commentary on Avicenna*. This, he ventured, would enable one to find out how much air such bodies contained. However, instead of amplifying this idea, Sanctorius referred to Archimedes' the famous experiment to determine the gold content of a crown. In the *De statica medicina,* in one of the aphorisms in the section on the non-natural pair air and water, Sanctorius likewise addressed the principle of specific gravity. He explained that the weight of water could be easily determined by weighing a heavy body in water. the deeper the body sank, the lighter and therefore healthier was the water; conversely, if the body sank only a little, the water was heavier and unhealthier. Thus, contrary to Cusanus, Sanctorius used the Archimedean principle here to measure the density of the water, and not of the body immersed in it. Yet, in another passage of the *De staticis experimentis*, Cusanus wrote about the weight of water, drawing on the Roman architect Vitruvius (fl. first century BCE), to assert that light

waters were healthier than heavy waters. The similarities between these two treatises pertain to issues that were common knowledge among scholars both in the
fifteenth and the seventeenth century, so their relevance should not be overestimated—but nor should their existence be neglected (Sanctorius 1614: 21r–21v;
1625: 152 f.).[24]

Many of the static experiments mentioned by Cusanus were based not on the
balance, but on the water-clock, which the ancient Greeks used to measure specific
intervals of time. It is one of the oldest time-measuring instruments and consisted,
in its basic form, of a vessel with a small opening near the bottom. A measured
amount of water was poured into it, which then flowed out through the hole. Given
the consistent use of the same instrument and the same quantity of water the time it
took for the vessel to empty was always the same. Cusanus suggested a slightly different use of the instrument in order to compare the pulse in healthy and in sick
adolescents, and in young and in elderly people. Instead of pouring a fixed amount
of water into the vessel, he proposed to weigh the water that traversed the clock during the time of one hundred pulse beats. Recording the different weights of water,
he believed, would make it possible to establish the respective weights of different
illnesses. According to Cusanus, the same method could be used with regard to
respiration; and he explained, thus, that if a person had fever, the physician should
measure the respiration by means of the water-clock during "the sudden episodes of
feeling hot and of feeling cold," in order to gain more precise knowledge of the
gravity of the disease and of the right moment to administer medication (Cusanus
and Hopkins 2001: 609). What is more, this would also help the physician to better
judge the course of the disease, so Cusanus.[25] As will be seen below (Sect. 7.2),
Sanctorius, too, engaged in attempts to measure the pulse and respiration. However,
instead of using a water-clock to determine changes in his patients' rates of pulse
and respiration, he devised several instruments of his own, most of which were
based on the swing of a pendulum. Thus, Cusanus and Sanctorius both proposed to
measure the pulse and respiration with instruments whose fundamental property
was to record equal intervals of time. In fact, Sanctorius's *pulsilogia* served him also
as a timekeeper. Furthermore, both scholars related their methods to medical practice, with the aim of helping the physician conduct diagnosis, prognosis, and
therapy.

Two further aspects have to be noted with regard to the *De staticis experimentis*.
While exploring possible means to measure the "weight of air," Cusanus suggested

---

[24] For more information on ancient hydrostatics and pneumatics, see: Valleriani 2016.

[25] Before Cusanus, the Alexandrian physician Herophilus is said to have used a water-clock in the
early third century BCE, to measure his patients' pulse. Drawing on his own experience, he had
determined which natural pulse rate for persons of different age groups should occur during the
time period measured by his water-clock. The amount by which his patients' pulse beats exceeded
or fell below the normal rate for their respective age group indicated the gravity of their disease.
As there is only one reference to Herophilus's use of the water-clock in the treatise *De pulsibus* (On
the Pulse, date uncertain) published (probably in the second century) by the otherwise unknown
physician Marcellinus, it can be assumed that Cusanus was unaware of it. See: Landels 1979: 32
f., Von Staden 1989: 354, Lewis 2015: 197 f., 200 f.

putting desiccated wool on one side of a pair of scales and stones on the other side, as a counterbalance.[26] If the balance was located outdoors in a temperate location, the weight of the wool would come to indicate the humidity or dryness of the air: growing heavier, if the air were moist, the wool would increase in weight: if the air were dry, the wool would become lighter. Writing about the weight of the air in the *De statica medicina*, Sanctorius stated:

> The weight of the air can be gathered first from the bigger or smaller weight of the sediment of alum, which is first dried in the sun and then exposed to nocturnal air (Sanctorius 1614: 20v–21r).[27]

Hence, here again Cusanus and Sanctorius put forward similar methods, in this case to determine the weight of air by measuring its humidity. Yet, while for Sanctorius, the physician, this was clearly related to the influence on insensible perspiration of the non-natural factor air, Cusanus did not relate his *experimentum* to medicine. Furthermore, as will be shown below (Sect. 7.4), Sanctorius dealt with the issue much more extensively than Cusanus and proposed also three other ways of determining the humidity of the air. In the *Commentary on Avicenna*, he even depicted two instruments for this purpose (Figs. 7.19 and 7.20) (Sanctorius 1625: 23, 215).

The last interesting point of comparison between Cusanus and Sanctorius under consideration here refers to their measurement of the impetus of wind. In the *De staticis experimentis*, Cusanus mentioned the possibility of investigating "the strength of winds … from experiments done with weight-scales" (Cusanus and Hopkins 2001: 617). He gave no further description of how these procedures should be conducted and made no reference to any medical application. However, he did correlate determining the strength of a wind and that of a man, stating that the latter could be ascertained by having a man lift a weight sufficient to bring a balance into equilibrium. In the *Commentary on Avicenna*, Sanctorius presented a special balance to measure the impetus of winds (Fig. 7.1). This was important, so Sanctorius, because of the different effects that rainy and windy air could have on the body. Noisy wind, for example, sometimes hindered sleep and sometimes induced it. What is more, the instrument helped predict sea storms and thus minimize the risks of flooding. Without going into the details of Sanctorius's apparatus and its operation, it should have become clear by now, that there are similarities not only between the *De staticis experimentis* and the *De statica medicina*, but also between the respective authors' quantitative endeavors, as indeed Sanctorius did mention in his other works (Sanctorius 1625: 246 f.).

---

[26] The idea that air has weight was much debated toward the end of the sixteenth century and was a topic of interest also for Sanctorius. The notion of determining the weight of the air by measuring its humidity must be seen in the light of the Aristotelian doctrine of the interconvertibility of air and water. For more information, see: Middleton 1964: 4, Middleton 1969: 3, 81. For an analysis of Sanctorius's concept of the weight of the air, see Bigotti (2018), who has argued that Sanctorius already recognized atmospheric pressure, an interpretation that I, however, do not share.

[27] "Quanta sit aeris ponderositas, colligitur primo ex maiori, vel minori gravitate aluminis faecum prius exiccati in sole, & deinde aeri nocturno expositi." See: Sanctorius 1614: 20v–21r.

The Accusation of Plagiarism In fact, these similarities did not go unnoticed by Sanctorius's contemporaries. In his attack on the *De statica medicina* (Sect. 3.1, fn. 2), Ippolito Obizzi (b. second half of the sixteenth century), a physician and philosopher from Ferrara, accused Sanctorius of plagiarism. After claiming that Sanctorius had learned about "static reasoning" from Cardinal Cusanus, he concluded that the *De statica medicina* was "deceptive and by no means a truthful experiment and cannot be called an original work" (Obizzi 1615: 71). Besides these general denunciations, Obizzi gave a rather detailed account of those arguments in Cusanus's treatise which he considered similar to those employed by Sanctorius, and asserted that Sanctorius had derived his *pulsilogium* from Cusanus's report on the use of the water-clock (Obizzi 1615: 71 f., 81 ff., 86).

What did Sanctorius say in his defense? What was his reaction to the grave allegations? The answer is: very little. Only in 1615, ten years after Obizzi first cast doubt on the originality of the *De statica medicina*, did Sanctorius comment on his possible debt to Cusanus. In the *Commentary on Avicenna*, he stated:

> … he [Ippolito Obizzi] suggested that our static [medicine] was taken from the static experiments of the Cardinal Cusanus, from which, as everyone can see, not a word is taken. For Cusanus never discusses that weighing of the insensible perspiration of the human body with which all of our aphorisms deal (Sanctorius 1625: 81).[28]

Thus, Sanctorius did not deny his knowledge of Cusanus's *De staticis experimentis*, nor did he explain in any detail how his work differed from that of the cardinal, except for the focus on insensible perspiration. Instead, in the next sentence, he directed the reader, first, to his earlier diatribe against Ippolito, which especially concerned the latter's inclination to astrology, and, secondly, to his defense of his own *De statica medicina*, which Sanctorius added as an eighth chapter to the revised edition, under the title *Ad Staticomasticem* (To the Scourge of Statics). This piece of the seventeen aphorisms comprising this defense not one made mention of Cusanus or of Ippolito's allegations of plagiarism. Sanctorius evidently regarded his examination of *perspiratio insensibilis* as unique and original work and accordingly saw no further need to distinguish his *De statica medicina* from Cusanus's treatise (Sanctorius 1625: 81; 1634: 69r–71v).

Sanctorius's meager reference to Cusanus makes it difficult to assess the relation between the *De staticis experimentis* and the *De statica medicina* and, more generally, Sanctorius's quantitative approach to physiology. Since the *De staticis experimentis* appeared in several editions in the sixteenth century, among them a popular edition of Vitruvius's *De architectura* (On Architecture, ca. 30–15 BCE), and since Sanctorius in his *Commentary on Avicenna* did not deny knowledge of Cusanus's treatise,it is likely that Sanctorius had read the work. What is more, Cusanus's

---

[28] "… protulit nostram staticam à staticis experimentis Cardinalis Cusani fuisse desumptam, à quibus, ut omnes videre possunt, nec verbulum desumptum est: nunquam enim Cusanus aegit de ponderatione insensibilis perspirationis humani corporis, de qua sunt omnes nostri aphorismi." See: Sanctorius 1625: 81. It is interesting to note that Sanctorius never mentioned Ippolito Obizzi by name, but referred to him for example as *Belluni*, or *Astrologus Magnus* (ibid.).

mathematical thoughts were known to a considerable number of Renaissance scholars, among them Girolamo Cardano, an author whom Sanctorius mentioned frequently in his commentaries. Still, these are not certain proofs of a simple and direct
connection between Cusanus and Sanctorius. Indeed, besides the many similarities
outlined above, there are also many differences between the two authors and their
treatises (Nagel 1984: 108; Rudolph 1996: 124).

Differences Between Cusanus and Sanctorius First of all, in the *De staticis
experimentis* there is no suggestion that the proposed experiments were actually
performed. Contrary to the *De statica medicina*, no measuring results are given and
the dialogue ends with the orator's promise to seek to realize the aforementioned
weighing procedures. Hence, Cusanus most probably presented thought experiments without any direct practical application. Whether the measuring instruments
were actually ever used must be asked also with regard to Sanctorius, although his
written work does contain much stronger indications that they were. Not only did
Sanctorius present some of his findings in the form of numerical values, but also his
terminology implies that he actually performed experiments in something like the
modern sense (Sect. 6.2.5). On the assumption that Sanctorius did use his instruments, it will be shown below that the path is long, from the intellectual conception
use of an instrument and its operation to its actual application in research and practice (Sect. 7.5). Accordingly, the question of plagiarism concerns here only the mental processes, the ideas behind the quantitative undertakings, and not their practical
and material dimensions. In this respect it must be noted also that Sanctorius put
forward a much wider range of measuring instruments than Cusanus did, the latter
having limited his static experiments to the use of the balance and the water-clock.
Sanctorius, by contrast, drew on very recent technologies when developing his measuring instruments, and this in itself does often raise the question of Sanctorius's
role in their invention, asin the case of the *pulsilogium* or the thermoscope, for
example (Sects. 7.2 and 7.3) (Hoff 1964: 113 f.).

A second major difference between the quantitative approaches of Cusanus and
Sanctorius is the context in which they appeared. As mentioned earlier, the *De staticis experimentis* is only one among four papers published by Cusanus in his book
Idiota and one should be careful not to consider it independently of the other papers.
Paula Pico Estrada has argued that the *De staticis experimentis* needs to be read as
an analysis, from a philosophical viewpoint, of the workings of the human mind
with regard to its knowledge of the natural world. The text does not express the
belief that reality has a mathematical structure apprehensible to the human mind.
Rather, it refers to the mind's action of "measuring" whatever it encounters, which
it conceives of as a creative action by which the power of the mind approximates
God's own creative power, so Estrada. Without going into the details of Cusanus's
philosophical notions underlying this, it is apparent that Sanctorius's *De statica
medicina* follows an entirely different goal. While Cusanus examined the human
mind and its rapport to truth, Sanctorius reinterpreted the doctrine of the six non-
natural things according to his concept and observation of insensible perspiration.
Even though Sanctorius, in his commentaries, related his quantitative approach to

his wish to attain more certainty in medicine, he thereby pursued different epistemic notions than Cusanus (Sect. 6.2).[29] He composed the *De statica medicina* explicitly as a medical treatise aimed at improving and facilitating the work of the practicing physician. Cusanus, on the contrary, did not focus exclusively on medicine in his *De staticis experimentis*, but dealt more generally with ideas about using quantitative procedures in the investigation of nature. The different orientations of these two works are reflected also in their titles: Cusanus referred to static *experiments* (*De staticis experimentis*) and Sanctorius to static *medicine* (*De statica medicina*) (Pico Estrada 2008: 137, 144).

In conclusion, after weighing up the differences and similarities between Sanctorius and Cusanus, I must say that this is more than just a "genial coincidence." In my opinion, it is significant that Cusanus had already conceptualized many of the quantitative measurements which Sanctorius later claimed to have realized and, moreover, had published them in a work with a title so similar to Sanctorius's *De statica medicina*. While not sharing the same epistemic goals as Cusanus, who, for his part, was no medical practitioner, Sanctorius was able to find in the *De staticis experimentis* much to inspire his own quantitative approach to physiology. Even though Cusanus's work makes no reference to the doctrine of the six non-natural things, it includes common Hippocratic-Galenic notions of the influence on the human body of external factors, such as the climate of the geographic region in which a person lives. What is more, it expresses a desire to put quantitative procedures in the service of medicine, in order to enhance the discipline's certainty. The fact that Sanctorius denied any connection between the *De staticis experimentis* and the *De statica medicina* by highlighting his measurement of insensible perspiration underlines that he considered this the original aspect of his work; and while Sanctorius was certainly right to do so, it is notable that he said nothing about Cusanus's possible influence on his other quantification efforts, not even to challenge Obizzi's remark regarding the similarity of their respective methods to measure the pulse. This is interesting, as *perspiratio insensibilis* played no part in these. It has been shown earlier that insensible perspiration and its quantification were not pivotal to Sanctorius's other publications (Sect. 3.3.7) and, as will be further elaborated below, nor were they pivotal to Sanctorius's other measuring instruments (Sect. 6.1.2). Hence, in this context, Sanctorius's appeal to the weighing of insensible perspiration as a distinguishing criterion is to no effect. In view of the evidence at hand, it is thus highly probable that the *De staticis experimentis* did

---

[29] A central aspect of Cusanus's epistemology was the idea of the impossibility of attainment of certain or complete knowledge on this earth. According to him, mathematics was the only measure by which the human mind could gradually approach knowledge of nature without ever fully achieving it. For more information, see: Nagel 1984: 1–85, Pico Estrada 2008. Contrary to this, Sanctorius still adhered to Aristotelian logic and conceived of medical knowledge as conjectural due to medicine's standing as an art (*ars*). However, departing from tradition, Sanctorius thought that uncertainty in medicine could be eliminated, or at least reduced by means of his measuring instruments (Sect. 6.2) (Siraisi 1987: 235–8).

indeed serve, in a more general sense, to inspire Sanctorius's quantitative approach to medicine and his development of measuring instruments.

5.4   Three Instances of Quantitative Physiological Reasoning

Another form in which quantification pervaded Galenic medicine was its use as a mode of argumentation in the discussion of physiological problems. Owsei Temkin and Jerome Bylebyl have drawn attention to instances of quantitative physiological reasoning in Galen's work as well as in the work of Renaissance scholars. In the next paragraphs, I will briefly describe these efforts and analyze their possible relation to Sanctorius (Temkin 1961; Bylebyl 1977).

5.4.1   Galen and the Quantification of Urine

In one and the same passage of his work *De naturalibus facultatibus* (On the Natural Faculties), Galen both mentioned Erasistratus's approach to insensible perspiration (Sect. 5.3.1) and tried to confute the view that the kidneys produced urine merely as a residue of their own nutrition. Galen believed rather, that the kidneys had a special faculty to attract for their nourishment only the thin and watery parts of the venous blood, generated during the process of digestion, and wouldexcrete the rest as urine. This was confirmed, so Galen, by the observation that the daily amount of urine corresponded to the daily amount of ingested drinks (Sect. 5.3.1) and could therefore be quite copious. If the urinary output was merely residue of the kidneys' nutritional matter, the absurd consequence, as Galen explained, would be that all the other body parts would produce similarly large amounts of residual fluid, proportionate to their size. And thus, he wrote:

> Now it is agreed that all parts which are undergoing nutrition produce a certain amount of residue, but it is neither agreed nor is it likely, that the kidneys alone, small bodies as they are, could hold four whole *congii*, and sometimes even more, of residual matter.[30] For this surplus must necessarily be greater in quantity in each of the larger viscera; thus, for example, that of the lung, if it corresponds in amount to the size of the viscus, will obviously be many times more than that in the kidneys, and thus the whole of the thorax will become filled, and the animal will be at once suffocated. But if it be said that the residual matter is equal in amount in each of the other parts, where are the *bladders*, one may ask, through which it is excreted? For, if the kidneys produce in drinkers three and sometimes four *congii* of superfluous matter, that of each of the other viscera will be much more, and thus an

---

[30] *Congius* was a Roman measure for liquids and corresponds to about six English pints, or 3.48 liters. The amount of urine that Galen specified, four *congii*, is thus about twenty-four English pints, or 13.92 liters. This is nearly five times as much as the average daily urinary output and could only be excreted if a very large amount of wine was drunk. See: Smith 1848a, Galen and Brock 1916: 111, fn. 2.

enormous barrel will be needed to contain the waste products of them all. Yet one often urinates practically the same quantity as one has drunk, which would show that the whole of what one drinks goes to the kidneys (Galen and Brock 1916: 111 ff.).

Hence, Galen pointed here to two difficulties: firstly, that the lungs were not able to eliminate residual fluid, which would consequently simply accumulate in the thorax and cause suffocation; and secondly, that there apparently was no surplus to supply the much larger amounts, which should be eliminated by the other parts, as the kidneys alone quickly eliminated a quantity of fluid nearly equal to that ingested. Without going into the details of Galen's argumentation, or discussing its conclusiveness, what is of interest here is that Galen put forward a numerical value for the urinary output to support his notion of the attractive faculty of the kidneys.[31] Remarkably, just as Sanctorius would do, more than a millennium later, in his measurement of insensible perspiration, Galen quantified a physiological process, the production of urine by the kidneys, by referring to the equilibrium between ingestion and excretion (Temkin 1961: 472–4; Bylebyl 1977: 374 f.).

Galen's argumentation suggests that he measured the amount of urine excreted by people who drank large amounts of wine (*drinkers*), and whose urinary output was therefore much larger than normal. It is of course questionable whether he really collected the urine of others, who might well have relieved themselves more than once while drinking large amounts of wine. Yet, the fact that Galen indicated the quantity of urine in *congius*, a measure which was often used for wine, gives cause to assume that he was directly comparing the consumption of wine with the excretion of urine. But regardless of whether or not Galen actually measured urine, the possibility and importance of quantifying excretions was conceptually formulated in his work. What is more, when comparing the intake of fluids to the output of urine in human bodies, Galen also already paid heed to the loss possibly caused by other excretions—feces, sweat, and insensible perspiration, for example—and therefore tried to reduce these to a minimum; which is why he proposed conducting his measurements in the wintertime, on people who rested and drank a lot in a short period of time (Sect. 5.3.1).[32] Thus, Sanctorius could draw on earlier works regarding not only the practice of quantifying food and drink, but also the quantification of excretions.

From references in his books, it is clear that Sanctorius was familiar with Galen's work *De naturalibus facultatibus* (e.g., Sanctorius 1625: 162; 1629a: 51, 514; 1629b: 137). While he did not discuss Galen's quantitative argumentation regarding urinary output, in the *Commentary on Galen* Sanctorius related the *De statica*

---

[31] According to Owsei Temkin and Rudolph Siegel, Galen's theory of urine formation is somewhat ambiguous and contradictory (Temkin 1961: 474, fn. 27, Siegel 1968: 131). For more information on this theory and on Galen's doctrine of kidney function, see: ibid.: 126–34.

[32] I assume that Galen proposed to compare the amount of ingested drink with the amount of excreted urine in winter because he thought that sweat and insensible perspiration were less profuse during this season (e.g., Galen and Johnston 2018: 363). However, Sanctorius held that in robust bodies insensible perspiration was greater in winter than in summer and claimed that this was also confirmed by Galen (Sect. 3.3.1) (Sanctorius 1629a: 382 f.).

medicina explicitly to Galen's statement, quoted above, that the whole amount of consumed drink became urine, except for that which was excreted as feces, sweat, or insensible perspiration (Sect. 5.3.1). In the context of a discussion on reduced urinary output as a sign of imminent disease, Sanctorius explained:

> But regarding the way in which insensible and free perspiration diminishes urine, as we explained so exactly in our book *De statica medicina*, nobody truly understands it without appeal to the principles of the *statica*. But that urine is often dissolved by means of sweat, or by means of invisible perspiration, Galen easily explains in the first book of *On the Natural Faculties*, ... (Sanctorius 1612a: 756).[33]

Thus, there is a direct link between Galen's quantitative reasoning and Sanctorius's measurement of insensible perspiration. Although Sanctorius did not refer in the *De statica medicina* to Galen's calculation of urinal output, but instead arrived at his own results, it seems that he accepted Galen's argumentation overall, as he was convinced of the kidneys' selective attraction of mattter. Due to the fact that Sanctorius on other occasions explicitly connected his weighing procedures to Hippocrates's work *De flatibus* (Breaths) and to various books by Galen, in particular his *De tuenda sanitate* (Hygiene) and *Methodus medendi* (Method of Healing), it is difficult to assess the relation between the quantitative reasoning in Galen's *De naturalibus facultatibus* and Sanctorius's static medicine. What is more, despite the striking parallels between the two quantitative approaches, there are also important differences (Sanctorius 1614: 13v; 1625: 21–4, 556, 569; 1629a: 23 f., 70 f.; 1902).

Galen used a quantitative argument, the high amount of urinary output, to defend his physiological theory, according to which the kidneys possessed a faculty to attract the matter they required to function. Sanctorius used a quantitative argument, the high amount of insensible perspiration, to show that this physiological phenomenon strongly influenced the state of health and that it was therefore necessary to systematically observe its occurrence. Hence, in the case of Galen, it was not important for the physician to personally observe for himself the quantity of urine, while in the case of Sanctorius there was a direct relation to medical practice. According to him, the monitoring of the *perspiratio insensibilis* by means of systematic weighing was fundamental to the preservation of health. Right at the beginning of the *De statica medicina*, he stated:

> If the physician, who is responsible for the health of others, takes care only of the sensible additions and evacuations and does not know their daily amount of insensible perspiration, he deceives them [his patients] and will not cure them (Sanctorius 1614: 1v).[34]

---

[33] "... quomodo verò insensibilis, & libera perspiratio minuat urinam: nos in lib.de statica medicina adeò exactè declaravimus, ut nemo sanè percipiet, nisi ad statica principia confugiat. Quod verò urina saepè resolvatur per sudorem, vel per invisibilem perspirationem Galenus facilè declarat I. de facul. naturalibus, ..." See: Sanctorius 1612a: 756.

[34] "Si medicus, qui praeest aliorum sanitati, sit solum capax additionis, vel evacuationis sensibilis, & nesciat quanta quotidie illorum sit perspiratio insensibilis, illos decipit, & non medetur." See: Sanctorius 1614: 1v.

This shows that Sanctorius's quantification of insensible perspiration had an immediate bearing on therapy—a cure was only possible if the amount of insensible perspiration in patients was known for certain. Accordingly, while Galen's quantitative physiological reasoning was purely conceptional, as a means to confirm a theory, Sanctorius's quantification of insensible perspiration was directly related to medical practice. As has been demonstrated in Chap. 3, the Galenic concepts of the internal physiological processes that led to the formation of insensible perspiration were not called into question by Sanctorius. His interest lay in insensible perspiration *per se* and in its systematic quantification by means of firsthand observation. It is, however, unclear whether Sanctorius really envisaged that his colleagues and readers might imitate his weighing chair and conduct weighing procedures themselves, or whether he considered it enough that they follow the rules he had laid down in the *De statica medicina*, based on his own measurements.[35]

Another difference between Galen and Sanctorius is their measuring methods. Urine could be measured directly, just like any other liquid.[36] Contrary to this, insensible perspiration could be quantified only indirectly, by inference, namely by comparing the quantities of substances ingested respectively of the substances excreted—be these sweat, urine, or feces.[37] Accordingly, Galen expressed the quantity of urine in a measure for liquids that was common in his time, while Sanctorius referred to insensible perspiration by weight, as it was a steelyard which enabled him, in the first place, to determine its amount. While Galen could simply use a prefabricated vessel to measure both, the volume of wine ingested and the volume of urine excreted, to determine the quantity of insensible perspiration Sanctorius had to develop, as will be shown below, a method and an instrument of his own.

All in all, the similarities between Galen's quantitative reasoning and Sanctorius's static medicine as well as Sanctorius's reference to Galen's work *De naturalibus facultatibus* imply that there was a relation between Galen's quantification of urine and Sanctorius's measurement of insensible perspiration. However, as has been mentioned, also other works of Galen may well have inspired Sanctorius's novel approach to physiology; and it was still a big step from Galen's observation of drinkers' urinary output in to Sanctorius's weighing procedures to indirectly quantify an invisible phenomenon. Moreover, Sanctorius related his static aphorisms

---

[35] In the *De statica medicina*, Sanctorius alternated between different perspectives. Often, he used an impersonal, objective style, e.g., "one discovers that …" (*deprehendatur*), or "it is demonstrated that …" (*patet*). But sometimes he directly addressed the reader by writing, for example, "if you then observe from the weighing that …" ("*si ex ponderatione videris, …*"), or "if you have transpired at night more than usual …" ("*si magis solito noctu paerspiraveris, …*"). Thus, occasionally it seems as if Sanctorius was inviting his readers to perform the weighing procedures themselves (ibid.: 3v, 10r, 12r, 14r–14v).

[36] Galen does not seem to have considered the different densities of wine and urine.

[37] In the *De statica medicina*, Sanctorius also referred to sweat, in certain cases, but only in more general terms, without stating exact quantities. This implies that he did not differentiate between sweat and insensible perspiration in his weighing procedures (Sect. 7.5.1) (Sanctorius 1614: e.g., 4r, 5v, 10r,14r–14v).

directly to medical practice, whereas in Galen's clinical practice there was no desire to quantify—an aspect that will be examined more closely below.

5.4.2   Leonardo Botallo and the Production of Blood

However, one does not have to go as far back as antiquity to encounter instances of quantitative physiological reasoning. In fact, such efforts can be found also in the Renaissance, chronologically close to Sanctorius. In the sixteenth century, the Italian physician Leonardo Botallo tried to determine the amount of blood generated daily in the human body. This endeavor was driven by practical medical considerations, as Botallo promoted a more liberal use of phlebotomy (drawing blood) than hitherto usual.

"Bloodletting"—as the evacuation of blood for a cure or prevention was known—was frequently recommended by physicians in order to mitigate the harmful effects of an abnormally large volume of blood in the body. In the words of Sanctorius, an "excess or plethora of blood" (*polyaemia*) might, for example, fill the veins to such an extent that neither spirits nor blood could pass, which would led to a sudden corruption of blood and then death (Sanctorius 1629b: 96 f.). Specific quantities of how much blood should be let had been already defined by Galen, who regarded three cotyles as a moderate evacuation. This implies that Galen measured the amount of blood that he removed from his patients—hence, a further form of quantification already practiced in ancient medicine, yet which shall be mentioned here only as a side note (Brain 1986: 133).[38]

In 1577, Botallo published his work *De curatione per sanguinis missionem* (On Healing by Phlebotomy), in which he put forward his concept of the human body as a siphon. On the basis of the Galenic physiology of nutrition, he argued that the body constantly lost substance through insensible perspiration and steadily compensated this loss by taking up fresh blood from the veins (Sect. 3.2.5). In order to determine the amount of blood to be extracted in phlebotomy, Botallo attempted to estimate the quantity of the liver's daily output of fresh blood to the veins. He admitted that no certain measure could be given, as the rate would differ greatly from one individual to another as well as from day to day, influenced as it was also by a person's activities, amount of nutrition ingested, and overall state of health. Still, he suggested that in a healthy, well-nourished body of moderate size, around ten to eight ounces of fresh blood were generated per day. In a later edition of his work, published in 1583, Botallo even increased this estimate to one pound (Botallo 1577: 11, 163 f.; 1583: 174).

Medical experience, rather than mathematical calculations, was the basis of Botallo's quantification of blood production. As a military surgeon, he had often

---

[38] For further passages in which Galen referred to specific quantities regarding venesection, see: Brain 1986: 31, 87, 89, 92.

treated patients who almost bled to death, but whose bodies were sufficiently replenished with fresh blood within three to four days. Probably, it was observations like these that made him conclude that the healthy body contained about twelve to fifteen pounds of blood in total. Botallo knew people who regularly had from eight to twelve ounces of blood let twice per month for many years without suffering from a blood deficiency. In his view, this confirmed that any blood lost was quickly replaced, since these people would otherwise be bled dry within one year. Another experience to which Botallo referred was the large loss of blood by patients with hemorrhoids or other forms of chronic bleeding, which could amount to fourteen or more pounds per month. In order to sustain such chronic losses, Botallo argued, the liver must produce at least eight to ten ounces of blood per day (Botallo 1577: 159–64).

But what happened in a body that was not subject to unusual losses of blood? How could it be capable of holding such large amounts of blood? Insensible perspiration was the solution. It has already been mentioned that Botallo thought that blood production was proportional to the excretion of insensible perspiration. By emphasizing the persistence and abundance of this invisible loss, he tried to show that the body was able to remove large quantities of blood in the course of normal, daily nutrition. Interestingly, in this context, he proposed to perform a weighing procedure, either in thought or in deed: weigh a piece of moist clay, then put it aside, and measure its weight again the next day to find out how much moisture it has lost. From this experience, it could be inferred, so Botallo, that even the healthy human body could daily dispose of eight to ten ounces of blood, if one considered its great size and the fact that it was subject to internal and external heat. It is unclear, however, whether Botallo himself carried out the weighing procedure he suggested, since he did not specify any quantitative outcome (Botallo 1577: 164–7).

Jerome Bylebyl already pointed out the striking similarities to Sanctorius's work. Both physicians emphasized the importance of insensible perspiration and posited its considerable occurrence on the basis of weighing procedures. Admittedly, Botallo's observation of the weight of moist clay can be hardly compared to Sanctorius's systematic weighing of human bodies and yet, the basic approach to quantifying invisible losses indirectly by means of a pair of scales was formulated already in Botallo's work. However, contrary to Sanctorius, Botallo was interested in quantifying insensible perspiration only insofar as this might support his view of the daily copious production of blood. His ultimate goal was to promote a liberal use of phlebotomy, and not the observation and quantification of invisible losses. In a way, this was the exact opposite of what Sanctorius did in the *De statica medicina*. According to him, insensible perspiration, as the main determinant of health and disease, could be regulated by dietetics with no need for phlebotomy. Thus, Botallo and Sanctorius advocated different forms of therapy, bloodletting and dietetics, but simultaneously shared a common concern for medical practice. Both based their quantitative reasoning on experience and observation. However, Botallo specified numerical values only for blood production and not for insensible perspiration. For the latter, he simply relied on the knowledge he had gained during his medical practice and, too, able to draw on the long tradition of phlebotomy, which had included a quantitative dimension—the measurement of blood removed from the patient—at the least since Galen's day. Sanctorius, on the contrary, had to break new ground for the measurement of insensible perspiration in the human body.

It is interesting, in this regard, to note that Sanctorius's static observations suggested that the body's normal rate of turnover of substance was much higher than had been previously assumed. While Botallo considered his estimate of the daily blood production, eight to ten ounces, already as immensely large, Sanctorius determined that the daily amount of insensible perspiration was far greater. In the *De statica medicina*, he stated that around five pounds of insensible perspiration were excreted per day. One can only imagine the astonishment of the readers, confronted with Sanctorius's claim that such large bodily losses occur daily yet go unnoticed (Sanctorius 1614: 2v).[39]

Perusal of Sanctorius's commentaries reveals that he was acquainted with the work of Botallo, as he mentioned him in the discussion of the production of the vital spirits. Moreover, Botallo's book, *De curatione per sanguinis missionem*, was published in several editions during Sanctorius's lifetime (1577, 1580, 1583). This, together with the fact that Botallo was of Italian origin and maintained relations with the famous Medici family, enjoying the favor of Catherine de' Medici (1519–1589), implies that Sanctorius knew Botallo's treatise on phlebotomy.[40] However, Sanctorius nowhere referred to Botallo's estimate of blood production nor to the work *De curatione per sanguinis missionem*. Therefore, the discussion of a possible relation between Botallo's quantitative reasoning and Sanctorius's measurement of insensible perspiration remains pure speculation. What is beyond all speculation, however, is that Sanctorius was familiar with Galen's treatises on phlebotomy and at times specifically quoted Galen with regard to the amounts of blood that should be evacuated in phlebotomy (Sanctorius 1612a: 260; 1625: 746; 1629a: 468, 517 f.; 1629b: 161).

---

[39] Writing in France in the sixteenth century, it can be assumed that Botallo used the Roman *libra* (pound) as a unit of measurement. It was equal to 327 g and divided into twelve *unciae* (ounces: one ounce = 27.25 g) (Cardarelli 2003: 73 f.). In Venice, the *oncia grossa* (39.5 g) and the *oncia sottile* (25 g) were in use. It is not known to which *oncia* Sanctorius was referring in the aphorisms of the *De statica medicina*. In the introduction to his edition of the *De statica medicina*, Giuseppe Ongaro referred to the *oncia sottile* without explaining the choice (Sanctorius and Ongaro 2001: 46). Given the precision of Sanctorius's quantitative statements and his aim to quantify the *perspiratio insensibilis*, this choice seems reasonable. Hence, Botallo's and Sanctorius's quantitative values cannot be compared directly. On the assumption that they used the units of measurement just mentioned, the five pounds to which Sanctorius referred are equal to 1500 g and Botallo's eight to ten ounces corresponds to 218–72.5 g.

[40] For bio-bibliographical information on Leonardo Botallo, see: Taccari 1971.

5.4.3   Cesare Cremonini and the Quantity of Arterial Blood

The last instance of quantitative physiological reasoning that shall be considered here is Cesare Cremonini's dispute of the view that arterial blood was excluded from nutrition.[41] According to Galenic medicine, the venous and arterial systems were completely distinct. While the veins carried blood and provided nutrition, the arteries contained a mixture of spirits and blood and served the dissemination of vital spirits (Sects. 3.2.5 and 3.2.6). This opinion was opposed by Aristotelians, who held that the heart was the main source of nutriment and that both, veins and arteries were involved in the process of nutrition. As one of the leading Aristotelian philosophers of the late sixteenth and the early seventeenth century in Italy, Cremonini wrote:

> Galen wants the venous blood to be, of its own substance, suitable for nutrition, so that from it all the members are nourished. I wish that Galen would tell me what becomes of the arterial blood? For it is continually generated, and once generated is diffused in great quantity to the entire body through the arteries. What then becomes of it, if it is always generated but is not consumed as nutriment? It would grow to infinity, because he says that an immense amount is always generated, but none of it is consumed" (Cremonini 1627: 338).[42]

Hence, Cremonini developed a quantitative argument, the danger of an impossibly large aggregation of blood in the arteries, in order to support his position that the arterial blood was consumed by the body as a major source of nutriment. Notwithstanding that Cremonini's quantification remained here on a rather general level and was purely conceptual—he did not mention any specific figures, nor did his reasoning include any form of measurement—it is still interesting for the following reasons. Cremonini was professor of natural philosophy at the University of Padua and the passage just quoted is from a transcript of his academic lectures, which he published in 1627 under the title *Apologia dictorum Aristotelis de origine, et principatu membrorum adversus Galenum* (Apology of Aristotle's opinions about the origin and the primacy of the members against Galen). As one of the three first ordinary professors of the arts and medicine faculty, he had many medical students in his audience. What is more, he was a direct colleague of Sanctorius during his tenure as first ordinary professor of theoretical medicine (Chap. 2). Sanctorius did not mention in his books the passage by Cremonini quoted here, but he did discuss the philosopher's opinion with regard to innate heat, mostly dismissively, in the

---

[41] In his article, *Nutrition, Quantification and Circulation*, Jerome Bylebyl pointed out two further examples of quantitative physiological reasoning that are contemporary to Sanctorius. These are the quantitative arguments of the Venetian physician Emelio Parigiano (1567–1643) and of Caspar Hofmann, a German professor of theoretical medicine (Sect. 2.5, fn. 46). As I could not find any reference to them in Sanctorius's works, I do not consider their quantitative reasonings here. For more information, see: Bylebyl 1977: 378–85.

[42] "Desiderarem, ut Galenus mihi diceret, quid fiat ex sanguine arteriali; nam continuè generatur, & generatum in multa quantitate diffunditur per totum corpus per arterias. Quidnam fit ab ipso, si semper generatur, & non absumitur in nutrimentum? Crescet in infinitum, quia semper generari nihil absumi dicit immensum." See: Cremonini 1627: 338. The English translation is taken from: Bylebyl 1977: 381.

Commentary on Hippocrates.[43] Here, too, it is unclear whether Sanctorius was aware of his colleague's quantitative argument regarding the nutritive process, so its influence should not be overestimated. Nonetheless, Cremonini's quantitative physiological reasoning, however basic it may be, attests that such notions were alive and kicking in medical circles at the University of Padua, and were used to refute rival theories. Even in the absence of specific figures, its existence alone proves that quantification was not unknown in the field of physiology, but was indeed put forward in arguments in the very milieu in which Sanctorius lived and worked (Sanctorius 1629a: 307 f., 329, 338 f.; Siraisi 1987: 222).

5.4.4   Quantification—A Growing Trend

To sum up, the picture painted by these last pages reveals that quantitative elements increasingly came to pervade Galenic medicine, both in theory and in practice. Certain forms of quantification can be identified in Galen's own work as well as that of Renaissance scholars. As has been shown, physicians, pharmacists, dieticians, and also laymen recognized the importance of measuring nutrition, excretions, and remedies and, in some cases, put this into practice. Cusanus proposed that the physician use a balance for quantification, in order to more accurately determine his patient's state of health. Galen and, much later, Botallo each pointed out the relevance of the amount of insensible perspiration. Like Botallo, the Alexandrian physician Erasistratus had put forward the idea of indirectly measuring invisible losses by means of a balance. In their discussion of physiological problems, Galen and Cremonini both used quantification as a mode of argument to defend one theory and refute another. In retrospect, Sanctorius's quantitative approach to physiology thus seems like a plausible evolution of these forms of quantification. However, Sanctorius placed himself explicitly in the tradition of Hippocrates and Galen, yet remained silent on, or—as with regard to Cusanus—altogether refutedany contemporary scholars' influence on his work. And nonetheless, the sometimes striking similarities between Sanctorius's endeavors and those of his contemporaries make it likely that their undertakings were to some extent related, even though the sources currently available here do not permit more than speculation. Hence, it is time now to finally consider Sanctorius's own quantification efforts and his development and use of measuring instruments in order to further uncover the path that led to his innovative approach to physiology.

---

[43] As Cremonini published his work *Apologia dictorum Aristotelis de origine, et principatu membrorum adversus Galenum* only in 1627, it is hardly surprising that Sanctorius did not refer to it in his works *Methodi vitandorum errorum*, *Commentary on Galen*, *De statica medicina*, and *Commentary on Avicenna*, which had all been published earlier. The *Commentary on Hippocrates* and the *De remediorum inventione* are the only works that Sanctorius published after 1627. It is of course possible that Sanctorius heard about Cremonini's quantitative argument while they both were teaching at the University of Padua.

References

Adamson, Melitta Weiss. 1995. *Medieval Dietetics: Food and Drink in Regimen Sanitatis Literature from 800 to 1400*. Frankfurt/Berlin: Peter Lang.

Albala, Ken. 2002. *Eating Right in the Renaissance*. Oakland CA: University of California Press.

Anagnostou, Sabine. 2008. "Qui bene purgat, bene curat!" Vom antiken Purgans zum modernen Laxans. *Pharmazie in unserer Zeit* 37: 121–129.

Anonymus, Londinensis, and William Henry Samuel Jones. 1968. *The Medical Writings of Anonymus Londinensis*. Amsterdam: Hakkert.

Bigotti, Fabrizio. 2016a. *Fisiologia dell'anima. Mente, corpo e materia nella tradizione galenica tardo rinascimentale (1550–1630)*. Unpublished manuscript, University of Exeter.

———. 2016b. Mathematica Medica. Santorio and the Quest for Certainty in Medicine. *Journal of Healthcare Communications* 1: 1–8.

———. 2018. The Weight of the Air: Santorio's Thermometers and the Early History of Medical Quantification Reconsidered. *Journal of Early Modern Studies* 7: 73–103.

Bigotti, Fabrizio, and David Taylor. 2017. The Pulsilogium of Santorio: New Light on Technology and Measurement in Early Modern Medicine. *Society and Politics* 11: 55–114.

Botallo, Leonardo. 1577. *De curatione per sanguinis missionem. De incidendae venae, cutis scarificandae, & hirudinum applicandarum modo*. Lyons: Apud Ioannem Huguetan.

———. 1580. *De curatione per sanguinis missionem. De incidendae venae cutis scarificandae, & hirudinum amplicandarum modo*. Lugduni: Apud Joan. Hugueta.

———. 1583. *De curatione per sanguinis missionem liber. De incidendae venae, cutis scarificandae, & hirudinum affigendarum modo*. Antwerp: Ex officina Christophori Plantini.

Brain, Peter. 1986. *Galen on Bloodletting: A Study of the Origins, Development and Validity of his Opinions, with a Translation of the Three Works*. Cambridge: Cambridge University Press.

Bylebyl, Jerome J. 1977. Nutrition, Quantification and Circulation. *Bulletin of the History of Medicine* 51: 369–385.

Cardarelli, François. 2003. *Encyclopaedia of Scientific Units, Weights and Measures: Their SI Equivalences and Origins*. London: Springer.

Castiglioni, Arturo. 1931. The Life and Work of Santorio Santorio (1561–1636). *Medical Life* 135: 725–786.

Cornaro, Luigi. 1591. *Discorsi della vita sobria*. Padua: Appresso Paolo Miglietti.

Cremonini, Cesare. 1627. *Apologia dictorum Aristotelis de origine, et principatu membrorum adversus Galenum*. Venice: Apud Hieronymum Piutum ad Signum Parnasi.

Cusanus, Nicolaus. 1983. *Nicolai de Cusa opera omnia*. Vol. 5. *Idiota de sapientia. Idiota de mente. Idiota de staticis experimentis*, ed. Ludovicus Baur and Renata Steiger with two appendices by Raymundus Klibansky. Hamburg: Meiner. http://www.cusanus-portal.de/content/werke.php?id=DeStatExper_161. Accessed 7 Oct 2019.

Cusanus, Nicolaus, and Jasper Hopkins. 2001. *Complete Philosophical and Theological Treatises of Nicholas of Cusa*, Vol. 1. Minneapolis: The Arthur J. Banning Press. https://urts99.uni-trier.de/cusanus/content/fw.php?werk=30&lid=43127&ids=&ln=hopkins. Accessed 7 Oct 2019.

Da Monte, Giambattista. 1556. *In artem parvam Galeni explanationes*. Lyons: Apud Ioannem Frellonium.

Del Gaizo, Modestino. 1889. *Ricerche Storiche intorno a Santorio Santorio ed alla Medicina Statica. Memoria letta nella R. Accademia Medico-Chirurgica di Napoli il dì 14 Aprile 1889*. Naples: A. Tocco.

Dictionary.com. 2020. static. https://www.dictionary.com/browse/static. Accessed 26 Mar 2020.

Ettari, Lieta Stella, and Mario Procopio. 1968. *Santorio Santorio: la vita e le opere*. Rome: Istituto nazionale della nutrizione.

Galen, and Arthur John Brock. 1916. *On the Natural Faculties*. Cambridge, MA/London: Harvard University Press.

Galen, and Ian Johnston. 2018. *Hygiene: Books 1–4*. Cambridge, MA/London: Harvard University Press.

Grmek, Mirko D. 1990. *La première révolution biologique: Réflexions sur la physiologie et la médecine du XVIIe siècle*. Paris: Payot.

Gruman, Gerald J. 1961. The Rise and Fall of Prolongevity Hygiene 1558–1873. *Bulletin of the History of Medicine* 35: 221–229.

Hippocrates, and Paul Potter. 1988. *Hippocrates,* Vol. VI. Cambridge, MA/London: Harvard University Press.

Hoff, Hebbel E. 1964. Nicolaus of Cusa, van Helmont, and Boyle: The First Experiment of the Renaissance in Quantitative Biology and Medicine. *Journal of the History of Medicine and Allied Sciences* 19: 99–117.

Joutsivuo, Timo. 1999. *Scholastic Tradition and Humanist Innovation: The Concept of Neutrum in Renaissance Medicine*. Helsinki: Academia Scientiarum Fennica.

Landels, John G. 1979. Water-clocks and Time Measurement in Classical Antiquity. *Endeavour* 3: 32–37.

Leong, Elaine, and Alisha Rankin. 2017. Testing Drugs and Trying Cures: Experiment and Medicine in Medieval and Early Modern Europe. *Bulletin of the History of Medicine* 91: 157–182.

Lessius, Leonardus. 1613. *Hygiasticon seu vera ratio valetudinis bonae et vitae una cum sensuum, iudicii, & memoriae integritate ad extremam senectutem conservandae*. Antwerp: Apud Viduam & Filios Io. Moreti.

Lewis, Orly. 2015. Marcellinus' *De pulsibus*: A Neglected Treatise on the Ancient "Art of the Pulse". *Scripta Classica Israelica* 34: 195–214.

Lohr, Charles H. 1988. Metaphysics. In *The Cambridge History of Renaissance Philosophy*, ed. Charles B. Schmitt and Quentin Skinner, 537–638. Cambridge: Cambridge University Press.

Maclean, Ian. 2002. *Logic, Signs, and Nature in the Renaissance: The Case of Learned Medicine*. Cambridge/New York: Cambridge University Press.

McVaugh, Michael. 1969. Quantified Medical Theory and Practice at Fourteenth-Century Montpellier. *Bulletin of the History of Medicine* 43: 397–413.

Middleton, W.E. Knowles. 1964. *The History of the Barometer*. Baltimore: John Hopkins Press.

———. 1969. *Invention of the Meteorological Instruments*. Baltimore: John Hopkins Press.

Mikkeli, Heikki. 1999. *Hygiene in the Early Modern Medical Tradition*. Helsinki: Academia Scientiarum Fennica.

Milani, Marisa. 2014. Introduction to Cornaro. In *Writings on the Sober Life: The Art and Grace of Living Long*, ed. Marisa Milani, Greg Critser, and Hiroko Fudemoto, 3–71. Toronto: University of Toronto Press.

Miller, Clyde Lee. 2017. Cusanus, Nicolaus [Nicolas of Cusa]. In *The Stanford Encyclopedia of Philosophy*, ed. Edward N. Zalta. https://plato.stanford.edu/archives/sum2017/entries/cusanus/. Accessed 7 Oct 2019.

Nagel, Fritz. 1984. *Nicolaus Cusanus und die Entstehung der exakten Wissenschaften*. Münster: Aschendorff.

Neumann, Caspar. 1752. *Chymiae medicae dogmatico-experimentalis, tomi secundi pars tertia*. Züllichau: Johann Jacob Dendeler.

Obizzi, Ippolito. 1615. *Staticomastix sive Staticae Medicinae demolitio*. Ferrara: Apud Victorium Baldinum.

Ottosson, Per-Gunnar. 1984. *Scholastic Medicine and Philosophy: A Study of Commentaries on Galen's Tegni (ca. 1300–1450)*. Naples: Bibliopolis.

Palmer, Richard. 1985. Pharmacy in the Republic of Venice in the Sixteenth Century. In *The Medical Renaissance of the Sixteenth Century*, ed. A. Wear, R.K. French, and Iain M. Lonie, 100–117. Cambridge/New York: Cambridge University Press.

Parrish, Sean David. 2015. *Marketing Nature: Apothecaries, Medicinal Retailing, and Scientific Culture in Early Modern Venice, 1565–1730*. Phd diss., Duke University.

Pico Estrada, Paula. 2008. Weight and Proportion in Nicholas of Cusa's *Idiota de Staticis Experimentis*. In *Nicolaus Cusanus: ein bewundernswerter historischer Brennpunkt:*

philosophische Tradition und wissenschaftliche Rezeption, ed. Klaus Reinhardt, Harald Schwaetzer, and Oleg Dushin, 135–146. Regensburg: Roderer.

Pontormo, Jacopo, and Salvatore Silvano Nigro. 1988. *Il libro mio: Aufzeichnungen 1554–1556*. Munich: Schirmer/Mosel.

Pugliano, Valentina. 2017. Pharmacy, Testing, and the Language of Truth in Renaissance Italy. *Bulletin of the History of Medicine* 91: 223–273.

Robens, Erich, Shanath Amarasiri A. Jayaweera, and Susanne Kiefer. 2014. *Balances: Instruments, Manufacturers, History*. Heidelberg: Springer.

Rudolph, Gerhard. 1996. La misurazione e l'esperimento. In *Storia del pensiero medico occidentale*. Vol. 2: *Dal Rinascimento all'Inizio dell'Ottocento*, ed. Jole Agrimi, R.A. Bernabeo, Mirko Drazen Grmek, et al., 93–154. Rome: Laterza.

Sanctorius, Sanctorius. 1603. *Methodi vitandorum errorum omnium, qui in arte medica contingunt, libri quindecim*. Venice: Apud Franciscum Barilettum.

———. 1612a. *Commentaria in Artem medicinalem Galeni*, Vol. I. Venice: Apud Franciscum Somascum.

———. 1612b. *Commentaria in Artem medicinalem Galeni*, Vol. II. Venice: Apud Franciscum Somascum.

———. 1614. *Ars Sanctorii Sanctorii Iustinopolitani de statica medicina, aphorismorum sectionibus septem comprehensa*. Venice: Apud Nicolaum Polum.

———. 1625. *Commentaria in primam Fen primi libri Canonis Avicennae*. Venice: Apud Iacobum Sarcinam.

———. 1629a. *Commentaria in primam sectionem Aphorismorum Hippocratis, &c. … De remediorum inventione*. Venice: Apud Marcum Antonium Brogiollum.

———. 1629b. *De remediorum inventione*. Venice: Apud Marcum Antonium Brogiollum.

———. 1634. *Ars Sanctorii Sanctorii de statica medicina et de responsione ad Staticomasticem*. Venice: Apud Marcum Antonium Brogiollum.

———. 1902 [1615]. Santorio Santorio a Galilei Galileo, 9 febbraio 1615. In *Le Opere di Galileo Galilei*, ed. Galileo Galilei, 140–142. Florence: Barbera. http://teca.bncf.firenze.sbn.it/ImageViewer/servlet/ImageViewer?idr=BNCF0003605126#page/1/mode/2up. Accessed 5 Nov 2015.

Sanctorius, Sanctorius, and Giuseppe Ongaro. 2001. *La medicina statica*. Florence: Giunti.

Siegel, Rudolph E. 1968. *Galen's System of Physiology and Medicine*. Basel/New York: S. Karger.

Siraisi, Nancy. 1987. *Avicenna in Renaissance Italy: The Canon and Medical Teaching in Italian Universities After 1500*. Princeton: Princeton University Press.

———. 1990. *Medieval & Early Renaissance Medicine: An Introduction to Knowledge and Practice*. Chicago: University of Chicago Press.

Smith, Philip. 1848a. Congius. In *Dictionary of Greek and Roman Antiquities*, ed. William Smith, 351. London: Taylor, Walton and Maberly.

———. 1848b. Cotyla. In *Dictionary of Greek and Roman Antiquities*, ed. William Smith, 367. London: Taylor, Walton and Maberly.

Storey, Tessa. 2017. English and Italian Health Advice: Protestant and Catholic Bodies. In *Conserving Health in Early Modern Culture: Bodies and Environments in Italy and England*, ed. Sandra Cavallo and Tessa Storey, 210–234. Manchester: Manchester University Press.

Sylla, Edith D. 1973. Medieval Concepts of the Latitude of Forms: The Oxford Calculators. *Archives d'histoire doctrinale et littéraire du Moyen Age* 40: 223–283.

———. 2011. Oxford Calculators. In *Encyclopedia of Medieval Philosophy*, ed. Henrik Lagerlund, 903–908. Dordrecht: Springer.

Taccari, Egisto. 1971. Botallo, Leonardo. In *Dizionario biografico degli italiani*. Vol. 13. Rome: Istituto dell'Enciclopedia Italiana. http://www.treccani.it/enciclopedia/leonardo-botallo_(Dizionario-Biografico)/. Accessed 30 Oct 2019.

Temkin, Owsei. 1961. A Galenic Model for Quantitative Physiological Reasoning? *Bulletin of the History of Medicine* 35: 470–475.

————. 1973. *Galenism: Rise and Decline of a Medical Philosophy*. Ithaca/London: Cornell University Press.

Trzeciok, Stefan Paul. 2016. *Alvarus Thomas und sein Liber de triplici motu:* Band I: *Naturphilosophie an der Pariser Artistenfakultät*. Berlin: Max-Planck-Gesellschaft zur Förderung der Wissenschaften. https://cdition-open sources.org/sources/7/index.html. Accessed 8 May 2022.

Valleriani, Matteo. 2016. Hydrostatics and Pneumatics in Antiquity. In *A Companion to Science, Technology, and Medicine in Ancient Greece and Rome,* Vol. 1, ed. Georgia L. Irby, 145–160. Chichester: Wiley Blackwell.

Vescovini, Graziella Federici. 2002. Cusanus und das wissenschaftliche Studium in Padua zu Beginn des 15. Jahrhunderts. In *Nicolaus Cusanus zwischen Deutschland und Italien. Beiträge eines deutsch-italienischen Symposiums in der Villa Vigoni*, ed. Martin Thurner. Berlin: Akademie Verlag.

Villanova, Arnaldus de, Michael McVaugh, and Luis García-Ballester. 1992. *Aphorismi de gradibus*. Barcelona: Public de la University de Barcelona.

Vogt, Sabine. 2008. Drugs and Pharmacology. In *The Cambridge Companion to Galen*, ed. R.J. Hankinson, 304–322. Cambridge/New York: Cambridge University Press.

Von Staden, Heinrich. 1989. *Herophilus: The Art of Medicine in Early Alexandria*. Cambridge/New York: Cambridge University Press.

Walker, William B. 1954. Luigi Cornaro, a Renaissance Writer on Personal Hygiene. *Bulletin of the History of Medicine* 28: 525–534.

Wear, Andrew. 1973. *Contingency and Logic in Renaissance Anatomy and Physiology*. Phd diss., Imperial College, London.

Chapter 6
Quantification and Certainty

Abstract In this chapter, I examine the context in which Sanctorius presented his measuring instruments in his publications. In difference to previous studies, which often have focused solely on the *Commentary on Avicenna*, this being the only work in which Sanctorius included illustrations of his instruments, I analyze the measuring instruments in the light of all that Sanctorius published. Furthermore, I scrutinize the relation of the various instruments to each other and discuss Sanctorius's possible complementary use of them. Of particular interest in this context is the role of the *De statica medicina*, which has become a keyword for Sanctorius's quantitative approach to physiology. These considerations will serve as an introduction to an in-depth study of Sanctorius's measuring instruments in Chap. 7, and reveal the agenda behind his quantification efforts—to enhance certainty in medicine. Given that the degree of conjecture in medicine was a contested issue in traditional introductory discussions in contemporary works on medicine, I examine Sanctorius's claim to enhance certainty through quantification, measurements, and instruments against the backdrop of the prevailing discourse(s) therein. While it is immediately obvious that Sanctorius departed from traditional views by introducing new quantitative procedures into medicine, the investigation of the roles that he assigned to logical reasoning, on the one hand, and to experience, empirical knowledge, and his new methods of quantification, on the other, draws a more complex picture regarding the combination of theory and practice in his works.

Keywords Certainty · Experience · Experiment · Quantification

Before analyzing Sanctorius's individual measuring instruments in more detail, in the following chapter, I will examine, more generally, the context in which Sanctorius presented these devices in his works. Since he published illustrations of them (and indeed of all of his instruments) exclusively in the *Commentary on Avicenna,* and only occasionally and superficially referred to them in his other books (Sect. 4.2), previous studies on Sanctorius's measuring instruments have

© The Author(s) 2023

T. Hollerbach, *Sanctorius Sanctorius and the Origins of Health Measurement*,
https://doi.org/10.1007/978-3-031-30118-6_6

often focused on this work.[1] In contrast, I shall consider the measuring instruments in the light of all of Sanctorius's published work, noting not only their mention, but also their omission. Moreover, I will scrutinize the relation of the various instruments to each other as well as Sanctorius's possible complementary use of them. In this context, the role of the *De statica medicina*, having become a keyword for Sanctorius's quantitative approach to physiology, is of particular interest. These considerations will serve as an introduction to an in-depth study of Sanctorius's measuring instruments in Chap. 7 and reveal the agenda behind his quantification efforts—to enhance certainty in medicine.

6.1 Measuring the Quantity of Diseases: Four Instruments

The four of Sanctorius's instruments to have received the most scholarly attention are: the *pulsilogium*, the thermoscope, the hygrometer, and the weighing chair that Sanctorius used to observe insensible perspiration.[2] In the secondary literature, they are often mentioned in connection with the sixth question (*quaestio*) of Sanctorius's *Commentary on Avicenna*, which discusses why the medical art is conjectural.[3] Sanctorius stated:

> The medical art is conjectural because of the quantity of diseases, of remedies, of virtues, because of idiosyncrasies or properties of nature and because of the individual conditions [of patients]. The reason why the quantity is conjectural is because in the first book of *Ad Glauconem*, at the beginning, and in the third chapter of the third book of *Methodi* Galen says *that the quantity of each thing can neither be written nor said*. With regard to the quantity of diseases, Galen states in the fifteenth chapter of the ninth book of *Methodi* that *in order to apply a remedy, not only the type of the disease must be known, but also its quantity*, which, according to the fourteenth chapter of the ninth book of Galen's *Methodi* *is a certain measure of the quantity of the deviation (recessus) from the natural state and this quantity can only be known by conjecture*. We have pondered for a long time, how that quantity of diseases can sometimes be partially known. We have invented four instruments (Sanctorius 1625: 21).[4]

---

[1] See e.g., Mitchell 1892, Miessen 1940, Guidone and Zurlini 2002: 129–133. Important exceptions are Bigotti and Taylor 2017, Bigotti 2018.

[2] Sanctorius usually refers to methods of measuring the humidity of air rather than to the two hygrometers he devised (Sanctorius 1612b: 105, 229 f., Sanctorius 1614: 20v–21r, Sanctorius 1625: 7, 522, Sanctorius 1629a: 24). The fact that he first mentions and illustrates these devices in the *Commentary on Avicenna* implies that he developed them in the period between his publication of the *De statica medicina* in 1614, and of the *Commentary on Avicenna* in 1625 (Sanctorius 1625: 23 f., 144, 215, 305). For the sake of simplicity, I subsume under the term "hygrometer" both the methods of measuring air humidity and the two instruments Sanctorius developed for this purpose. Distinctions between the instruments and the methods as well as their relation to each other are explored in Sect. 7.4.

[3] E.g., Ettari and Procopio 1968: 88, Grmek and Gourevitch 2001: 2010 f., Sanctorius and Ongaro 2001: 24 f., Bigotti 2016: 4 f.

[4] "Ars medica est coniecturalis ratione quantitatis morborum, remediorum, virtutis, ratione idiosyncrisiae, i. proprietatis naturae, & ratione conditionum individuantium. Ratione quantitatis est

From this citation it seems that the four measuring instruments constituted a coherent program of measurements, which were developed in response to Galen's assertion that it was impossible to detect the quantity of a disease, that is, the measure of divergence in a body from its natural state. Accordingly, historians have interpreted them as interdependent devices, used complementarily by Sanctorius.[5] Yet, perusal of all the passages in which Sanctorius referred to his measuring instruments reveals a more ambiguous relation between the instruments and their use. Around two hundred pages after the quoted citation, Sanctorius mentioned only the *pulsilogium*, the thermoscope, and the hygrometer, in a discussion that touched on the same aspects as the sixth question referred to above. Sanctorius here again explained that he had invented instruments to determine the "certain measures of the affections of the body" (Sanctorius 1625: 214 ff.). Moreover, a hundred pages later still, Sanctorius referred only to two instruments: the thermoscope and the hygrometer. By means of these two devices, he explained, it was possible to discern a balanced and an imbalanced complexion. Interestingly, in his description of the weighing chair, with which he conducted his static observations, Sanctorius made no mention of any other device (Sanctorius 1625: 304 f., 555–8).

In the *Commentary on Hippocrates*, he grouped the four instruments together differently again. While he dealt with the question of why the medical art was conjectural in almost the same manner as in the *Commentary on Avicenna*, referring therefore to all four of the measuring instruments, in a later passage of *Hippocrates*, Sanctorius spoke of the *pulsilogium*, the thermoscope, and the weighing chair as the three instruments that served his pursuit of a "certain knowledge of the quantity of the vital virtue" (Sanctorius 1629a: 23 ff., 136 f.). A look into the *Commentary on Galen* shows that here, too, Sanctorius discussed the measuring instruments in various combinations. Published in 1612, two years before the *De statica medicina*, 13 years before the *Commentary on Avicenna*, and 17 years before the *Commentary on Hippocrates*, the work already contains all of the four instruments and yet does not present them together as a group.[6] In the discussion of the possibility of measuring deviations from the balanced complexion, which Galen had considered impossible, Sanctorius suggested three instruments: the thermoscope, the *pulsilogium*, and the hygrometer. A few chapters later, however, writing on this same topic, he referred to the thermoscope, the *pulsilogium*, and the weighing chair, but not to the

---

coniecturalis: quia Galenus primo ad Glauconem in principio, & 3. meth. 3. dicit, *quod nec scribi, nec dici potest de unoquoque, illud esse quantum.* Ratione quantitatis morborum: Galenus enim 9. Meth. 15. dicit, *ut verum exhibeatur remedium, non solum oportet cognoscere morbi speciem, sed etiam eius quantitatem, quae ex Gal. 9. Meth. 14. est certa mensura quantitatis recessus à naturali statu, quae quantitas solum coniectura haberi potest.* Nos diu cogitavimus, quomodo illud quantum morborum aliqua ex parte aliquando cognosci possit. Excogitavimus quatuor instrumenta." See: Sanctorius 1625: 21. Original emphasis.

[5] E.g., Ettari and Procopio 1968: 88, Grmek and Gourevitch 2001: 2010, Bigotti 2016: 5.

[6] In Sanctorius's first published work, *Methodi vitandorum errorum*, reference is made only to the *pulsilogium* and it can be assumed that Sanctorius had not yet developed either the thermoscope or the hygrometer. But according to his own testimony, he must already have been engaged in the weighing procedures whose results he later published in the *De statica medicina* (Sanctorius 1603: 109r–109v) (Sect. 2.2).

hygrometer (Sanctorius 1612b: 229 f., 374 ff.). Contrary to this, in the *De statica medicina*, published 2 years after the *Commentary on Galen*, Sanctorius mentioned the hygrometer and the thermoscope, but did not allude to the *pulsilogium* (Sanctorius 1614: 20v–21r).

Thus, Sanctorius's varying grouping of the four measuring instruments calls into question whether he really conceived of them as complementary parts of an overall program geared to the quantification of physiological parameters and fundamental to his novel doctrine of static medicine—the *Ars de statica medicina*. As has become apparent, the different combinations do not stem from the chronological development of his instruments, or medical theory, as different combinations can be found in the same work, even in Sanctorius's last book, the *Commentary on Hippocrates*, published in 1629.[7] In the following, a closer examination of the theoretical context in which Sanctorius presented the measuring instruments in his different books will shed more light on their relation to each other and, more generally, on Sanctorius's quantification efforts.

6.1.1   Galen's Latitude of Health Quantified: The Measurement of Disease, Virtue, and Humors

Sanctorius presented his measuring instruments in his published works mostly in relation to the question of the quantity of diseases, which was taken to mean a deviation (*recessus*) from the natural state of a body, i.e., from its temperate, balanced complexion. Sanctorius presented his measuring instruments as a direct advancement of Galenic medicine, as a solution to a problem that Galen had been unable to resolve. According to the Greek physician, so Sanctorius, it was impossible to determine quantity in medicine and medicine therefore had a conjectural character. In Chap. 5, it has been outlined that Galen tried to classify the complexion of drugs and of human bodies along ranges, or latitudes, which permitted of degrees. These degrees were, however, not expressed in numerical values and so remained conjectural, as, according to Galen, the intensity of a drug and likewise the complexion of a patient could be detected only by touch (Sect. 5.2). This is where Sanctorius stepped in. Sanctorius thought it possible to establish the norm for individuals, i.e., their natural state, and to measure deviations from that norm by measuring various parameters, such as the pulse and respiration, the heat of the body, and its parts, as well as the surrounding air, perspiration loss, and the humidity and dryness of the air. Doing so would enable him to put numerical values to the gradual differences that occurred in health and disease in the complexions of bodies. How Sanctorius conceived of this in detail and how he put this into practice with regard to the respective measuring instruments will be considered below.

---

[7] As mentioned earlier, Sanctorius published the *Commentary on Hippocrates* together with another work entitled *De remediorum inventione*, which, however, contains no reference to his measuring instruments (Sect. 2.6 and 5.2.4).

Browsing through the different passages in which Sanctorius mentioned his measuring instruments, the theoretical context is more or less identical to the one just portrayed. Hence, the differing groupings of the four devices are not connected to the specific text in which they appear. Yet, two passages deserve further mention, as they diverge slightly from the others. As stated above, in the *Commentary on Hippocrates*, Sanctorius mentioned the *pulsilogium*, the thermoscope, and the weighing chair in relation to his efforts to determine the quantity of the vital virtue and not, as was otherwise the case, with regard to the natural state of a body and the deviation (*recessus*) from it (Sect. 6.1). According to Galenic medicine, the vital spirits conveyed vital virtue, a power which ensured that life itself was maintained. This vital virtue manifested itself in the rhythms of heartbeat, pulse, and respiration. Thus, its relation to Sanctorius's *pulsilogia*, which served to measure pulse and respiration, is clear. Furthermore, the principal product of respiration was thought to be heat, generated and distributed by the heart as well as the arteries. Given that the arteries contained blood mixed with vital spirit, which in turn conveyed vital virtue, it is understandable why Sanctorius employed the thermoscope, too, in order to measure the vital virtue of a body.[8] He explained, for example, that, in acute diseases, a large increase in heat over a period of a few days indicated that the vital power was steady. In addition, heat was crucial for the digestive process. As insensible perspiration resulted from the processes of respiration and digestion, the connection between the quantification of the vital virtue and Sanctorius's observations with the weighing chair is also clear.[9] However, here again, the question remains: Why did Sanctorius, in omitting the hygrometer, fail to mention all four of his measuring instruments? Since the vital spirits were generated from inhaled air, the humidity or dryness of the air must have been important to the vital virtue. Moreover, as shown above (Sect. 3.3.1), air was thought to be the most important factor of the six non-natural things and the quality of air had a considerable effect on insensible perspiration. Therefore, Sanctorius's reasons for excluding the hygrometer, when seeking to determine the quantity of the vital virtue, remain obscure (Sanctorius 1629a: 136 f.; Siegel 1968: 163; Siraisi 1990: 107 ff.).

Taking all the aspects into consideration, it seems that Sanctorius considered the power of the vital spirits, the vital virtue, to be one of the various parameters that indicated how much a body deviated from the normal, healthy condition. Earlier in the *Commentary on Hippocrates*, he wrote that the vital virtue was robust only in those whose four humors were "in symmetry" and whose body parts were "optimally uniform." Hence, measuring the robustness of the vital power enabled one to determine the humoral balance, i.e., the health of a patient (Sanctorius 1629a: 86).

The second passage that differs a little from Sanctorius's usual presentation of his measuring instruments is the description of the weighing chair in the *Commentary on Avicenna*. Instead of referring to the measurement of the natural state of a body,

---

[8] The connection between Sanctorius's use of thermoscopes and Galenic concepts of fever will be considered below, see: Sect. 7.3.2.

[9] For more information on the generation of the vital spirits and the processes of respiration and digestion, see Sects. 3.2.5 and 3.2.6.

he included the explanation of his large steelyard in a discussion of the signs that indicated the quantity of humors necessary to preserve health. Sanctorius explained that, according to Avicenna, not only the proportion of the four humors in the body was important, in order to preserve health, but also their quantity. However, Avicenna did not teach how to determine this quantity and Galen held it impossible to gain knowledge of it, per Sanctorius. He continued: "But after having thought about this for a very long time, we invented static medicine, in which we declared when the quantity and proportion of the humors can be found in our body" (Sanctorius 1625: 555).[10] With reference to Galen's teachings, Sanctorius argued that the quantity of the humors could be determined by measuring the ratio between ingestion and excretion. If the intake of nutrition corresponded to the output of sensible and insensible evacuations, the humors would be balanced, quantitatively and proportionally, and health would be preserved. It is to be recalled here that, according to Galenic physiology, the humors were generated during the digestive process from the ingested nutrition. Thus, Sanctorius measured the quantity of the humors only indirectly, by observing the equilibrium between the substances the organism consumed and those it rejected. A lack of equilibrium in this regard indicated an imbalance in the humors and hence a deviation from the natural healthy state (Sects. 3.1.2 and 3.2). Accordingly, like the robustness of the vital virtue, the quantity or, rather, the balance of the humors, measured via intake and output, was another parameter that helped Sanctorius determine the natural state of a body (Sanctorius 1625: 555 f.).

In this context, it is interesting to note that Sanctorius referred in his discussion of how to determine the quantity of humors not only to Avicenna and Galen, but also to another physician: Jacques Despars (ca. 1380–1458). The French doctor was famous for his commentary on Avicenna's *Canon* and this is the work that Sanctorius mentioned here. He explained that Despars, in dealing with the same issue, had written that the quantity of the humors must be equivalent to the release (*resolutio*) from the body parts and the spirits. But according to Despars, so Sanctorius, these daily evacuations could not be quantified, since a lot of them were released insensibly. Hence, Despars had already related the determination of the quantity of the humors to the measurement of insensible evacuations, but considered the latter impossible. Furthermore, he had already noted the great quantity of these excretions. This implies that Despars's commentary on Avicenna's *Canon* was a source of inspiration for Sanctorius's proposal to determine the balance of the humors by measuring insensible perspiration (Sanctorius 1625: 555 f.; Jacquart 1980).

As indicated above, Sanctorius did not point to any of the other measuring instruments in his description of the weighing chair in the *Commentary on Avicenna*. Even though the humors were directly linked to the primary qualities and to the individual complexion of a body and thus to its natural state, Sanctorius specifically and exclusively related the determination of the quantity of the humors to the weighing chair and to his measurement of insensible perspiration. This further calls into

---

[10] "Nos autem hoc diutissimè excogitando adinvenimus staticam medicinam, in qua declaravimus, quando in corpore nostro, & quantitas, & humorum proportio reperiatur." See: Sanctorius 1625: 555.

question the role of the *De statica medicina* as an overall framework for Sanctorius's quantification of physiological parameters by means of his four measuring devices. The fact that Sanctorius related his weighing procedures to the preservation of health, namely to the question of how to find the quantity of humors that was needed to preserve health, underscores the orientation of the *De statica medicina* toward individual hygiene and its dietetic handbook character (Sect. 4.1). Yet, in view of the strong connection of his other measuring instruments to the six non-natural things, such as the thermoscope and the hygrometer, the context of the preservation of health does not set the *De statica medicina* apart. Why then did Sanctorius not mention here the other measuring devices? And why did he point in only one of the many aphorisms of the *De statica medicina* to the thermoscope and the hygrometer, and not at all to the *pulsilogium*? To explore these questions, I shall take a closer look at Sanctorius's representation of the *De statica medicina* in his published works. Consideration of the different contexts in which Sanctorius mentioned the work will enhance understanding of the role that Sanctorius assigned to the *De statica medicina* and its relation to the other measuring instruments.

6.1.2   The Relation of the De statica medicina to the Measuring Instruments

Sanctorius first mentioned the *De statica medicina* in the *Commentary on Galen*, which was published 2 years earlier than the aphoristic treatise. In most instances, Sanctorius referred to the *De statica medicina* in the context of the measurement of insensible perspiration, and more generally, of bodily evacuations, sometimes in connection with the digestive process. Moreover, he included references to his weighing procedures in discussions of the non-natural things and at times highlighted their importance for the preservation of health. The same picture emerges with regard to his other two commentaries—the *Commentary on Avicenna* and the *Commentary on Hippocrates*.[11] In the few passages in which Sanctorius grouped the *De statica medicina* together with other measuring instruments, the context is, as outlined above, the determination of the deviation of a body from its natural healthy state and the measurement of the quantity of the vital virtue.[12] Hence, Sanctorius usually presented his weighing procedures in isolation from his other quantification efforts. If he did mention the *De statica medicina*, i.e., the weighing chair, in con-

---

[11] In his two other published books, *Methodi vitandorum errorum* and *De remediorum inventione*, Sanctorius does not refer to the *De statica medicina*. For Sanctorius's references to the *De statica medicina* in his three commentaries, see: Sanctorius 1612a: 139, 348, 352 f., 358, 447, 496, 756 f., 759, 761, Sanctorius 1612b: 4, 40, 48, 71 f., 84 f., 87, 95, 198, 342, 357, 374 f., Sanctorius 1625: Ad lectorem, 7, 23 f., 27 f., 60, 68, 81, 157, 161, 264, 373, 375, 394, 522, 555 ff., Sanctorius 1629a: 23 f., 70 f., 78, 137, 204, 207, 210, 276, 291, 300 f., 309 f., 367, 381 ff., 429, 469.

[12] The text passages in which Sanctorius connects the *De statica medicina* with other measuring instruments are Sanctorius 1612b: 374 ff., Sanctorius 1625: 21–5, Sanctorius 1629a: 23 ff., 136 f.

nection with some of his other measuring instruments, then not in isolation, but ranked among the other devices. This reinforces the impression that static medicine was not a sort of superstructure for Sanctorius's physiological measurements, but rather served to determine certain parameters that, together with his other quantitative examinations, indicated to him the quantity of disease. In the preface to the *Commentary on Avicenna*, when Sanctorius proclaimed his new approach to the teaching of theoretical medicine, he wrote of his "instruments and static experiments"—a statement that highlights the specificity of the weighing chair and its differentiation from Sanctorius's other instruments.

Interestingly, apart from the description of the weighing chair in the *Commentary on Avicenna*, Sanctorius never referred to the actual instrument with which he conducted his observations of insensible perspiration. Instead, he wrote of his "static medicine" (*statica medicina*), "static experiments, experiences, and observations" (*staticis experimentis, staticae experientiae, staticis observationibus*), or simply "our statics" (*staticis nostris*). At this point it should not be forgotten that Sanctorius expressly gave his aphorisms the title *Ars ... de statica medicina* and also repeatedly wrote in his commentaries about his "static art" (*statica ars*). As elucidated in Chap. 3, static medicine was about much more than a steelyard and other weighing measurements. It also comprised Sanctorius's new interpretation of the doctrine of the six non-natural things, which he considered apparently to be a whole new medical art, in which the instrument was but a means to achieve the ultimate goal: the exact measurement of insensible perspiration. Indeed, the *De statica medicina* contains neither an illustration nor a description of the weighing chair. Only to later editions published after Sanctorius's death in 1636 were an illustration and a description of the steelyard sometimes added, reproduced from the *Commentary on Avicenna*. Thus, the focus of static medicine was not the weighing chair and its use, but rather the results of the weighing procedures and the conclusions that Sanctorius drew from them.

While the further implications of Sanctorius's labelling of static medicine as an art are explored below, what is of interest here is that Sanctorius seemingly distinguished this static art from his use of other measuring devices. The equal ranking of the *De statica medicina* and his other instruments stands in stark contrast to the importance that Sanctorius occasionally assigned to static medicine, namely that healing and preserving health was impossible without a knowledge of the quantity of insensible perspiration in patients (Sect. 5.4.1). In view of this, it is curious that Sanctorius did not always mention the *De statica medicina*, i.e., the weighing chair, when dealing with the measurement of how much a body deviated from its natural state (Fig. 6.1). Moreover, in the passages in which he grouped the weighing chair together with other measuring devices, he did not emphasize the former's relevance, which suggests that all of the measurements to which Sanctorius referred were of equal importance to him. Contrary to the thermoscope and the *pulsilogium*, whose complementary use Sanctorius explicitly described in the *Commentary on Avicenna*, as will be shown later, Sanctorius did not allude to any similar interrelated use of different instruments with regard to his weighing chair and the observation of insensible perspiration. In the aphorism of the *De statica medicina*, in which he

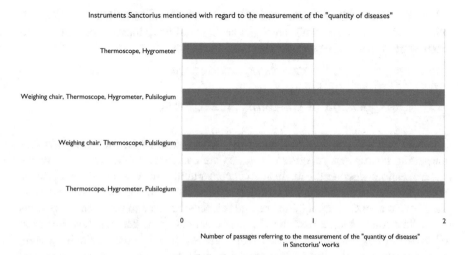

Fig. 6.1   Different combinations in which Sanctorius mentioned his instruments with regard to the measurement of the "quantity of diseases". (Sanctorius 1612b: 229 f., 374 ff.; 1625: 21–5, 214 ff., 304 f.; 1629a: 23 ff., 136 f.)

mentioned the hygrometer and the thermoscope, he did so in the section on the non-natural pair air and water, discussing how to determine the "weight of the air" (Sect. 5.3.2). The relation to insensible perspiration was thereby only indirect, since Sanctorius described the harmful effects of "heavy" air on insensible perspiration only in a later aphorism. Hence, the measurements that Sanctorius conducted with the thermoscope and the hygrometer seem not to have been included in his newly formulated rules of health that revolved around insensible perspiration and constituted for Sanctorius a new medical art. Concerning the *pulsilogium*, the fact that Galen's work *De pulsibus ad tirones* (On the Pulse for Beginners) was one of the sources for the doctrine of the six non-natural things, a work Sanctorius was very familiar with, and in which Galen used the expression "non-natural" when referring to the causes of alterations in the pulse, makes it even more obscure why Sanctorius did not refer to his *pulsilogia* in the *De statica medicina* (Sect. 3.1.1) (Sanctorius 1614: 20v–21v; 1625: 24, 76 ff., 219–22, 346).

Hence, it has become apparent that static medicine cannot be identified as an overall program of measurements conducted with various measuring instruments. Instead, the quantification of insensible perspiration by means of a weighing chair was only one of several means that helped Sanctorius quantify diseases, i.e., determine any deviation from the natural state of a body. Thus, there is a tension between the importance that Sanctorius ascribed to the *De statica medicina* and the rather minor role he gave to it when he mentioned the work together with the other measuring instruments. Furthermore, despite the strong relations between the six non-natural things and the thermoscope, the hygrometer, the *pulsilogia* and, as will be seen below, also the two steelyards that Sanctorius built to measure climatic conditions (Sect. 7.1), none of these instruments was integrated into the measurement of insensible perspiration. These conclusions mirror those formulated in Chap. 3, where it was stated that static medicine cannot readily be identified as the overall framework of Sanctorius's works (Sect. 3.3.7).

6.2   The Question of Certainty in Medicine

The only two text passages in Sanctorius's works in which he mentioned his four measuring instruments together as a group are connected to the discussion of the same question: whether the medical art is conjectural. This was a standard question which featured in the traditional introductory discussions of medical knowledge and the place of medicine among the arts and sciences included in the opening sections of medical commentaries as well as in general works on medicine. In his discussion of these topics, Sanctorius argued that certainty in medicine could be greatly enhanced through the use of his weighing chair, *pulsilogium*, thermoscope, and hygrometer. In order to better understand this important feature of Sanctorius's four measuring instruments and, more generally, his integration of quantitative methods and instruments into the discussions of the conjectural character of medicine and the related aim of enhancing certainty in medicine, I will briefly outline the main issues that were at play and Sanctorius's stance on them.

6.2.1   *Medicine*—ars *or* scientia?

The authoritative differentiation of art (*ars*) and science (*scientia*) with regard to disciplines derives from Aristotle. Thus, in the discussions of the status of medicine as either an art or a science, the basic understanding of terms was Aristotelian. As Nancy Siraisi aptly summarized:

> *Scientia* is usually assumed to offer certain knowledge about universal truths arrived at by demonstration (that is, syllogistic reasoning) from generally accepted principles, and to be pursued for the sake of truth. Different *scientiae* are distinguished by their subject matter. … And *ars* is a rationally organized and transmitted body of knowledge or skill resulting in a product (not necessarily a material one) (Siraisi 1987: 226).

Hence, *scientia* was understood as a theoretical discipline concerned with the knowledge of universal causes that were hidden from the senses and could be perceived only by the mind. *Ars*, on the contrary, referred to practical skill and ordered knowledge. It was associated with empirical and individual aspects, with particulars perceived by the senses. While *scientia* offered certain knowledge, *ars* always involved conjecture. Medicine with its ambiguous position, swaying between the university classroom and the sickbed, was not easy to fit into either of the categories. In the Middle Ages and the Renaissance, academically educated physicians usually wanted at least some aspects of medicine to qualify as science, not least in order to guarantee its high status as a core university subject. However, they also readily admitted that much of it belonged to the kind of knowledge identified as art. Without delving into the depths of the topic, which have been explored elsewhere, I shall focus here on Sanctorius's answer to the question.[13] His answer was clear and unambiguous: medicine is an art and not a science. In his three commentaries, he put forward several reasons why "medicine could by no means be *scientia*" (Sanctorius 1612a: 74). He thereby refuted not only the teaching of Avicenna—who considered medicine to be both a science and an art—and the academic medical convention of the fourteenth and fifteenth century, which upheld the latter's view, but also the opinions of those sixteenth-century medical commentators who claimed medicine for *scientia* alone.[14] Contrary to these authors, Sanctorius, in identifying medicine as an *ars*, saw himself in the tradition of Hippocrates, Aristotle, Galen, and Averroes. While a detailed analysis of Sanctorius's argument lies beyond the scope of the present work, certain statements deserve further consideration, as they help us appreciate Sanctorius's conception of medicine and medical knowledge.[15]

In the *Commentary on Galen*, Sanctorius explained that Aristotle had described arts as productive sciences (*scientias effectivas*). He thereby referred to the Aristotelian tripartite division of human knowledge, which was oriented to the purposes this knowledge served: speculative (i.e., theoretical), practical (related to leading a good and useful life), or "factive" (related to the production of things in the arts and trades). According to Sanctorius, it was clear that medicine was not a science, as the purpose of *scientia* was knowing (*scire*), while the purpose of medicine was operating (*operari*). However, this did not mean that medicine, as an art, concerned solely practical aspects or necessarily entailed the habitual practice of it. The habitus of an art could be acquired either from repeated activities (*iteratis actibus*) or from a master, explained Sanctorius in the *Commentary on Avicenna*, and it was hence possible to speak of "excellent theoretical and practical physicians [*medici theorici* and *practici*] who never exercised the art" (Sanctorius 1625).[16] Thus, on the one hand Sanctorius's clear identification of medicine as an art suggests that he highlighted the practical dimensions of medicine related to the senses and to utilitarian knowledge. On the other hand, he did not dismiss the intellectual dimensions of medicine, but, quite on the contrary, considered them of integral importance to the art. This ambiguous attitude toward the role of the senses and the

---

[13] For an account of the debates concerning the status of medicine as an art or a science, see: Siraisi 1981: 118–37, Ottosson 1984: 68–74, Siraisi 1987: 226–38.

[14] "… medicina nullo modo potest esse scientia, …." See: Sanctorius 1612a: 74.

[15] In the *Commentary on Avicenna* and in the *Commentary on Hippocrates*, Sanctorius discusses the question of whether medicine is an art or a science in separate *quaestiones*. See: Sanctorius 1625: 28–37, Sanctorius 1629a: 18–23. In the *Commentary on Galen*, Sanctorius refers to the issue in discussions about both the subject of Galen's *Ars medica* and definitions of medicine. See: Sanctorius 1612a: 9–15, 63–7. For an account of Sanctorius's arguments in the *Commentary on Avicenna*, see: Siraisi 1987: 236 f.

[16] "Respondemus dari duplicem habitum, vel acquisitum ex iteratis actibus, vel à magistro: hac enim ratione possunt dari optimi medici theorici, & practici, qui nunquam artem exercuerint, …." See: Sanctorius 1625: 29. The English translation is based on Siraisi 1987: 236.

role of the mind in gaining medical knowledge relates to the ambiguous relation between theory and practice that is found in Sanctorius's works. As was mentioned earlier, Sanctorius rejected the division of medicine into theory and practice in the university curricula, on the grounds that medicine, contrary to theory and practice, was a "factive" or operative art, meaning that its purpose was neither truth, as in the case of theory, nor action, as in the case of practice, but instead, operation, i.e., the preservation and restoration of health (Sect. 4.3) (Sanctorius 1612a: 64; Park and Daston 2006: 6).

   With his strong emphasis on the nature of medicine as an art, Sanctorius was in line with a general trend at the beginning of the seventeenth century, namely to pay far greater attention to the status of medicine as an art. In conjunction with the revaluation of practical medicine, starting from the fifteenth century, the practical and social dimension of medicine was increasingly stressed, a development that, according to Ian Maclean, can be associated with the rising value attributed to therapeutics, to clinical precepting, and to the design of hospitals at this time (Maclean 2002: 70–5). Yet, there is one feature that distinguishes Sanctorius's concept of the "art of medicine" considerably from the conventions of Latin academic medicine and this was his quantitative approach to medicine and attendant use of instrumentation.

6.2.2   Enhancing Certainty in Medicine through Quantification

Contrary to the common Aristotelian understanding of *ars* as knowledge that offered no prospect of certitude, Sanctorius thought it possible to enhance or even to gain certainty. In a letter to his friend Senatore Settala (life dates unknown), in 1625, the year he published the *Commentary on Avicenna*, he stated:

> I send his Lordship the two books on Avicenna's text, as He wrote me, and I pray His Lordship to read them carefully, because He will read new thoughts, which are, however, based on the authorities of Hippocrates and Galen with regard to practice and experience. … Besides, He will frequently see the advantages which one can gain from the use of the static, invented by me, which one can certainly call mathematical medicine (*mathematica medica*) as it adds so much certainty to medical things (Castellani 1958: 5).[17]

Hence, according to Sanctorius, the *De statica medicina* increased the certainty of medicine to such an extent that it could be termed "mathematical" medicine. While the reference to mathematics certainly pointed to the quantitative method on which Sanctorius allegedly based his aphorisms, it also had other connotations. Based on

---

[17] "Mando a V.S. li 2 libri sopra la parte di Avicenna secondo mi ha scritto et prego V.S. che li lega con diligenza perchè legerà pensieri nuovi fondati però nella autorità di Hippocrate et Galeno, nella pratica et nella esperienza. … Di più vedrà spesso li benefitij che cavar si può dal uso della statica inventata da me la qual certo si può chiamar mathematica medica tanto ci fa certi nelle cose di medicina." See: Castellani 1958: 5. The translation is based on Bigotti 2016: 1.

the authority of Aristotle and of Averroes (1126–1198), mathematics was tradition-ally considered as the demonstrative science (*scientia*) par excellence, which thus provided knowledge with the highest degree of certainty.[18] By comparing his static medicine to mathematics, Sanctorius therefore made a very strong statement for the certainty of his newly invented art. In doing so, he claimed that it was possible for an art to achieve a degree of certainty comparable to that accomplished in the sci-ences and indicated that this certainty was attained by using the subject of mathe-matics, namely quantification. Along the same lines, in the *Commentary on Galen*, Sanctorius termed his aphoristic treatise "static theorems" (*staticis theorematibus*) and explained that his weighing procedures (*staticis experimentis*) were in the first degree of certainty. It seems, thus, that he thought that from the *De statica medic-ina*, understood as an art, knowledge of universal causes could be gained and that it was therefore possible to achieve certain knowledge from particulars by means that did not refer to deductive reasoning but to the senses—to the observations and expe-riences (*experimenta*) that he made with his weighing chair. This interpretation is reinforced by the fact that in the *Commentary on Hippocrates*, Sanctorius explicitly stated that the theorems (*theoremata*) of medicine were most certain (*certissima*), since, following Galen, the universal precepts of medicine had most certain and eternal truth (Sanctorius 1612b: 71, 95; 1629a: 23).

However, Sanctorius's claim to certainty in medicine was not limited to the *De statica medicina*, but also included some of his other measuring instruments. Coming back to the citation quoted at the beginning of this chapter (Sect. 6.1), in the discussion of the conjectural character of medicine, Sanctorius made it clear that elements of uncertainty could be greatly reduced not only by static medicine, but also through the use of his *pulsilogium*, thermoscope and hygrometer. In Sect. 6.1, I have outlined that Sanctorius usually presented these instruments, in varying com-binations, as a solution to one aspect that made the medical art conjectural: the quantity of diseases. In this connection, he also frequently stressed the certainty that the use of his measuring devices provided. In the *Commentary on Galen*, he explained for example that the *pulsilogium* enabled one to know, not by conjecture but with the most certain knowledge (*scientia*), how much the movements of the pulse of a patient deviated from its natural state. Around a hundred pages later, Sanctorius similarly proclaimed that his four measuring instruments ascertained (*reddimur certi*) the quantity of the deviation from the natural state. In the *Commentary on Avicenna*, he wrote: "But we find out the quantities or certain mea-sures of the affections with various instruments" (Sanctorius 1625: 215).[19] By means of the thermoscope and the hygrometer, Sanctorius maintained in the *Commentary on Galen*, one could exactly perceive the furthest degrees of active and

---

[18] The issue of the certainty of mathematics gained considerable attention in the second half of the sixteenth century, when a dispute arose over the question of the causes and foundations of this certainty and the way in which it was interpreted. For more information on Renaissance debates on the *certitudine mathematicorum*, see: De Pace 1993.

[19] "Nos verò instrumentis varijs adinvenimus quantitates sive certas affectuum mensuras ...." See: Sanctorius 1625: 215.

passive qualities. He thereby alluded to the four primary qualities, which, according to Galenic medicine, could be divided into active qualities (hot and cold) and passive qualities (wet and dry) in accordance with the Aristotelian distinction between the active and the passive pair amongst the four elements. As the mixture of these primary qualities in the body, the so-called complexion, was the decisive factor for the body's state of health, measuring them was crucial, per Sanctorius, in order that deviations from the balanced complexion, i.e., from the natural, healthy state of a body, could be determined. The thermoscope and the hygrometer not only enabled him to *exactly* measure the primary qualities, but also to determine their "furthest degrees," as he explained. Referring to the Galenic concept of the latitude of qualities, this implies that by using the two instruments, extreme deviations from a healthy state could be measured (Sanctorius 1612b: 105, 229, 374).

In order to guarantee that the measurements provided certainty, some other factors had to be considered, too. The instruments needed to be used repeatedly in sickness and in health, the measuring results had to be carefully recorded, and even minor variations noted. In the *Methodi vitandorum errorum*, Sanctorius stated:

> … only from this comparison [of the pulse of the previous attack of disease and the present pulse] can we obtain a certain and infallible judgement on whether the patient is in a better or worse condition (Sanctorius 1603: 109r).[20]

In the same vein, Sanctorius described in the *Commentary on Avicenna* that the use of his thermoscope allowed one to compare febrile heat from 1 day to another, or from one paroxysm to another. On this basis, the physician could infer with certainty, so Sanctorius, whether the febrile heat increased, or decreased, and to what degree. An important point in this regard was that the instruments aided the physician's memory. According to Sanctorius, "no physician is provided with such ingenuity and memory as to be able, without the *pulsilogium*, to keep in mind the minimal differences of the movement and rest of the artery" (Sanctorius 1625: 222).[21] Therefore, Sanctorius continued, other physicians determined the pulse by conjecture, whereas he, by using his *pulsilogium*, could instead gain infallible knowledge (*cognitionem infallibilem*) of it. Hence, the measuring instruments not only served to quantify and to record a patient's state of health, but also helped the physician compile accurate data sourced from medical practice. Memorizing by heart the details of patients' histories also assured greater certainty in diagnosis. It is interesting to recall here that Sanctorius's choice of the *De statica medicina*'s form and structure was likewise informed by the wish to makes its content easier to memorize (Sect. 4.1.1). This testifies again to Sanctorius's strong concern for medical practice and his awareness of the pitfalls that a physician daily encountered at

---

[20] "… solum ex hac collatione certum & infallibile iudicium colligemus, an aeger sit in meliori, vel deteriori statu; …." See: Sanctorius 1603: 109r. The English translation is taken from: Bigotti and Taylor 2017: 87.

[21] "… nullus Medicus sit tam faelici ingenio, & memoria, qui posset sine pulsilogio tenere memoria minimas differentias motus, & quietis arteriae: …." See: Sanctorius 1625: 222. The English translation is taken from: Bigotti and Taylor 2017: 94.

the bedside. In this spirit, Sanctorius intended his measuring instruments, just as his other devices, to facilitate the work of practicing physicians and to improve therapeutics. What set the *pulsilogium*, the thermoscope, the hygrometer, and the *De statica medicina* apart was their ability to enhance the certainty of medical knowledge and thereby to improve the physician's judgment, his diagnosis. Notwithstanding that Sanctorius still adhered to the Aristotelian definition of *scientia* and thus placed his measuring instruments at the service of *ars*, he claimed that he could bring to the medical art a new precision which would, if not achieve absolute certainty, then in any case approximate it in a way never before believed possible. An epitome of Sanctorius's ambiguous concept of the status of medical knowledge can be found in the preface to the *Commentary on Avicenna*, where he explained that "through the long use and trial of all these things [healing, experiments, instruments, and static art], I found out that they can make this medical philosophy clear and manifest" (Sanctorius 1625: Ad lectorem).[22] Similarly to his description of the *De statica medicina* as a mathematical medicine, Sanctorius seemingly contradicted here his clear identification of medicine as an art. In naming the subject matter of medical theory "philosophy," he implied that he conceived of it as having the same status as philosophy, which was commonly assumed to be a science. What is more, he maintained that his medical approach, which was based on experience, observation, and the use of instruments, could enhance the clarity of this "philosophy" (Sanctorius 1625: 222; Siraisi 1987: 237 f.).

However, this is but one side of the coin. Along with the insistent claims as to the certainty of medicine, brought about by his new approach to the art, Sanctorius also repeatedly qualified his statements. An example of this can be seen in the citation quoted above, when he declared that he had "pondered for a long time, how that quantity of diseases can sometimes be partially known" (Sect. 6.1). In the *Commentary on Galen*, after having presented his *pulsilogium*, thermoscope, and hygrometer, he explained that these instruments enabled him to approximate the quantity of diseases to the greatest possible extent (*quammaxime*). This implies that, according to Sanctorius, a true, mathematical knowledge of this quantity could not be gained. In a later passage of the same work, Sanctorius made this even more explicit. He stated:

> … along with Galen at the start of the first book of [*De methodo medendi*] *ad Glauconem*, however, I shall admit that it is impossible that the ultimate and specific quantity will be fathomed by the physician, and so Galen rightly states that: "if I knew that quantity of action, I would consider myself to be as people say Asclepius was" (Sanctorius 1612b: 376).[23]

---

[22] "Hippocrates enim 2. Aphorismorum 17. vult, quod sanatio indicet morbum: Ego quoque Divini Senis imitation dico, quod & sanatio, & experimenta, necnon etiam instrumenta, & statica ars; quae omnia longo usu, & periclitatione adinveni, hanc medicam philosophiam reddere possint claram, & manifestam." See: Sanctorius 1625: Ad lectorem.

[23] "… quamvis fatear cum Galen 1. ad Glauc. in principio esse impossibile, ut illud ultimum & specificum quantum à medico penetretur: meritoque ibi dicit: si ego scirem illud quantum agendum, talem me reputarem, qualem fuisse ferunt Aesculapium." See: Sanctorius 1612b: 376. The English translation is based on: Bigotti 2018: 97.

Hence, in striking contrast to his insistent claims as to the certainty of medicine and his ability to achieve this certainty, Sanctorius apparently also had his doubts. In his last published commentary, the *Commentary on Hippocrates*, he cautiously wrote about the "quantity of diseases" which "might be occasionally perceived," and that he had invented four instruments for this purpose. Consequently, Sanctorius questioned the certainty with which the quantity of diseases could be determined in all three of his commentaries. In my opinion, it is significant that he did so in the two passages of his commentaries in which he explicitly dealt with the question of whether the medical art is conjectural—in the *Commentary on Avicenna* and in the *Commentary on Hippocrates*. Interestingly, these are also the only two instances in which he presented all four of his measuring instruments together as a group (Sanctorius 1612b: 230; 1625: 21–5; 1629a: 23–6).

To further blur the picture, when Sanctorius discussed the conjectural character of medicine, he did not only refer to the quantity of diseases, but also to other quantities that, following the Galenic teachings, made medicine uncertain: the quantity of remedies and the quantity of virtues. Moreover, idiosyncrasies and individual conditions of patients also added to the uncertainty of medicine, so Sanctorius (Sect. 6.1). Remarkably, while describing in some detail how the quantity of diseases could be ascertained by means of his instruments, Sanctorius offered hardly any solutions as to how to make these other conjectural factors more certain. Concerning the quantity of remedies, Sanctorius simply quoted various writings of Galen that relate to the latter's pharmacological theory and to the concept of the latitude of qualities (Sect. 5.2.2). From these, Sanctorius concluded that it was impossible to know with absolute certainty the strength of a remedy, i.e., its degree of intensity. In the *Commentary on Galen*, he explained that he used the weighing chair, the thermoscope, and the *pulsilogium* to determine the dosage of remedies, but this statement was followed by the assertion quoted above, that it was impossible to know the "ultimate and specific quantity." With regard to the quantity of virtues, matters are more ambiguous. I have shown above that Sanctorius held that he could gain certain knowledge of the quantity of the vital virtue by using three of his measuring instruments (Sects. 6.1 and 6.1.1). However, in discussing the question of the conjectural character of medicine, he made no reference to this solution, but briefly explained that it was necessary to know the quantity of the virtue in order to determine the quantity of remedies, both of which quantities remained conjectural, according to him. Thus, from today's standpoint, Sanctorius is once again equivocal, leaving one to wonder about his actual concept of medical knowledge and the status he assigned to his instruments and quantitative observations (Sanctorius 1625: 24, 215 f.; 1629a: 24 f.).

In contrast, on the question of the indeterminable nature of idiosyncrasies and individual conditions Sanctorius was clear: it was impossible to ascertain these two factors. Referring to Galen, he explained that it was necessary for the physician to know not only the common nature, but also individual natures, since there were, for example, people who had an idiosyncrasy that made them suffer so much from the smell of roses, or from eating cheese, as to fall at times into syncope (*lipothymia*). However, these properties of nature were as diverse as individuals and so innumerable as to be hidden (*occultae*) from the physician. Even in the *De statica medicina*, described elsewhere by Sanctorius as a mathematical medicine, he included an aphorism that says:

> The quantity of insensible perspiration varies according to the differences of natural properties, of regions, of seasons, of ages, of diseases, of food, and of the other non-natural things (Sanctorius 1614: 2v).[24]

Thus, according to Sanctorius, the influence of individuals' peculiar nature, or constitution on their state of health could not be determined with any certainty, and this made individuals incomprehensible to the physician. However, he did tone down this element of uncertainty in medicine a little, by maintaining that the task of the physician was not to treat individuals but to treat specific diseases. Accordingly, he understood an effective medicine to be one that cured the same disease in any number of different people. This added a universal aspect to therapy and weakened the argument that medicine was conjectural because it dealt so largely with particulars and thus did not arrive at general truths (Sanctorius 1625: 25, 214 f.; 1629a: 25 f.).

What to make now of these noticeable ambiguities in Sanctorius's work? All things considered, it seems Sanctorius was convinced that his instruments provided certainty, since he often referred to the values gained with them as being "most certain" or even as having "mathematical certainty." However, when it came to determining quantity in medicine and, more generally, to those five factors that made the medical art conjectural—the quantity of diseases, remedies, and virtues, as well as idiosyncrasies and individual conditions—Sanctorius was no longer so sure. While he was often quite confident about reducing, or even eliminating conjecture with regard to the quantity of diseases, he was strikingly reluctant to suggest solutions to making the other factors more certain. He appears to have been of the opinion that not all quantities in medicine could be determined and that, owing to the individuality of patients, medicine always would include a conjectural element. While his measuring instruments, when used alone, provided reliable and certain findings, their combined use in the quantification of disease might still leave room for uncertainty and provide only an estimate of the patient's state of health. Sanctorius's doubts in this regard might also explain the ambiguous relation between the measuring instruments and his varying grouping of them, analyzed above. And yet, despite all the equivocations, it is important to stress that Sanctorius's conception of the medical art as being able to approximate certainty, and his recourse to instruments and quantitative observation in order to enhance this certainty, clearly demonstrate his marked departure from tradition. From today's perspective, Sanctorius was at the threshold of a new understanding of medical knowledge and, more generally of *scientia*, according to which certainty would lie in the observation and experience of material things rather than in causal first principles. In this period of transition, Sanctorius proposed a specific approach: quantitative observation by means of instruments.

---

[24] "Quantitas perspirationis insensibilis aliquam varietatem patitur pro varietate naturae, regionis, temporis, aetatis, morborum, ciborum, & aliarum rerum non naturalium." See: Sanctorius 1614: 2v.

6.2.3   The Role of Reasoning and the Method of the Six **Fontes**

The preceding paragraphs disclosed the complex constellation in Sanctorius's works of traditional ideas on medical knowledge, his reinterpretation of them, and his introduction of new procedures based on quantification and instrumentation. However, these procedures were not the only means by which Sanctorius claimed to enhance certainty in medicine. Notwithstanding that Sanctorius's identification of medicine as an art stressed its practical and empirical dimensions, reasoning still played an important part for him in the purview of medicine. As mentioned above, Sanctorius did not consider it strictly necessary for a physician to actually exercise the art, which, he felt, could also be learned from a master alone, by using the mind rather than the senses (Sect. 6.2.1). Moreover, Sanctorius argued that anatomists could obtain mathematical certainty in their inquiry into disease and its causes by emphasizing that anatomical studies were not based on the senses alone, but involved reasoning, too (Sect. 4.2.1). This implies that, in the case of anatomy, it was the intellectual activities involved rather than anatomical practice and experience which made this field of medicine certain for him. In fact, already in his first publication, Sanctorius presented his doctrine of six *fontes* (sources), based on Aristotelian syllogistic logic, as the most certain of the, as the title says, "Methods to avoid all errors occurring in medical art" (*Methodi vitandorum errorum omnium qui in arte medica contingunt*). Without going into the details of this method, which have been outlined elsewhere, I will only briefly summarize its main features.[25]

Sanctorius's six *fontes* method was based on the collection of signs or symptoms (*per syndromen signorum*) and their progressive analysis. He identified six sources (*fontes*) of diagnostic signs that he considered would suffice to remove all ambiguity and uncertainty from diagnostic conclusions.[26] These were: external (procatarctic) causes, like bitter foods or remedies, the disposition of the patient; internal efficient causes, like bitter humors, symptoms, affected parts; and those things which aggravate or alleviate the condition.[27] According to Sanctorius,, the physician following this method could overcome the problem of the idiosyncrasy of patients as well as the problem of diseases having contrary symptoms but same cause, or,

---

[25] For accounts of Sanctorius's method of six *fontes*, see: Wear 1973: 173 ff., 214 f., 238 f., 243, Maclean 2002: 162, 285, 288, 300 f., 336 f., Poma 2012: 222 ff.

[26] Ian Maclean argues that Sanctorius's determination of six as a sufficient number of sources is a mathematical and not a logical claim and can be related to the trend toward computation mentioned in Sect. 5.2.1 (Maclean 2002: 162).

[27] The term "procatarctic causes" refers to a specific Galenic doctrine of causes based on Galen's treatise *De causis procatarcticis* (On Procatarctic Causes), which differentiates between *causa continens*, usually taken to mean "sustaining," "internal," "material," "remote," or "occult," and *causa procatarctica* which can be described as "preliminary," "external," "material," "proximate," or "efficient" and involved the six non-natural things. For more information on this doctrine and on Renaissance debates about the issue, see: ibid.: 146 f., 262–5, Galen and Johnston 2016: xxxv–xxxvi. Efficient causes form part of the Aristotelian doctrine of the four causes and are described by Aristotle as "the primary sources of the change or rest." See: Falcon 2019.

at an early stage, almost indistinguishable symptoms. In brief, Sanctorius suggested the six *fontes* as a means to apply to the fundamentals of medicine, i.e., to the established universal causes or categories, those particulars encountered in medical practice and perceived by the senses. In developing his own sign theory, Sanctorius was following a trend toward the reorganization of the medical field of semiology, which had begun in the late sixteenth century. As Ian Maclean has argued, the doctrine of signs grew in importance at this time, and Renaissance physicians put forward very different versions of sign theory. While Sanctorius's interest in semiology was thus in line with contemporary tendencies, his claim to have identified *the infallible* method that guaranteed a certain diagnosis was remarkable.[28] It shows that, according to Sanctorius, the means by which the physician could solve the problems and uncertainties that occurred in medical practice by no means related only to the senses, to experience, instrumentation, and quantitative observation, but also to mental procedures in the form of a logical methodology focused on categories and causes as well as on theories and reason. Hence, in the quest for certainty in medicine, Sanctorius did not only suggest his novel quantitative approach, but also drew on traditional sign theory (Sanctorius 1603: esp. 8v–9v; Maclean 2002).

To get a clearer picture of the significance and status that Sanctorius assigned to his two methods for enhancing certainty in medicine—the logical method set out in the *Methodi vitandorum errorum* and the practical and quantitative procedures set out mainly in the *De statica medicina* and the *Commentary on Avicenna*—it is instructive to compare how he referred to them in his other published works. Contrary to the instruments and quantitative measurements which, as was stated above, Sanctorius repeatedly mentioned in his three commentaries, he rarely mentioned the six *fontes* method in these works.[29] However, in discussing the second part of Galen's work *Ars medica*, which deals with semiology, Sanctorius, in his *Commentary on Galen*, frequently emphasized the importance of sign theory in diagnosis and the necessity of detecting a syndrome of signs. In this context he often mentioned the work *Methodi vitandorum errorum* (Sanctorius 1612a: e.g., 322, 335–9, 344 f., 499 f., 634). What is more, in 1630 Sanctorius published a revised version of this book, which implies that he still considered its content and the six *fontes* method significant.[30] In the same year, he also released a second edition of his *Commentary on Galen*, to which he added, among other things, a fairly lengthy passage outlining Galen's sign theory in more detail (Sanctorius 1630a: 854–67). This shows that, late in life, he still saw semiology based on logic and reasoning as a topic worthy of further discussion. Besides all this, as noted earlier, Sanctorius's work *De remediorum inventione*, that dealt with finding the correct remedies, was

---

[28] For more information on medical semiology, see: Maclean 2002: esp. 276–332.

[29] The only references by Sanctorius to his six *fontes* method that I can find in his commentaries are in the *Commentary on Galen*, see: Sanctorius 1612a: 170 f. [erroneously paginated 174 instead of 170], 308.

[30] I did not check all the revisions that Sanctorius made for the second edition of the *Methodi vitandorum errorum*, but only looked at the passage in which he presented his six *fontes* method, which remained unchanged. See: Sanctorius 1630b: 33–8.

based on syllogistic logic and focused on a method for identifying the specific differences between diseases which Sanctorius had presented in the *Methodi vitandorum errorum* (Sect. 5.2.4). Directly at the start of the work, Sanctorius stated: "The reason why physicians very rarely find the proper and particular remedy is their ignorance of the art of medicine, of philosophy, and of logic" (Sanctorius 1629b: 1).[31] Thus, even though Sanctorius propounded the use of instrumentation and measurements in order to improve the work of the physician, he still held that logic and philosophy were essential foundations of medicine, as was common among contemporary learned physicians.

All in all, given the minor role that the method of the six *fontes* plays throughout the whole of Sanctorius's works, it appears that this approach was less important to him than his instruments and measurements, which he mentioned more often. In the 1630 edition of the *Commentary on Galen*, Sanctorius not only dwelled longer on Galen's sign theory, but also included references to some of the devices that he had presented 5 years earlier in the *Commentary on Avicenna*.[32] In view of this, it is conceivable that Sanctorius developed and presented the six *fontes* method in his first publication, the *Methodi vitandorum errorum*, due to strategic considerations. Since sign theory was very popular at the time, he might have seen this as a way to promote his career. Being aware of its lack of originality, he later no longer emphasized his six *fontes* method. And yet, even after he had become professor at the University of Padua, Sanctorius still held that sign theory, more generally, and likewise syllogistic reasoning were highly relevant for gaining medical knowledge and for the success practice of medicine. And so, he did not tire of repeating that, in order to determine the complexion of a patient, it was necessary to consider a collection, or syndrome of signs. It is striking that Sanctorius did not weigh the two procedures against each other, but dealt separately in his works with sign theory and logical method on the one hand, and instruments and measurements on the other. Despite the eminent practical orientation of two procedures that ultimately fulfilled the same purpose, namely to determine the complexion of a patient, Sanctorius never sought to systematically merge them. In this context, it is interesting to note that Sanctorius considered semiology as an important means to aid the memory of the physician—a function fulfilled, too, as we have seen, by his *pulsilogium* and the *De statica medicina* (Sect. 6.2.2). This notwithstanding, Sanctorius connected his quantitative approach to physiology to the Galenic concept of the latitude of health and to Galenic pharmacological theory, and not to sign theory. It remains thus an open question how, for example, Sanctorius's strong emphasis on the importance of insensible perspiration in diagnosis and therapy relates to his declared necessity of always observing a collection of signs, or symptoms when making a diagnosis. A manifestation of the rather independent existence of the two procedures in

---

[31] "Causa, cur medici admodum rarò verum & proprium remedium inveniant, est artis medica, Philosophiae, & Logicae imperitia ...." See: Sanctorius 1629b: 1.

[32] To his second edition of the *Commentary on Galen*, Sanctorius added references to the thermoscope, the hygrometer, the *pulsilogium*, the clyster (*mitrenchyta*) and to the instrument to quench the thirst of fever patients. See: Sanctorius 1630a: 262 f., 594, 693, 762, 807 f.

Sanctorius's works and probably also in his concept of medicine is his last publication, *De remediorum inventione*, which focuses on a logical method for finding the correct remedies without any reference to measuring instruments or quantification. More pointedly, Sanctorius's concurrent but independent use of mental and sensuous, or experiential procedures with the aim of enhancing the certainty of medicine and of improving diagnosis and treatment reflects the complex relations between the empirical and rational parts of the discipline of medicine and coincides with the ambiguous relation between theory and practice found in his works (Sect. 4.3). In order to better understand these relations, it is pertinent to now take a look at the other side of the spectrum and to examine more closely Sanctorius's notions of the role of experience and empirical knowledge in medicine (Sanctorius 1612a: 335 f.; 1630a: 854; Poma 2012: 218).

6.2.4   The Role of Experience and Empirical Knowledge

In his first publication, *Methodi vitandorum errorum*, released in 1603, Sanctorius wrote:

> From this nature of tastes and colors, as explained so far, those mixtures of the humors, which are manifest to the senses, can be gathered. In order to fully know, however, the [mixtures] which are in the most inner parts of the body, where neither the tongue, nor the eyes can go, there are methods proposed in the sixth book that can teach every of the predominant humors and consequently any of their mixtures (Sanctorius 1603: 149r).[33]

Hence, Sanctorius pointed here clearly to the limits of experience and the use of the senses. According to him, the only way for the physician to penetrate into the depths of the body was to use his mind and thereby apply the method of a syndrome of signs that he had presented in the sixth book of his *Methodi vitandorum errorum*. Accordingly, sign theory was the means by which the physician could gain knowledge of things that were not accessible to the senses. In another passage of the same work, Sanctorius stated that not even "millions of thousands of particulars" (*milliona millia particularia*) could produce a universal. According to him, no universal cause could be derived from the experience of single events and he argued that one would need an infinite number of instances in order to logically produce a universal from particulars—an undertaking that was impossible for mortal man. Consequently, neither experience nor experiments—Sanctorius used both words indiscriminately in this context—could ever provide certain knowledge, since they were concerned only with particulars. As medicine, in the words of Sanctorius, "centered on universals and not particulars," the physician needed to know universal causes and thus

---

[33] "Ex hac natura saporum, & colorum hactenus explicata illae humorum misturae, quae sensibus sunt manifestae, colligi possunt: Quomodo verò pernoscantur, quae sint in penitissimis corporis partibus, in quas neque lingua, vel oculi penetrare possunt, traditae sunt in 6.lib. Methodi, quae possunt docere omnes praedominantes humores, & per consequens quamlibet eorum miscellam: ...." See: Sanctorius 1603: 149r.

use other means than experience and the senses to gain this knowledge, namely Aristotelian syllogistic reasoning and sign theory (Sanctorius 1603: 188v; 1612a: 90).

Nonetheless, there still was a connection for Sanctorius between particulars and universals and in this regard, experience could be useful. He explained:

> We do not deny, however, that induction or experiments can contribute toward knowing a universal; because as Boethius said in [his commentary on Aristotle's] *Categories*, experience is the collection of examples, and after the collection, the intellect is urged on by its own light to separate the natural universals from the individual, for the whole universal nature is in any individual (Sanctorius 1603: 189v–190r).[34]

From this citation, it is clear that Sanctorius held, even while admitting that experience might help the physician arrive at universal truths, that there was no infallible way or method to proceed from personal experience to universals, and that it was ultimately the mind that gleaned universal truths from appearances, "by its own light." The perception of particulars triggered the mind to identify the correct cause of the perception. Thus, Sanctorius adhered to the Aristotelian theory of knowledge and its division into sensory experience and intellection (Sect. 6.2.1). His association of Galenic semiology with Aristotelian scientific methodology in order to explore the possibilities of induction, i.e., the methodological derivation of knowledge from particulars, reflects a development that Per-Gunnar Ottosson detected in medieval and Renaissance commentaries on Galen's *Ars medica*. Moreover, as Andrew Wear has shown, Sanctorius's view of the role of experience in medicine was influenced also by contemporary discussions on medical method, which stressed an a priori type of knowledge, according to which theory preceded action, explained the action, and gave it its sense.[35] Notwithstanding that the physician first examined the patient by looking for symptoms and relating these to the possible cause of the disease, the investigation of symptoms and signs would have been pointless, had the causes of the signs not previously been known. As Sanctorius explained in the preface to the *Commentary on Avicenna*, when discussing the division of medicine into theory and practice, the physician first explored the truth and then directed it to action, that is, to the preservation or restoration of health. Along the same lines, Sanctorius's new approach to the teaching of medical theory aimed to confirm theory a posteriori, by means of practice, and to corroborate practice a priori, by means of theory (Sect. 4.3) (Ottosson 1984: 196).

Thus, Sanctorius's critical opinion of sensory experience and his emphasis on reason as the preferable means to gain knowledge about universal truths, show him to be very traditional and conform with contemporary views. In light of this,

---

[34] "Non tamen negamus inductionem, vel experimenta conferre posse ad cognoscendum universale: quia, ut dicit Boetius in praedicamentis, experientia est exemplorum collectio, post quam collectionem intellectus à proprio lumine excitatur ad separandam naturam universalem ab individuali; tota enim natura universalis est in quolibet individuo, ..." See: ibid.: 189v–190r. The English translation is taken from: Wear 1981: 255.

[35] For accounts of Sanctorius's views on medical methods, see: Wear 1973: esp. 210–56, Wear 1981, Poma 2012. See also Sect. 4.1.1, fn. 2.

Sanctorius seems far removed from the figure of the ingenious innovator who pioneered a new medical science. Yet, as one might guess after having read the previous chapters, things are not always quite as simple as they seem at first glance. Eleven years after the publication of the *Methodi vitandorum errorum*, Sanctorius wrote in the preface to the *De statica medicina* that "not only do the mind and the intellect perceive sincere and pure truth, but also the eyes and the hands virtually palpate it" (Sanctorius 1614: Ad lectorem).[36] This fits with Sanctorius's description of the work as "mathematical medicine" or "static theorems" (Sect. 6.2.2) and implies that he did, after all, believe it possible that knowledge of universal causes could be gained from particulars by means of the observations and experiences that he made with his weighing chair. The contrast with his statements in the *Methodi vitandorum errorum*, outlined above, is immediate and striking. Curiously, in the first edition of this work, a chapter title stated that "induction gives sufficient proof" (*probatur inductione sufficientissima*), which suggests that Sanctorius was not as convinced of the impossibility of induction as it might seem from his other statements in the book. But in the second edition of the *Methodi vitandorum errorum*, published 27 years later, Sanctorius deleted the "sufficient" from the chapter's title. Hence, despite his bold claim in the *De statica medicina* that eyes and hands could feel truth, he seems still to have been in doubt about the possibility of induction as late as 1630.[37] Similarly, a year earlier, Sanctorius wrote in his book *De remediorum inventione* that "without reason and the advice of Galen or Hippocrates, experience cannot be trusted" (Sanctorius 1629b: 11).[38] In fact, as Elaine Leong has pointed out, such a pairing of experience with reason was ubiquitous and enduring in medieval and Renaissance learned medical writings. It served to distinguish the Hippocratic-Galenic medical sect (usually referred to as dogmatic, or rational sect) against the rival empirical sect, which Galen had so fiercely attacked in his works and whose members relied, according to the Greek physician, on experience alone. By emphasizing the need to always couple experience with reason, learned physicians tried to distance themselves from the practices of unlearned healers, and invoked a picture of an acceptable empiricism that was backed up by medical learning. A loyal Galenist, Sanctorius's ambiguous attitude toward the role of experience and empirical knowledge was certainly influenced by Galen's dislike of the empirical sect and by the anxiety of being perceived as an adherent of this medical school. The statement quoted above is preceded by Sanctorius's warning that one should not listen to

---

[36] "… veritatem ipsam sinceram ac puram putam non solum animo & intellectu percipiant, sed oculis etiam ac ipsis quasi manibus palpent, …." See: Sanctorius 1614: Ad lectorem.

[37] Ian Maclean pointed out this change in the chapter title in Sanctorius's *Methodi vitandorum errorum*. However, Maclean did not consult the first edition of the work, and therefore assumed—having referred to the second edition published in 1630—that the adjective *sufficientissima* was added only to the 1631 edition of the book. In fact, the 1631 edition of the *Methodi vitandorum errorum*, published in Geneva, was a copy of the original edition of the book from 1603. See: Sanctorius 1631: 162, Maclean 2002: 169, fn. 87.

[38] "Nos verò experientiam, esse concedendam putamus, sed sine ratione, & Galeni seu Hippocratis consilio, credimus experientiae non esse fidendum." See: Sanctorius 1629b: 11.

the Empiricists (*Empirici*), who rejected reason and authority and said that experience was worth more than the philosophies of Hippocrates and Galen. In the same way, Sanctorius frequently attacked present-day Empiricists in the *Methodi vitandorum errorum* and complained in the *De remediorum inventione* about unlearned physicians and surgeons (*medici* and *chirurgi plebei*), who did not properly follow the Galenic teachings. All things considered, Sanctorius undoubtedly assigned an important role to experience and the senses, but was at the same time careful to acknowledge their limitations. According to him, the physician was a "*sensatus philosophus*," who used his mind to derive universal knowledge. However, sometimes he also was a "*sensatus artifex*," who rather used experience and practical skill for the same purpose. Since Sanctorius switched in his works between these two ideas, he appears to have considered the physician to be both—a "sensible" philosopher and a "sensible" artisan (Sanctorius 1603: 7v, 16v–18v, 61r, 170r–170v; 1612a: 107, 117, 123; 1629b: 11, 39, 66; 1630b: 258; French 1994: 322; Maclean 2002: 169–98; Leong and Rankin 2017: 168, 170).

The preceding passages have shown that the question of certainty in medicine was, to be sure, not easily answered. Sanctorius put forward two methods that he believed would make the work of the physician more certain and so reduce the errors committed in medical practice. Whether he really believed that conjecture could be completely eliminated from medicine and absolute certainty achieved remains an open question. The method of the six *fontes*, or more generally, of a syndrome of signs, tied in with contemporary attempts to reorganize the medical field of semiology and adhered to traditional views of the role and limits of experience and the senses. Contrary to this, Sanctorius broke new ground by using instruments in order to enhance the physician's perceptions and so make the medical art more certain. Especially with the *De statica medicina*, Sanctorius attempted to overcome the division made between sensory experience and intellection. The very idea of using a mechanical instrument to render visible an internal and invisible bodily process which was completely hidden from the senses and thereby lay claim to mathematical certainty shows that Sanctorius was prepared to think what was, by earlier Aristotelian-Galenic standards, the unthinkable: namely, that experience and quantification could provide knowledge about universal causes. In doing so, Sanctorius walked a tightrope between the traditional Galenic position, accepted and cultivated at the universities, and the empiricist position, deemed by the learned medical community to be inferior, arbitrary, and even dangerous. Sanctorius left no doubt as to which camp he belonged In. His firm commitment to Galenic medicine can then explain how Sanctorius's attitude toward experience, empirical knowledge, and the use of the senses, which sounds ambiguous and contradictory today, was no contradiction for Sanctorius himself. In his attempt to improve Galenic medicine, he reconsidered the relation between the empirical and the rational parts of the discipline without, however, abandoning the fundamental principles upon which the whole discipline of medicine rested. Sanctorius's thoughts on the roles of experience and reason in medicine also elucidate something about the way in which he conceptually integrated his weighing procedures, and, too, his experiences with the other devices, into a traditional Galenic framework. It should now be clear that, for

Sanctorius, theoretical medical concepts, such as dietetics and the doctrine of the six non-natural things, had necessarily to be the starting point for any inquiry into the uncertainties involved in the medical art. These uncertainties were strongly felt by Sanctorius, who, as a diligent practitioner, was eager to avoid errors in diagnosis and treatment and, more generally, aimed to improve the day-to-day work of the physician. To further investigate Sanctorius's understanding and use of experience, empirical knowledge, and practice, it is pertinent to take a look at the terminology he used to describe these factors in his works.

6.2.5   Experience or Experiment?

In the Middle Ages and the Renaissance, the Latin word *experimentum* was closely aligned to the word *experientia* (experience) and both were usually used indiscriminately, with no systematic distinction between them. Generally, they simply referred to experience of some kind and included a whole range of empirical practices, such as drug testing or dissections. Furthermore, neither *experientia* nor *experimentum* had to result from firsthand experience, but might well be based on others' reports.[39] Perusal of Sanctorius's works suggests that he, too, employed the two words interchangeably, although a systematic analysis of his use of the terms would be needed to confirm this hypothesis, and that is not feasible here. Rather, I want to draw attention to another related Latin term, *periculum*, which can be translated as "trial" or "test" and began to be used in the sixteenth century to designate the deliberate execution of a trial, as in: *periculum facere*, "to put to the test." As Roger French has argued, this phrase alluded to the famous first of Hippocrates's *Aphorisms* that says: "Life is short, the Art long, opportunity fleeting, experiment treacherous, judgment difficult" (Hippocrates and Jones 1931: 99).[40] According to French's research, the phrase *experimentum periculosum* (treacherous experiment) was used consistently in the various Latin translations of the originally Greek aphorism. He concluded that this expression could not signify passive experience, since *periculum* also meant an attempt or trial, including the attendant risks. In his opinion, the Renaissance translators qualified the noun *experimentum* with an adjective derived from *periculum* in order to highlight that what was meant was an active attempt with

---

[39] The historical development of "experiment" is complex and difficult to pin down, since the roles and functions this notion has had in different contexts and times are manifold. For accounts of early modern understandings of the term, see e.g., Schmitt 1969, Dear 1995, Dear 2006, Leong and Rankin 2017, Steinle et al. 2019. For a study of the various uses of "experiment" in research processes and the understanding of experiment as a means for empirical research, see: Steinle 2005.

[40] Evan Ragland has shown that sixteenth-century writers also referred to other precedents for using the phrase *periculum facere* to mean the conduct of a trial or test. These were taken from classical Latin literature, such as Cicero and the plays of Terence and Plautus. See: Ragland 2017: 511.

uncertain outcome, a clinical or medical trial (French 1994: 320–33; Dear 1995: 13; 2006: 106; Leong and Rankin 2017: 162–70; Ragland 2017: 512).

Interestingly, Sanctorius used the phrase *periculum facere* in the preface to the *De statica medicina*. He wrote: "But I am the first to make the trial [*periculum feci*], and unless I am mistaken I have by reasoning and by the experience [*experientia*] of 30 years brought this art to perfection …" (Sanctorius 1614: Ad lectorem).[41] Hence, following French's interpretation of the expression, this implies that Sanctorius wanted to stress here that he was the first to make a *deliberate* test in order to determine the quantity of insensible perspiration. He evidently considered it important to inform his readers that he gained his information not from passive observation, but from an active, contrived event. He appears, thus, to have had some notion of "experiment" according to which the "experimenter" consciously, and with forethought, attempted to test a particular hypothesis by devising a specific observational situation by which to resolve the question. This understanding of experiment, or rather "putting to the test," is of course very different from modern experimental methods and randomized clinical trials. However, Sanctorius's use of the phrase *periculum fecit* in the preface to the *De statica medicina* shows that he was aware he was presenting a new and different approach to a medical problem, based on a specific empirical practice that might best be described as controlled and deliberate observation. It appears then, that it was this procedure that, according to Sanctorius, enabled the eyes and hands of the physician to feel truth. Remarkably, Sanctorius used the phrase *periculum facere* also in a passage of the *Commentary on Avicenna*, when describing his thermoscopes. He explained that by means of these instruments he "put to the test" whether the heat in children and adolescents was the same (Sanctorius 1625: 357; Schmitt 1969: 105–21).

But there is also another dimension to this. Using the expression *periculum facere* instead of *experimentum* or *experientia* in reference to Hippocrates's first aphorism might simply mean that Sanctorius did not want to risk being regarded as an empiricist. Hippocrates served here as a model for the empirical observer, recording case histories and justifying "experiment" with regard to the patient, and impartially recording empirical data. As mentioned earlier, Sanctorius not only used the phrase *periculum facere* in the preface to the *De statica medicina*, but also presented himself as a follower of Hippocrates, especially regarding the use of the aphoristic form (Sect. 4.1.1). Thus, it is very probable that Sanctorius invoked the Physician of Kos strategically, in order to legitimize his new approach to physiology as an acceptable empiricism. Indeed, this might even have been the reason why he chose to present the results of his weighing procedures in aphorisms. Note that in the preface to the *De statica medicina* Sanctorius again paired experience with reason. What is more, in the preface to the *Commentary on Avicenna* Sanctorius introduced his novel way of teaching medical theory, which, as a direct continuation of the Hippocratic teachings, was based on the use of "experiments [*experimenta*],

---

[41] "… ego verò primus periculum feci, & (nisi me fallat genius) artem ratione & triginta annorum experientia ad perfectionem deduxi, …." See: Sanctorius 1614: Ad lectorem. The English translation is based on: Foster 1924: 145.

instruments, and static art." Here Sanctorius did not use the phrase *periculum fecit*, but rather the Latin word *periclitatio*, which, like *periculum,* can be translated as "test" or "trial," but also as "danger," "risk," or "hazard." The fact that Sanctorius drew so heavily on the authority of Hippocrates in the introductions to the two publications, in which he mainly set out the practical and quantitative procedures aimed at enhancing certainty in medicine, strongly suggests that he struggled to distance himself from the empirical sect and to emphasize that his novel methods were based on learned medical knowledge. It is easy to understand the importance of this to Sanctorius, if one considers that he was still working as a university professor of medical theory at least at the time when he published the *De statica medicina* (Sanctorius 1625: Ad lectorem; Ramminger; Lewis and Short 1879).

Having said all this, it must be noted nonetheless that what Sanctorius actually practiced might have differed from the words he used and from the methods that he recommended in his books to enhance certainty in medicine. Similarly, the use of the word *periculum*, like that of *experimentum* and *experientia*, does not necessarily imply that Sanctorius performed actual experimental procedures as opposed to hypothetical "thought experiments." It is therefore necessary to finally take a closer look at his measuring instruments in order to further examine Sanctorius's making and doing: his use of experience, observation, quantification, and experimentation in medical practice. In the process, not only will the material dimensions of his endeavors come to the fore, but also the ways in which contemporary technology and craftmanship played a part in Sanctorius's concept of medicine as an art that could, if not attain, then at least approximate certainty (Maclean 2002: 296).

References

Bigotti, Fabrizio. 2016. Mathematica Medica. Santorio and the Quest for Certainty in Medicine. *Journal of Healthcare Communications* 1: 1–8.
———. 2018. The Weight of the Air: Santorio's Thermometers and the Early History of Medical Quantification Reconsidered. *Journal of Early Modern Studies* 7: 73–103.
Bigotti, Fabrizio, and David Taylor. 2017. The Pulsilogium of Santorio: New Light on Technology and Measurement in Early Modern Medicine. *Society and Politics* 11: 55–114.
Castellani, Carlo. 1958. Alcune lettere di Santorio Santorio a Senatore Settala. *Castalia* 1: 3–7.
De Pace, Anna. 1993. *Le matematiche e il mondo: Ricerche su un dibattito in Italia nella seconda metà del Cinquecento*. Milan: Franco Angeli.
Dear, Peter. 1995. *Discipline & Experience: The Mathematical Way in the Scientific Revolution*. Chicago: University of Chicago Press.
———. 2006. The Meanings of Experience. In *The Cambridge History of Science,* Vol. 3: *Early Modern Science*, ed. Katharine Park and Lorraine Daston, 106–131. Cambridge\New York: Cambridge University Press.
Ettari, Lieta Stella, and Mario Procopio. 1968. *Santorio Santorio: la vita e le opere*. Rome: Istituto nazionale della nutrizione.
Falcon, Andrea. 2019. Aristotle on Causality. In *The Stanford Encyclopedia of Philosophy*, ed. Edward N. Zalta. https://plato.stanford.edu/archives/spr2019/entries/aristotle-causality/. Accessed 12 Dec 2019.

Foster, Michael. 1924. *Lectures on the History of Physiology during the Sixteenth, Seventeenth and Eighteenth Centuries*. Cambridge: Cambridge University Press.

French, Roger K. 1994. *William Harvey's Natural Philosophy*. Cambridge: Cambridge University Press.

Galen and Ian Johnston. 2016. *On the Constitution of the Art of Medicine; The Art of Medicine; A Method of Medicine to Glaucon*. Cambridge, MA\London: Harvard University Press.

Grmek, Mirko D., and Danielle Gourevitch. 2001. La maladie mesurée: l'apport de Santorio Santorio. *La Revue du Praticien* 51: 2009–2012.

Guidone, Mario and Fabiola Zurlini. 2002. L'introduzione dell'esperienza quantitativa nelle scienze biologiche ed in medicina Santorio Santorio. In *Atti della XXXVI tornata dello Studio firmano per la storia dell'arte medica e della scienza, Fermo, 16–17–18 maggio 2002,* ed. Studio firmano per la storia dell'arte medica e della scienza, 117–137. Fermo: A. Livi.

Hippocrates, and William Henry Samuel Jones. 1931. *Hippocrates,* Vol. IV. Cambridge, MA: Harvard University Press.

Jacquart, Danielle. 1980. Le regard d'un médecin sur son temps: Jacques Despars (1380?–1458). *Bibliothèque de l'École des chartes* 138: 35–86.

Leong, Elaine, and Alisha Rankin. 2017. Testing Drugs and Trying Cures: Experiment and Medicine in Medieval and Early Modern Europe. *Bulletin of the History of Medicine* 91: 157–182.

Lewis, Charlton T. and Charles Short. 1879. periclitatio. In *A Latin Dictionary*. Oxford: Clarendon Press. http://www.perseus.tufts.edu/hopper/text?doc=Perseus:text:1999.04.0059:entry=periclitatio. Accessed 23 Dec 2019.

Maclean, Ian. 2002. *Logic, Signs, and Nature in the Renaissance: The Case of Learned Medicine*. Cambridge\New York: Cambridge University Press.

Miessen, Hermann. 1940. Die Verdienste Sanctorii Sanctorii um die Einführung physikalischer Methoden in die Heilkunde. *Düsseldorfer Arbeiten zur Geschichte der Medizin* 20: 1–40.

Mitchell, S. Weir. 1892. *The Early History of Instrumental Precision in Medicine. An Address before the Second Congress of American Physicians and Surgeons, September 23rd, 1891*. New Haven: Tuttle, Morehouse & Taylor.

Ottosson, Per-Gunnar. 1984. *Scholastic Medicine and Philosophy: A Study of Commentaries on Galen's Tegni (ca. 1300–1450)*. Naples: Bibliopolis.

Park, Katharine, and Lorraine Daston. 2006. Introduction: The Age of the New. In *The Cambridge History of Science,* Vol. 3: *Early Modern Science*, ed. Katharine Park and Lorraine Daston, 1–17. Cambridge\New York: Cambridge University Press.

Poma, Roberto. 2012. Santorio Santorio et l'infallibilité médicale. In *Errors and Mistakes. A Cultural History of Fallibility*, ed. Mariacarla Gadebusch Bondio, Paravicini Bagliani, and Agostino, 213–225. Florence: SISMEL-Edizioni del Galluzzo.

Ragland, Evan R. 2017. "Making Trials" in Sixteenth- and Early Seventeenth-Century European Academic Medicine. *Isis* 108: 503–528.

Ramminger, J. periclitatio. In *Neulateinische Wortliste. Ein Wörterbuch des Lateinischen von Petrarca bis 1700.* www.neulatein.de/words/0/018468.htm. Accessed 23 Dec 2019.

Sanctorius, Sanctorius. 1603. *Methodi vitandorum errorum omnium, qui in arte medica contingunt, libri quindecim*. Venice: Apud Franciscum Barilettum.

———. 1612a. *Commentaria in Artem medicinalem Galeni,* Vol. I. Venice: Apud Franciscum Somascum.

———. 1612b. *Commentaria in Artem medicinalem Galeni,* Vol. II. Venice: Apud Franciscum Somascum.

———. 1614. *Ars Sanctorii Sanctorii Iustinopolitani de statica medicina, aphorismorum sectionibus septem comprehensa*. Venice: Apud Nicolaum Polum.

———. 1625. *Commentaria in primam Fen primi libri Canonis Avicennae.* Venice: Apud Iacobum Sarcinam.

———. 1629a. *Commentaria in primam sectionem Aphorismorum Hippocratis, & c. … De remediorum inventione*. Venice: Apud Marcum Antonium Brogiollum.

———. 1629b. *De remediorum inventione*. Venice: Apud Marcum Antonium Brogiollum.

———. 1630a. *Commentaria in Artem medicinalem Galeni*. Venice: Apud Marcum Antonium Brogiollum.

———. 1630b. *Methodi vitandorum errorum omnium, qui in arte medica contingunt, libri quindecim*. Venice: Apud Marcum Antonium Brogiollum.

———. 1631. *Methodi vitandorum errorum omnium, qui in arte medica contingunt, libri quindecim … Nunc primùm acceßit eiusdem Authoris De inventione remediorum liber*. Geneva: Apud Petrum Aubertum.

Sanctorius, Sanctorius, and Giuseppe Ongaro. 2001. *La medicina statica*. Florence: Giunti.

Schmitt, Charles B. 1969. Experience and Experiment: A Comparison of Zabarella's View With Galileo's in *De Motu*. *Studies in the Renaissance* 16: 80–138.

Siegel, Rudolph E. 1968. *Galen's System of Physiology and Medicine*. Basel\New York: S. Karger.

Siraisi, Nancy. 1981. *Taddeo Alderotti and His Pupils: Two Generations of Italian Medical Learning*. Princeton: Princeton University Press.

———. 1987. *Avicenna in Renaissance Italy: The Canon and Medical Teaching in Italian Universities After 1500*. Princeton: Princeton University Press.

———. 1990. *Medieval & Early Renaissance Medicine: An Introduction to Knowledge and Practice*. Chicago: University of Chicago Press.

Steinle, Friedrich. 2005. *Explorative Experimente: Ampère, Faraday und die Ursprünge der Elektrodynamik*. Stuttgart: Steiner.

Steinle, Friedrich, Cesare Pastorino, and Evan R. Ragland. 2019. Experiment in Renaissance Science. In *Encyclopedia of Renaissance Philosophy*, ed. Marco Sgarbi, 1–15. Cham: Springer.

Wear, Andrew. 1973. *Contingency and Logic in Renaissance Anatomy and Physiology*. Phd diss., Imperial College London.

———. 1981. Galen in the Renaissance. In *Galen: Problems and Prospects*, ed. Vivian Nutton, 229–262. London: The Wellcome Institute for the History of Medicine.

Chapter 7
The Measuring Instruments

Abstract As the title suggests, this chapter deals with the most famous of the devices which Sanctorius developed to measure and to quantify physiological change: pulsilogia, thermoscopes, hygrometers, and balances. Having attracted considerable scholarly attention, they form the backbone of the narrative that identifies Sanctorius as a great innovator, who founded a new medical science, a science to which mechanization, measurement, and numerical values were integral. The findings of the foregoing chapters allow us now to go beyond this selective account of Sanctorius and his works and to reevaluate his celebrated measuring instruments and their use from a closer perspective. To this end, I explore their design and basic functioning, the contexts in which they emerged, how Sanctorius possibly used them, and what precisely they measured. In this connection, I also analyze two steelyards for the measurement of climatic conditions which have hitherto been largely ignored, thereby covering the whole range of Sanctorius's measuring instruments. Moreover, I present the results of my reconstruction of the Sanctorian weighing chair and the attendant replication of his experimental practice, and thereby show how this approach opened up new perspectives on Sanctorius's works, his doctrine of static medicine, and the function and purpose of his weighing chair.

Keywords Material culture · Measuring instruments · Replication · Sanctorian chair · Weighing

The index of the *Commentary on Avicenna* contains sixteen items that can be subsumed in the following types of measuring instrument: *pulsilogia*, thermoscopes, hygrometers, and balances (Fig. 4.1). As already seen, this group comprises the most famous instruments devised by Sanctorius, which have already attracted considerable scholarly attention (Sect. 4.2, fn. 16). They form the backbone of the narrative that identifies Sanctorius as a great innovator, who founded a new medical science, a science to which mechanization, measurement, and numerical values were integral. I have pointed out earlier that this storyline omits some important points. It concentrates solely on those parts of Sanctorius's works that are, or appear to be innovative, isolating them from the context in which they emerged. The analyses of the preceding chapters allow me now to go beyond this selective account of

© The Author(s) 2023
T. Hollerbach, *Sanctorius Sanctorius and the Origins of Health Measurement*,
https://doi.org/10.1007/978-3-031-30118-6_7

Sanctorius and his works and to reevaluate his famous measuring instruments and their use from a broader perspective. Against this backdrop, it is possible to critically review the image of Sanctorius and to ask whether it is still tenable to label him the innovator of a new medical science.

In the following, I will analyze all of Sanctorius's measuring instruments, including two steelyards for the measurement of climatic conditions which have hitherto been largely ignored. Priority disputes are considered only insofar as they provide important insights into Sanctorius's social and intellectual milieu and thus allow some conclusions to be drawn about the way in which the physician developed his innovative ideas. Here, too, instead of focusing only on the *Commentary on Avicenna*, I will examine the measuring instruments with regard to all of Sanctorius's books. However, in order to fully grasp the material dimensions of Sanctorius's quantitative approach to physiology, it is necessary to look beyond the written and pictorial sources. Illustrations and descriptions of the instruments represent codified forms of the knowledge produced in the very process of their invention, from the first idea to their realization and use. They are the end products of active processes of knowledge making. The reconstruction of such instruments and the attendant replication of the experiments conducted with them is a means for the historian to gain insight into these active processes of knowledge making, and of knowledge in its uncodified form (Smith and Schmidt 2007: 3 f.; Smith 2017: 372 ff.). In my attempt to understand how Sanctorius developed his quantitative approach to physiology and to trace the mechanical and practical knowledge involved in his undertakings, I reconstructed his most famous instrument, the weighing chair, and sought to replicate his experimental practice. This opened up new perspectives on Sanctorius's works, his doctrine of static medicine, and the function and purpose of his weighing chair. But before addressing this, at the end of this chapter, I will begin my study of Sanctorius's measuring instruments by examining two other balances that the Venetian physician devised.

7.1   Two Balances to Measure Climatic Conditions

In addition to the famous weighing chair, Sanctorius developed two other balances, which enjoy far less renown: one, to measure the *impetus* of wind (Fig. 7.1); the other, to measure the *impetus* of water currents (Fig. 7.2).[1]

Sanctorius described the design of the two balances in the *Commentary on Avicenna* as follows:

---

[1] I use here the term *impetus*, because this is the term that Sanctorius always uses in his description of the two steelyards to measure climatic conditions. This term was highly relevant at the time and played an important part in Galileo Galilei's theory of motion. There is no standard translation of *impetus*, as its meaning has often changed over time and been further differentiated. Today, there is no direct equivalent for *impetus*. For more information on the term and concept *impetus*, see: Elazar 2011, Van Dyck and Malara 2019.

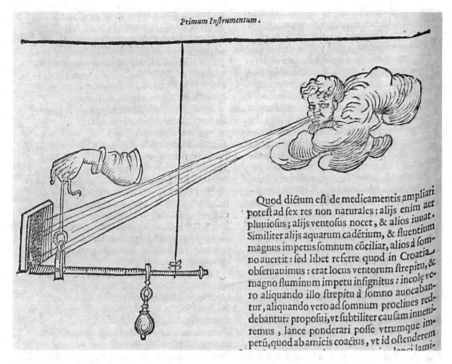

Primum Inſtrumentum .

Quod di&um eſt de medicamentis ampliari poteſt ad ſex res non naturales : alijs enim aer pluuioſus; alijs ventoſus nocet, & alios iuuat. Similiter alijs aquarum cadétium, & fluentium magnus impetus ſomnum côciliat, alios à ſomno auertit : ſed libet referre quod in Croatiæ obſeruauimus : erat locus ventorum ſtrepitu, & magno fluminum impetu inſignitus : incolę vero aliquando illo ſtrepitu à ſomno auocabantur, aliquando vero ad ſomnum procliues reddebantur: propoſui, vt ſubtiliter cauſam inueniremus , lance ponderari poſſe vtrumque impetû, quod ab amicis coa&us, vt id oſtenderem

Fig. 7.1 Balance to measure the *impetus* of wind (Sanctorius 1625: 246). (© British Library Board 542.h.11, 246)

I proposed … that both impetuses can be weighed with a scale pan and, encouraged by friends to show this, I provided two balances, the first for the impetus of wind, the second for the impetus of water and added to both scale pans an iron plate. With the one, in which the iron plate is above [the beam], we weigh the impetus of wind. … But by means of the other [balance], to which the same plate is attached, we discern how much the weight of the impetus of flowing water is (Sanctorius 1625: 246 f.).[2]

---

[2] "… proposui, …, lance ponderari posse utrumque impetum, quod ab amicis coactus, ut id ostenderem praestiti duobus stateris, per primam ventorum, per secundam vero aquae impetum, utrique; lanci laminam ferream apponendo: illa, in qua lamina ferrea supereminet, perpendimus ventorum impetum: …. Alia verò cui appensa est eadem lamina aquae currentis impetum dignoscimus quanti sit ponderis." See: Sanctorius 1625: 246 f. It is interesting to note that Sanctorius refers here to *the weight of the impetus* (*impetum … ponderis*). The physical concept of force as we use it today, did not yet exist, but contemporaries like Galileo Galilei used the term *force (forza)*. As with *impetus*, the concept behind this was in flux and cannot be mapped seamlessly onto the modern physical concept of force. The attempt to measure with a balance the *impulsive forces* (in Galileo: *forza della percossa*) then assumed to be proportional to the *impetus*, was nothing new at the time. The English mathematician and philosopher Thomas Harriot (1560–1621), for example, dropped balls from different heights onto the pan of a balance with equal arms. Similar to Sanctorius, Galileo tried to measure with a scale the *force* of an impinging water jet. But Galileo used an equal-armed balance and falling, not streaming water as Sanctorius did. The fact that Sanctorius wrote of the *weight of the impetus* is not surprising, since it is derived ad hoc from his experimental arrangement—a scale measures weights. See: Settle 1996, Schemmel 2008.

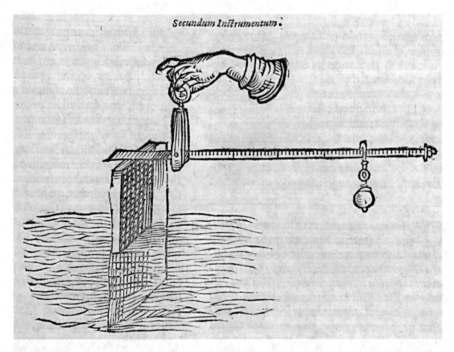

Fig. 7.2 Balance to measure the *impetus* of water currents (Sanctorius 1625: 247). (© British Library Board 542.h.11, 247)

These scant remarks, together with the illustrations, are the only information that Sanctorius gave about how the two balances work. Therefore, it is difficult to understand how he came to design and use these devices. That historical studies on the development of anemometry and hydraulics have mostly overlooked his devices further aggravates the problem.[3] Thus, a comprehensive analysis of Sanctorius's two balances is required. Notwithstanding that such an analysis goes beyond the scope of the present work, I will present a first step in this endeavor.

---

[3] In his study on the invention of meteorological instruments, W.E. Knowles Middleton describes Sanctorius's anemometer only in a few sentences and does not consider instruments for the measurement of moving water (Middleton 1969: 185, 187). Arthur Frazier's article on Sanctorius's "water current meter" does not discuss the design and functionality of the instrument (Frazier 1969) and is basically reproduced in Frazier's later study on water current meters, which contains some inaccuracies regarding Sanctorius and his works (Frazier 1974: 18–21). Other historical studies on the measurement of moving water ignore Sanctorius's steelyard for the measurement of the *impetus* of water currents completely, see: AWWA Meter Manual 1959, Maffioli 1994, Di Fidio and Gandolfi 2011.

7.1.1   Technical Interpretation of the Steelyards

The illustrations of the two instruments (Figs. 7.1 and 7.2) indicate that Sanctorius used Roman steelyards. Scales of this type were widely in use at the time, especially in a trading hub like Venice, Sanctorius's second home. Merchants and traders used steelyards the size of those depicted by Sanctorius to weigh small items of merchandise in ounces. Thus, it can be assumed that Sanctorius used the steelyards already in circulation for his measurement of the *impetus* of wind and water currents. This is also implicit in his statement that he "provided two balances" (Sect. 7.1). The Roman steelyard consists of a straight beam with arms of unequal length (Fig. 7.3). The beam is suspended from a defined pivot (C), which is flanked by two arms. The longer arm is graduated and incorporates a counterweight (A), which can be moved along the arm to counterbalance the object to be weighed, the load (B), hanging on the short arm. When the two arms are balanced in a horizontal position about the pivot, the weight of the load is indicated by the position of the counterweight on the graduated arm. Thus, the weight can either be read directly from the graduation marks or calculated according to the law of the lever (Robens et al. 2014: 169; Hollerbach 2018: 129).

In order to measure the *impetus* of wind and of water currents with a steelyard, Sanctorius had to adapt its design, as he himself explained in the quoted citation. He added an iron plate to the short arm, in the place where usually the load is positioned, and, depending on what he wanted to measure, placed the plate either below, or above the arm. From the illustrations, it seems that both plates are firmly mounted perpendicular to the beam. Under the influence of air or water flow, the plate is pushed to the side and the pressure thereby generated is transformed into a downward or upward movement, due to the plate affixed to the beam. This movement can be compensated by moving the counterweight until an equilibrium is gained, whereupon the weight can be read in the usual way described above. However, contrary to the weighing of a load, Sanctorius's measurements were complicated by the erratic character of air and water currents. Therefore, the arrangement of the iron plate was crucial, particularly in the case of the anemometer, as wind, even more than water, not only arrives from unforeseeable directions but also in irregular gusts. The rope attached to the long arm of the beam might have had a dual function: to better orient the instrument toward wind direction; and to (generally) enhance stability. Even

Fig. 7.3 Illustration of a
Roman steelyard
(Comstock 1836: 69)

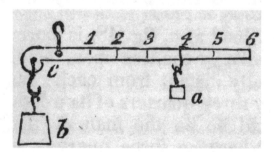

though the illustration does not show any device to determine wind direction, it is possible that Sanctorius used a wind vane for this purpose, as these simple devices had long been known, in his day, and were often attached to church towers in the Middle Ages. But generally, it is quite questionable how Sanctorius conducted a measurement with his anemometer in strong wind given that the latter affected the whole steelyard and not only the iron plate (Middleton 1969: 177, 185).

With regard to the measurement of the *impetus* of water currents, other questions arise.[4] Why did Sanctorius use a grid here, rather than a continuous plate (Fig. 7.2)? To guarantee the comparability of the measurements taken, the grid has always to be immersed in water to the same depth. How did Sanctorius achieve this, especially in strong currents and given the fact that both steelyards were operated by hand? It's easy to imagine how difficult it must have been to keep a steady hand and not inadvertently falsify the measurements, especially when the wind or water currents were strong. Moving the counterweight must have been a challenge, too, and even more so when strong currents of water or air were continuously pushing against the iron plate at the other end of the steelyard.[5] Further investigation is necessary, in order to better understand these difficulties and how they were possibly overcome. In the 1960s, the medical historian Loris Premuda made replicas of the two steelyards, but they are not fully functional, as one can see (Fig. 7.4): both are insufficiently stable for the plate to be mounted above the beam. Since the replicas were made in the context of an exhibition, I assume that they served purely illustrative purposes. New replicas of Sanctorius's two steelyards as well a reenactment of his measuring procedures would be necessary to shed more light on their respective design and use. This, however, lies beyond the scope of the present study.

The initial assessment, here, of Sanctorius's two steelyards implies that the practice of taking measurements was not impossible but certainly, very difficult. Although clearly identifiable graduations on the beam of each instrument suggest that reading and comparing measurements was possible, at least, Sanctorius made no mention of the numerical outcomes of his weighing procedures with the two steelyards. The only indication that the devices were ever put to use is Sanctorius's remark, that he demonstrated how they worked to his friends. Accordingly, there is much to suggest that Sanctorius conducted the weighing procedures with the two steelyards in thought only, and never in deed. The practical difficulties of using Sanctorius's anemometer might also explain why it was neither adopted nor advanced by other scholars and practitioners and has received little attention from historians. In fact, a look into the history of anemometry reveals that Sanctorius is the only scientist ever to have suggested using a steelyard to measure the *impetus* of wind (Sanctorius 1625: 246 f.).

---

[4] For more information on the larger topic of Renaissance hydraulics and the measurement of water flow, see: Maffioli 1994.

[5] I thank Jochen Büttner, Bernadette Lessel, and Markus Hollerbach for their help with the technical interpretation of Sanctorius's two steelyards for the measurement of the *impetus* of wind and of water currents.

Fig. 7.4 Replicas of Sanctorius's steelyards for measuring the *impetus* of wind and water currents. (These replicas were made by Loris Premuda for an exhibition held in 1961 at the University of Padua, where they can still be found today (Biblioteca medica 'Vincenzo Pinali antica' dell'Università degli Studi di Padova, © Philip Scupin))

7.1.2   The Technological Context

The swinging-plate instrument devised by the Italian scholar Leon Battista Alberti (1404–1472) is generally regarded as the first anemometer, followed by the wind plate of Leonardo da Vinci (1452–1519), which was most probably inspired by Alberti's device. Alberti described and illustrated his anemometer in the work *Ex ludis rerum mathematicarum* (On the Pleasures of Mathematics), which was completed sometime between 1450 and 1452. Alberti's anemometer was a little swinging board, directed into the wind by a vane, and equipped with an arc on which its degree of deflection could be read (Fig. 7.5). A sign swinging in the wind, or sheets drying on a clothes line may have given him the idea for the design of his anemometer.

The illustration of Alberti's anemometer shows that it is quite different from Sanctorius's instrument, the only similarity being the plate, whose deflection serves in both devices to indicate the strength of the wind. Without going into a detailed comparison of the two devices, it must be noted that Alberti's instrument was not operated by hand and was therefore not prone to the imprecision caused by irregular movements of the hand and arm. What is more, Alberti proposed that his anemometer be used in the context of sailing, while Sanctorius's device had a clear medical purpose. It is likely that Sanctorius was familiar with Alberti's *Ludi matematici*,

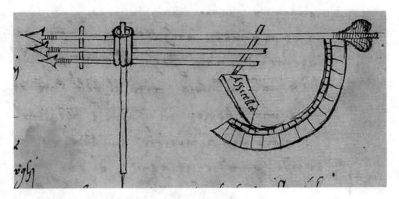

Fig. 7.5 Illustration of the anemometer by Leon Battista Alberti (Wassell 2010: 64)

whose popularity grew after its release in print in 1568. Thus, despite the many differences between the two anemometers, perhaps Alberti's illustration inspired Sanctorius to attach a plate to one end of a steelyard in order to determine the *impetus* of wind. What is more, as mentioned earlier, for the idea of using a pair of scales to measure the strength (*vis*) of wind, Sanctorius could draw on another work—the *De staticis experimentis* of Nicolaus Cusanus (Sect. 5.3.2) (Sanctorius 1625: 246 f.; Middleton 1969: 182 f.; Wassell 2010: 64–77).

A slightly different picture emerges with regard to Sanctorius's instrument for measuring the *impetus* of water currents. Even though I was unable, also in this case, to find any earlier device based on the manual steelyard mechanism described above, the so-called *hydraulic steelyards* presented in the eighteenth-century works of Jacob Leupold (1674–1727) and Francesco Michelotti (1710–1787) do bear similarities to Sanctorius's instrument (Leupold 1724: 150 f., tab. LIX, fig. 1; Michelotti 1771: 116 ff., tavola II). In his article, "Dr. Santorio's Water Current Meter, circa 1610," Arthur Frazier argued that Sanctorius's steelyard had started a vogue, and named further similar devices. However, in the absence of any reference to Sanctorius by the designers of these instruments, it is hard to say whether they knew of, or were influenced by Sanctorius's steelyard. Further research would be required to clarify this issue, for it lies far beyond the period under consideration here (Frazier 1969: 251 ff.).

In Sanctorius's direct context, sixteenth-century Italy, the investigation of moving water, especially the study of rivers and the engineering problems associated with river control, was a matter of deep and widespread concern, and could look back on a long tradition. It responded, both to the preservation of the Venetian lagoon and the very practical issue of flood prevention, especially along the river Reno, in the Bologna region. Finding solutions to such technological problems played a significant role in Renaissance hydraulics, which remained a largely empirical undertaking until the early eighteenth century. People involved in designing and supervising the construction of waterworks were therefore called architects (*architetti*), water experts (*periti delle acque*), foremen (*proti*, in Venice), or simply, engineers (*ingegneri*). One of the most famous representatives of this profession is Leonardo da Vinci, who, among other things, used rod floats to determine the velocity of river currents.[6] It can be assumed that Sanctorius was familiar with the practical hydraulics of the time, living and practicing frequently in Venice as he did. His development of an early form of a water current meter shows his interest in, and receptivity to contemporary practical technologies, which he endeavored to put at the service of medicine.

However, Sanctorius also considered other practical applications for his steelyard. In the *Commentary on Avicenna*, he referred to its great potential for milling (*molendis efficiendis*), which implies that he was thinking of the water milling technology of his day. What is more, as Frazier assumed, the design of his instrument might well have been provoked by his experience of rowing (or of being rowed), or more specifically, by his observation that an oar or paddle in water tends to be pushed backward by the current. Noteworthy, here, is that Sanctorius saw the greater potential for flood prevention, not in his steelyard for measuring the *impetus* of water currents, but in the anemometer. For this instrument could, he believed, determine an incipient increase in the *impetus* of wind and thus easily predict imminent sea storms and high tides (Sanctorius 1625: 247; Frazier 1974: 8, 18; Maffioli 1994: foreword, 6–25).

Another important detail proffered by Sanctorius regarding his two steelyards is that he designed and used them in Croatia. This, together with Sanctorius's reference to sea storms, led Mirko Grmek to conclude that Sanctorius developed the steelyards somewhere between Senj and Trsat, close to the north Adriatic coast. But Sanctorius's reference to Croatia is interesting also for another reason. Sanctorius spent time in Croatia as a practicing physician sometime between his graduation in 1582 and his appointment as professor of *theoria* at the University of Padua in 1611. This was also when he started his static observations of insensible perspiration, for which he used a special weighing chair suspended from one of the beams of a large balance—hence, a steelyard, here, too (Sect. 2.2). Thus, it seems likely that Sanctorius was simultaneously engaged in several studies with steelyards, which may well have been interrelated. And indeed, he connected his use of the steelyards for measuring climatic conditions with the doctrine of the six non-natural things (Sects. 3.1 and 3.3) (Sanctorius 1625: 246; Grmek 1952: 14, 48).

7.1.3   The Dietetic Context: The Six Non-Natural Things

In the *Commentary on Avicenna*, Sanctorius explained that just as the effect of a drug depended always on the complexion of the patient taking it, so, too, the effect of the six non-natural things had to be considered in relation to the human body.

---

[6] For more information on Renaissance hydraulics and the developing "science of waters," see: Maffioli 1994. For Leonardo da Vinci's use of rod floats to measure stream velocities, see: Frazier 1974: 8–11.

Rainy air was harmful to some people, while windy air made others suffer, and others again found both rainy and windy air beneficial, per Sanctorius. Similarly, falling and flowing water with a big *impetus* lulled some people to sleep, while keeping others awake. In a Croatian town with noisy winds and a river with a strong current (*magno impetu*), Sanctorius continued, he had observed that these factors at times hindered the inhabitants' sleep and, at other times, positively fostered it.[7] By means of his two devices, Sanctorius intended to measure variations in the *impetus* of wind and of water currents, which were, according to him, responsible for the various effects that these climatic conditions had on the Croatians' sleep. According to his own testimony, he investigated which *impetus* was healthy and which was harmful and, on this basis, why the larger or smaller *impetus*, or noise was sometimes the cause of health and sometimes the cause of disease (Sanctorius 1625: 246 f.).

It has been mentioned earlier that the non-natural pair air and water was thought to have a considerable impact on health and disease (Sect. 3.3.1). With his two steelyards, Sanctorius attempted to determine this impact quantitatively by measuring the *impetus* of wind and of water currents. According to him, these measurements were a means for the physician to identify the correlation between the external factors of air and water and the well-being of his patient. The two steelyards thus helped the physician make a correct diagnosis and identify general patterns or regularities regarding the effect of the *impetuses* of wind and of water currents on health and disease. In Sanctorius's opinion, such generalization based on repeated measurements enabled one to differentiate between healthy and harmful *impetuses*. Indeed, Sanctorius explained that he most certainly (*certo certiores*) could detect with his anemometer, whether the *impetus* of wind was beginning to increase or to decrease—and so was evidently convinced that his steelyards were capable of measuring such climatic conditions. Furthermore, this statement corresponds with his attempt to enhance certainty in medicine by means of his measuring instruments, as described in the previous chapter. Here again, Sanctorius put forward innovative ideas and integrated them into the traditional framework of Galenic dietetic medicine. Interestingly, in the *Commentary on Avicenna*, Sanctorius related the influence of the *impetus* of wind and of water currents to another non-natural thing, sleep, but remained completely silent on the effects these climatic conditions might have on insensible perspiration. Despite the strong relation of the two steelyards to the six non-natural things and the fact that Sanctorius also used a steelyard to measure insensible perspiration, there is no connection to the *De statica medicina*. Likewise, the static aphorisms bear no trace of the two devices (Sanctorius 1625: 246 f.).

---

[7] Carlo Zammattio suggests that the location to which Sanctorius refers here may have been at the Škocjan Caves (now in Slovenia), around twenty kilometers east of Trieste. There, a river disappears with a strong roar into a huge underground cavern. Moreover, the gale force bora wind sweeps the region (Frazier 1969: 251, 1974: 20).

7.1.4 The Context of Pharmacology

A last remark must be made concerning the embedding of the two steelyards in the pharmacological context. As outlined above, Sanctorius compared the effect of drugs with the effect of the six non-natural things, before launching into a description of his devices (Sect. 7.1.3). Remarkably, before ending that description, he resumed this comparison, asking:

> Wherefore, if the larger or smaller impetus, or noise is at one time a healthy cause and at another an unhealthy cause, how much more must the strengths of the ingested drugs weigh? (Sanctorius 1625: 247).[8]

Thus, Sanctorius seems here to ponder the possibility of measuring the strengths (*vires*) of drugs in relation to their effect on the body. His statement implies that he wondered whether it was feasible to differentiate between healthy and harmful strengths of drugs by means of weighing, in a way similar to that used for the measurement of the *impetuses* of wind and of water currents. This is further indicated by his use of the Latin verb *perpendo*, which he also employed in the description of the two steelyards (Sect. 7.1, fn. 2). But it remains unclear whether Sanctorius really considered it possible to quantitatively determine the degrees of intensity, or strength of Galenic pharmacological theory, described in Sect. 5.2.2. He formulated this idea only as a question and did not further explain how such a measurement or weighing procedure might be conducted. Instead, he resumed his commentary on a passage of Avicenna's *Canon* in a traditional manner, by discussing doubts (*dubitatio*). Even though these discussions concerned the complexion of drugs and their faculties, no further reference was made to quantification. Moreover, as stated earlier, in other passages of his works Sanctorius clearly concluded that it was impossible to know for certain the strength of a remedy (Sect. 6.2.2) (Sanctorius 1625: 247 ff.).

In conclusion, Sanctorius's presentation of the two steelyards to measure climatic conditions shows that he looked beyond the confines of medicine and was attentive to the practical technologies of the time. The Renaissance engineering tradition in Italy was the backdrop against which Sanctorius came up with novel methods to measure the *impetus* of wind and of water currents. Yet, regarding the measurement of the *impetus* of wind, Sanctorius did not use the contemporary technology of Alberti's anemometer, but came up with a different method that does not seem to have been oriented toward practical use. His familiarity with handling a steelyard, gained through the use of his weighing chair to quantify insensible perspiration, probably encouraged him to apply this technology to other areas, too. This notwithstanding and despite the fact that steelyards, anemometers, and instruments to examine water currents did not originate with Sanctorius, his dealings with such devices as a practicing physician were exceptional, as was his later inclusion of them in his university lectures on a traditional textbook, the *Commentary on*

---

[8] "Quare si maior, vel minor impetus, vel strepitus modo est causa salubris, modo insalubris: quanto magis erunt perpendendae vires medicamentorum quae intus sumuntur?" See: Sanctorius 1625: 247.

Avicenna, Sanctorius's strong interest in practical technologies, more specifically, mechanics, was anything but ordinary for a physician. While it is widely known that Sanctorius brought physiology and mechanics together in the *De statica medicina*, his use of steelyards to measure climatic conditions and to detect their influence on health and disease is largely unknown. Despite the unclear relation between his study of insensible perspiration and his examination of the *impetus* of wind and water currents, it is significant that Sanctorius worked with steelyards in both cases, thereby relating the same instrument to very different applications. Moreover, as touched on above, he found a further use for them in meteorological studies, such as weather forecasting. Traditional dietetic medicine, according to which the environment had an important influence on the health and disease of a body, provided the framework in which Sanctorius combined the quantification of meteorological factors with medical diagnosis and treatment. Sanctorius's efforts, albeit most probably not put into practice, illustrate how medicine contributed to the development of meteorology and spurred the use of quantification and measurement methods in this field.[9]

7.2   The *Pulsilogia*

Sanctorius presented an instrument that he described as a *pulsilogium* as early as 1603, in his first publication *Methodi vitandorum errorum*. However, he limited himself in this work to describing the function and purpose of his allegedly newly invented device, offering neither technical details nor an illustration. Nine years later, in the *Commentary on Galen*, he revealed that the instrument relied on the properties of the pendulum; and thirteen years after that, he published illustrations and descriptions of several types of *pulsilogia* in his *Commentary on Avicenna*. In what follows, I will outline the design, functioning, and use of these instruments, and consider the historical context in which they emerged. Since the reception of Sanctorius's *pulsilogia* has been dealt with in the secondary literature, I will refer to this in some detail, too.[10] Moreover, going beyond existing studies, I will provide new reflections on the actual practical application of Sanctorius's *pulsilogia* as well as on the relation between these instruments and the efforts of Nicolaus Cusanus and Girolamo Cardano, who were both engaged in studies of the pulse (Sanctorius 1603: 109r–109v; 1612b: 374; 1625: 21 f., 77 f., 219–22, 346, 364 f.).

[9] For further information on Renaissance meteorology, see: Martin 2011.

[10] In a recent paper, Fabrizio Bigotti and David Taylor have closely analyzed Sanctorius's *pulsilogia* and also considered their reception. Their study is not only based on written documents but also refers to insights that were gained by reconstructing and experimenting with one type of these instruments. My following account of Sanctorius's *pulsilogia* draws largely on this study. See: Bigotti and Taylor 2017.

7.2.1 The Use of the Pendulum: How Did the **Pulsilogia** Measure?

Sanctorius put forward five designs of *pulsilogia*, to all of which he ascribed two uses: to record pulse frequency and to measure time. As the illustrations in the *Commentary on Avicenna* show, these five designs, depending on their form and appearance, fall into three main types: the beam type (Figs. 7.6 and 7.7), the dial type (Figs. 7.8 and 7.9) and the pocket watch type (Fig. 7.10). At least four of the five *pulsilogia* designs are based on the properties of the *pendulum*.

The simplest and, according to Sanctorius, also handiest version consisted of a thread to which a lead ball was attached (Fig. 7.6). The physician used this handheld pendulum by synchronizing the swing of the pendulum with the patient's pulse at two pulse strokes per pendulum cycle. In order to do so, he adjusted the length of the pendulum cord until the swing matched the patient's pulse. The length of the cord was then measured with a measuring rod that was divided into eighty degrees. To make it easier to read the measurement, a vertical white line marked the circumference of the lead ball (letter C in Fig. 7.6). Although Sanctorius described this *pulsilogium* as easy to handle (*paratu facile*), from my perspective, the use of the instrument in medical practice required some dexterity, as the physician had to operate the pendulum with both hands and, at the same time, to determine the pulse of his patient by touch. During this process, the hand holding the pendulum had to be kept as still as possible so as not to falsify the measurements (Sanctorius 1625: 21 f.).

Maybe in response to these difficulties, Sanctorius presented a second, advanced version of the beam type *pulsilogium* (Fig. 7.7). Based on the same principle, the pendulum here was not handheld but attached to a horizontal beam which, in turn, was attached to a wall or a fixed vertical stand in order to guarantee stability. The length of the thread could be adjusted by means of a tapered peg mounted to the bottom right of the instrument. Another difference to the first beam type *pulsilogium* was the scale, which was divided here not into the range zero to eighty degrees, but into seventy unnumbered parts or degrees. Fixed to the thread over the scale was a knot or a little wooden bead (letter O in Fig. 7.7), which indicated the degree measured. Based on the empirical testing that Fabrizio Bigotti and David Taylor undertook with their replica of the device, made in the framework of their recent study of Sanctorius's *pulsilogia*, they argued that the beam was actually angled horizontally and not vertically as in Sanctorius's illustration.[11] This means that the broad face of the beam was laid flat with the scale uppermost. The contemporary depiction of a similar device in the frontispiece of the book *De proportione motuum* (On the Proportion of Motions, 1639) by the physician Jan Marek Marci (1595–1667) further corroborates this assumption. It shows a portable version of the *pulsilogium* with the beam angled horizontally (Fig. 7.11). Following this line of argument,

---

[11] For more information on the technical and empirical factors that led Bigotti and Taylor to assume that the beam of this type of *pulsilogium* was horizontally angled, see: ibid.: 78 ff.

Fig. 7.6 Simple beam type
pulsilogium (Sanctorius
1625: 22). (© British
Library
Board 542.h.11, 22.)

Fig. 7.7 Advanced beam
type *pulsilogium*
(Sanctorius 1625: 77 f.).
(© British Library Board
542.h.11, 77 f.)

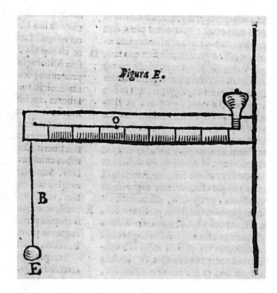

Fig. 7.8 First version of a
dial type *pulsilogium*
(Sanctorius 1625: 220).
(© British Library Board
542.h.11, 220)

Sanctorius presented the instrument in perspective in order to show the reader its
overall function. However, since the device was mounted to a wall, it is also conceivable that the physician used it horizontally when adjusting the length of the
pendulum cord and inclined it vertically afterwards, to facilitate reading the measurement while simultaneously taking his patient's pulse (Sanctorius 1625: 78;
Marci 1639: frontispiece; Bigotti and Taylor 2017: 70–82).

Fig. 7.9 Second version of
a dial type *pulsilogium*
(Sanctorius 1625: 364).
(© British Library Board
542.h.11, 364)

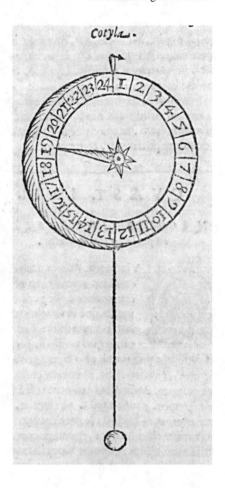

Hence, the design of the second beam type of *pulsilogium* improved the reliability of measurements by avoiding interferences that occurred in the first type due to
its manual operation. Moreover, it enabled the physician to adjust the swing rate and
to read the cord length while the pendulum was still in motion. Therefore, compared
to the simple handheld *pulsilogium*, it provided more reliable measurements and its
use must have been more convenient in medical practice.

With regard to the dial type of *pulsilogia*, the illustrations in the *Commentary on
Avicenna* suggest that they were likewise based on the use of a pendulum (Figs. 7.8
and 7.9). It seems that the pendulum cord could be wound around a pivot at the back
of the dial in order to adjust its length. If, as one may assume, this pivot and the hand
on the front of the device were connected, then winding the cord would move the
hand and so indicate on the dial the degree measured (Fig. 7.12). The number of
degrees into which the dial was divided differed in the two instruments, being
twelve in the one, and twenty-four in the other. Interestingly, the *pulsilogium* with
the twelve-degree dial seems to have not only a moving hand, but also a moveable
dial, as the latter is shown rotated clockwise in the illustration (Fig. 7.8). What is
more, Sanctorius occasionally described both devices as *cotyla*, which could be

Fig. 7.10 Pocket watch
type *pulsilogium*
(Sanctorius 1625: 78, 220,
346). (© British Library
Board 542.h.11, 78)

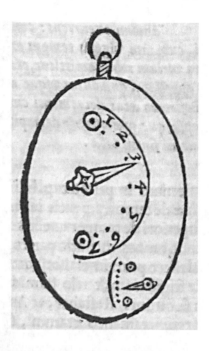

Fig. 7.11 Detail of the
frontispiece to Jan Marek
Marci's work *De
proportione motuum*
displaying a *pulsilogium*
(Marci 1639: frontispiece).
(Courtesy of
Niedersächsische
Staats- und
Universitätsbibliothek
Göttingen (SUB
Göttingen))

translated as "concave bowl." Against this background, the sketches respectively underneath and behind the dial can be interpreted as some kind of bulky boxes (Sanctorius 1625: 219–22, 364 f.).

Fig. 7.12 Replica of
Sanctorius's second
version of a dial type of
pulsilogium. (The replica
was made by Loris
Premuda for an exhibition
held in 1961 at the
University of Padua, where
it can still be found today.
The replica's winding
mechanism of is imperfect,
since it does not operate
smoothly and the
connected hand moves
through all the degrees on
the dial after winding up
only a very small length of
the cord (Biblioteca
medica 'Vincenzo Pinali
antica' dell'Università
degli Studi di Padova, ©
Philip Scupin))

In his descriptions of the dial *pulsilogia*, Sanctorius indicated that the instruments measured both time and pulse frequency. It must therefore be assumed that they provided comparisons of degrees. In view of this, the bulky boxes as well as the movable dial might have been part of a special mechanism that allowed the two values to be registered simultaneously—a hypothesis that must, however, be further investigated (Sanctorius 1625: 222).

In this context, it is important to note that Sanctorius intended both devices to measure, besides pulse frequency, also and particularly the respiration cycle. In doing so, he tried to evaluate the difference between the "diastolic" and "systolic" pulses. Whilst this distinction seems somewhat counterintuitive today, since the focus now is on examining the succession of pulse beats, in Galenic medicine the pauses between single pulse beats were thought to be important, too. These pauses were conceived of as the phases of arterial contraction and described as "systolic" pulses, whereas the "diastolic" pulses referred to arterial expansion—hence, the converse of modern terminology. Within a complicated body of theory, the pauses revealed qualitative features such as the pulse's "width" or "breadth," thereby alluding to the dimensions of the arteries. Since, according to Sanctorius, the systolic pulses, i.e., the pauses, were "not made apparent by touching the pulse with our fingers" (Sanctorius 1625: 364), he came up with another method to detect them: via respiration.[12] As explained in Sect. 3.2.6, in Galenic medicine, inspiration corresponded to diastole and expiration to systole. Accordingly, Sanctorius held that if expiration was faster than inspiration, the systolic pulse would be faster than the diastolic pulse, too. Likewise, faster inspiration indicated a faster diastolic pulse. In order to measure the duration of inspiration and expiration, respectively, Sanctorius synchronized the swing of the pendulum with the cycle of respiration. But confusingly, he detected the latter by "putting the hand over the heart" of the patient (Sanctorius 1625: 364).[13] This implies that he actually measured the heart and not the respiration cycle. Most probably, this differentiation was not important, to him, since the medical theory of his day held that these two processes were coincident. However, it did have major implications for his measurements (Sanctorius 1625: 364 f.; 1630: 594; Bedford 1951: 427; Bacalexi and Katouzian-Safadi 2019: 3).

As is known today, diastole and systole are of very brief duration, less than one second. Hence, it is not at all clear how Sanctorius managed to synchronize the swing of a pendulum with these processes. In a general sense, it was certainly possible to determine whether the duration of systole was shorter than the duration of diastole. But measuring the frequency of diastole and systole along a scale of twelve or even twenty-four degrees is questionable, at best. Moreover, it is difficult to understand why the systolic pulses apparently could not be identified by touching the wrist of a patient, but were detectible by feeling the beat of his heart.

Adding to the curiosity, the only measuring result that Sanctorius mentioned in this context was that he usually detected two or three pulses between inspiration and expiration. Hence, he observed here the quantitative relation between pulse and respiration without differentiating between diastolic and systolic pulses. Furthermore, the number of pulse beats to which he referred could not be measured when working with the heart cycle as an indicator of the respiration cycle, due to the problems outlined above. Perhaps his statement that he "put the hand over the heart" has to be interpreted differently. Sanctorius might have simply put his hand on his patient's chest, probably close to the heart, to determine its movement during respiration. In this manner, it would be possible to differentiate the frequency of inspiration and expiration according to different degrees, since respiration is much slower than the processes of systole and diastole. Yet, there remains the problem of how Sanctorius differentiated between diastolic and systolic pulses, since this would have required him to somehow simultaneously account for the pulse beats occurring within the time span of inspiration and expiration, respectively. Moreover, his explicit reference to the heart does not make much sense, if he in fact measured the movement of the chest. Given that Sanctorius did not provide any further details of

---

[12] "… systole digitis nostris pulsus tangentibus non occurrit …." See: Sanctorius 1625: 364. The English translation is taken from: Bigotti and Taylor 2017: 95.

[13] "… manu ad cor admota, …." See: Sanctorius 1625: 364. The English translation is taken from: Bigotti and Taylor 2017: 95.

these procedures, his alleged measurement of diastolic and systolic pulses via respiration leaves many questions open (Sanctorius 1625: 364).[14]

Still, the analysis of the dial *pulsilogia* did serve to reveal an important dimension of Sanctorius's *pulsilogia*: their close integration into Galenic pulse lore.[15] Indeed, since taking the pulse was, along with uroscopy, the physician's main diagnostic tool at the time, every physician was familiar, at least to some degree, with the pulse teachings of Galen. But even university-trained physicians mentioned that they struggled to understand the complexities of the Galenic ideas, according to which the pulse had many variations in almost innumerable combinations, each of either diagnostic or prognostic significance. Moreover, they doubted whether the Galenic doctrines could be implemented in medical practice, discussing, for example, whether analysis of the pauses between pulse beats, i.e., systolic pulses, was possible in practice. Hence, these contemporary issues seem to reflect, in some way, the difficulties one encounters when trying to interpret Sanctorius's dial *pulsilogia* today. In any case, it is within the intricacies of Galenic pulse lore that Sanctorius's *pulsilogia* have to be seen (Horine 1941: 219; Siraisi 1990: 58–127).

The fifth *pulsilogium*, classified as a pocket watch type, is the one that raises the most questions (Fig. 7.10). In his descriptions of the device, Sanctorius ascribed to it the same function as to the other *pulsilogia*, namely to measure pulse frequency and time. He especially used this type as a timekeeper, during the observations that he made with his thermoscope (Sects. 7.3.2 and 7.3.3). However, the illustrations of the *pulsilogium* do not show a pendulum and Sanctorius never wrote a word about how the device worked. Hence, it is unclear how he took measurements and how these related to the scale, which is arranged in this case in two semicircles. Each of the semicircles is divided into seven parts that represented, so Sanctorius, seven divisions (*differentiae*) and seven subdivisions (*minuta*) (Sanctorius 1625: 77 f., 219–22, 346).

The preceding paragraphs have demonstrated how Sanctorius based most of his different types of *pulsilogia* on the swing of a pendulum. This testifies that Sanctorius, like many others at the time, was familiar with this phenomenon and understood its most fundamental property, the production of equal intervals of time.[16] But given that he provided no mathematical details of his grasp of the properties of a pendulum and limited himself to rather general statements, it is difficult to assess the mechanical reasoning underpinning his *pulsilogia*. An attempt to do so has been made by Bigotti and Taylor, but shall not be discussed here, since such an

---

[14] A reconstruction of the dial *pulsilogia* and their use could help further clarify how Sanctorius might have measured diastolic and systolic pulses via respiration. This, however, lies beyond the scope of the present work.

[15] I use the term "pulse lore" to refer to the study and examination of the pulse, i.e., to the theories and practices connected with taking a person's pulse.

[16] With regard to pendulum motion, Jochen Büttner has aptly summarized: "A characteristic property of pendulum motion is its period, that is the time it takes the pendulum to complete one full oscillation. The assumption that this period does not depend on the initial displacement has become known as the 'isochronism' of the pendulum. The 'isochronism' of the pendulum holds, according to classical mechanics, only approximately. The full solution of the equation of motion of a pendulum, which requires the use of elliptic integrals, shows that the period does indeed depend on the displacement of the pendulum" (Büttner 2008: p. 227, fn. 11).

analysis lies beyond the scope of the present work.[17] Instead, I will focus in the following on the broader context in which Sanctorius's pendulum-based *pulsilogia* emerged and consider its possible influence on Sanctorius's undertakings.

7.2.2   The Pulsilogia *in Context*

Long before Sanctorius, scholars such as Nicole Oresme (1320–1382), Giovanni Marliani (1420–1483), Leonardo da Vinci, and Girolamo Cardano referred to observations made with the pendulum.[18] In Sanctorius's times, pendulums became a part of contemporary technology and were built into various machines serving different functions. There is even evidence that they were used as timekeepers in clocks as early as the sixteenth century. What is more, a whole group of intellectuals, including, for example, Isaac Beeckman (1588–1637), Niccolò Cabeo (1586–1650), and Marin Mersenne (1588–1648), tried to integrate the pendulum into their mechanical theories.[19] Hence, theoretical reflection on the properties of pendulum motion and the practical applications of pendulums occurred at the time when Sanctorius put forward his *pulsilogia*. Remarkably, two figures with whom Sanctorius was well acquainted also dealt with the issue: Paolo Sarpi and Galileo Galilei. Without going into analyses of their respective studies of the pendulum conducted elsewhere, it is enough to note that, once again, Sanctorius's network of friends in Venice, the *Ridotto Morosini*, was an important focal point, where topics of current scholarly interest were discussed.[20] Sarpi and Galileo both frequented the meetings in the house on the Grand Canal and it is therefore most certain that Sanctorius discussed and observed the phenomenon of the pendulum with the two scholars, the former,

---

[17] Bigotti and Taylor have argued that Sanctorius's theoretical mechanical explanation for the *pulsilogium* drew on an understanding of the Renaissance controversy on equilibrium, see: Bigotti and Taylor 2017: 60–3. For more information on the so-called *equilibrium controversy*, see: Renn and Damerow 2012.

[18] Marliani 1482: 4r, Cardano 1550: 50r–51r, Oresme et al. 1968: I.18, 30a–b, Da Vinci et al. 2018: 383–7, 515 ff.

[19] Illustrations in, for example, the work *Theatrum instrumentorum et machinarum* (1569) by the French engineer Jacques Besson (ca. 1540–1576), or in the work *Machinae novae* (1615) by Fausto Veranzio (1551–1617) show that pendulums were used as parts of different machines in the early modern period. See: Büttner 2008: 228. For more information on the use of pendulums as timekeepers in clocks in the sixteenth century, see: ibid.: 228, fn. 15. For a cursory account of Beeckman's attempt to integrate the phenomenon of the pendulum into his mechanical theories, see: ibid.: 232–5. Marin Mersenne corresponded, for example, with René Descartes (1596–1650) on questions regarding pendulum motion, see, e.g., letters written on October 8, November 13, and December 18, 1629 in: Mersenne et al. 1932–1988. For Niccolò Cabeo, see: Cabeo 1646: 93, 98 f.

[20] For more information on the roles that Paolo Sarpi and Galileo Galilei played in the invention of the *pulsilogium*, see: Bigotti and Taylor 2017: 56 ff. For an account of Galileo's studies on the pendulum, see e.g., Büttner 2019 and for Sarpi, see: Sarpi and Cozzi 1996: 111, 408 ff. There are also studies on the relation between Sanctorius and Galileo, in which the inventions of the *pulsilogium* and the thermometer have been of particular interest, see: Bizzarrini 1947, Grmek 1967, Ongaro 2009.

moreover, being his close friend. Yet, recent historical research suggests that it was Sanctorius who first applied the pendulum as a timing device to medicine and that his *pulsilogia* did not result from Galileo's studies of the pendulum, but were rather a source of inspiration for Galileo. Notwithstanding that physicians like Cardano had been interested in the pendulum before Sanctorius, they did not consider its application in a medical context (Büttner 2008: 227–32; 2019: esp. 91 f., fn. 32).

Just as the phenomenon of the pendulum was a topic of great contemporary interest, the counting of the pulse was also practiced at the time. Since the clocks that were then available did not allow brief intervals of time to be measured with any precision, scholars, especially astronomers, used the pulse for this purpose. The pulse beat was a tangible parameter, and hence a suitable measure able to be counted within the longer periods of time that could already be determined rather accurately by clocks, i.e., the period of one hour. In his work *De proportionibus* (On Proportions, 1570), Girolamo Cardano, for example, illustrated very fast movements in the heavens, like those of the moon, by converting the incredibly wide distance covered in one hour into the distance covered during one pulse beat. In this context, he tried to determine the number of pulse beats per hour and came to the fairly accurate number of four thousand pulse beats, which corresponds to sixty-seven beats per minute. Later, in 1618, Johannes Kepler counted the pulse in relation to minutes and assessed that the pulse of a healthy man at rest corresponded to an average of seventy beats per minute. Accordingly, his count could provide a rather reliable indication of the time elapsed in any given observation. However, these attempts did not aim to measure the pulse per se. Rather, they were informed by a general interest in the relation between the human pulse and time, or by the effort to improve the precision of the pulse as a timekeeper (Cardano 1570: 50; Kepler 1618: 278 f.).

But, besides these, there was also an effort to measure the pulse frequency related to medical practice. As was discussed in Sect. 5.3.2, in the fifteenth century, Nicolaus Cusanus already suggested using a water-clock to compare the pulse of different people, in health and in disease. This would help the physician, so Cusanus, in diagnosis, prognosis, and therapy. From the evidence at hand, it is highly probable that Sanctorius knew of his work and was inspired by it to pursue his quantitative approach to medicine. Thus, there is good reason to assume that Sanctorius took from Cusanus's work *De staticis experimentis* the idea of measuring the pulse with an instrument that could record equal intervals of time. As likewise mentioned, Cusanus also put forward the idea of measuring respiration by the same method, based on the water-clock. Interestingly, a good hundred years later, Cardano examined the quantitative relation between pulse and respiration. Moreover, contrary to his count of pulse beats per hour, he did so in a medical context, in his commentary on the Hippocratic treatise *Nutriment* (*Commentaria in librum Hippocratis de alimento*, 1574). He concluded that, independent of age and complexion, this relation would always be 3:1. Given that Cardano did not explain how he arrived at this ratio, whether he used an instrument to this end or not, it is difficult to assess the relation of his observation to Cusanus's proposed measurement of respiration using the water-clock. It is known that Cardano was familiar with the mathematical thoughts of Cusanus, but due to the fact that Cusanus, unlike Cardano, did not consider the relation between respiration and pulse, but suggested measuring both parameters with a timekeeping instrument, it is doubtful whether the two undertakings were in any way related. Yet, it seems significant that Sanctorius knew both authors and presented a method of his own to measure the respiration cycle in comparison with the pulse. By means of his dial *pulsilogia*, he allegedly observed that there were usually two or three pulses between inspiration and expiration. While it is most certain that Cusanus's *De staticis experimentis* stimulated Sanctorius in his measurement of respiration, it remains unclear whether Sanctorius was aware of Cardano's quantification of the relation between pulse and respiration. Notwithstanding that Sanctorius frequently mentioned him in his commentaries, I could not find any reference to Cardano's commentary on *Nutriment*. At any rate, at least in hindsight, Sanctorius's solution appears to be a combination of Cusanus's and Cardano's efforts (Cardano 1574: 230v; Sanctorius 1625: 364; Kümmel 1974: 4–12).

In summary, it can be said that the phenomenon of the pendulum, occasionally already applied as a timekeeper in clocks, was of interest to scholars, practitioners, and engineers both before and contemporary to Sanctorius. It was a part of contemporary technology as well as of intellectual reflection and discourse. Most likely, it was among the subjects that Sanctorius discussed with people like Sarpi and Galileo in the *Ridotto Morosini*. In a similar manner, the counting of the pulse was a current means to measure time, especially in astronomy. What is more, the importance of assessing the frequency of the pulse in a medical context had been recognized long before Sanctorius by Cusanus, who had suggested that respiration be measured, too. In the sixteenth century, Girolamo Cardano not only studied the motion of the pendulum, but also counted the pulse and compared the frequency of pulse with the frequency of respiration. However, it was Sanctorius who brought these different strands of contemporary interest and investigation together by devising a series of instruments called *pulsilogia*. Most importantly, he put his instruments at the service of medical practice and was thereby the first to *apply* the pendulum and the measurement of the frequency of the pulse to medical diagnosis, prognosis, and therapy.

The Reception of the *Pulsilogia* Following Sanctorius's description of the *pulsilogium* in the *Methodi vitandorum errorum*, many other physicians and scholars remarked on the device. And in fact, someone else had already announced it in writing, a year before Sanctorius first did. This was Eustachio Rudio (1548–1612), professor of practical medicine at the University of Padua and a member of the College of Physicians of Venice, who died shortly after Sanctorius, too, entered these two institutions (Facciolati 1757: 332 f.; BNMVe n.d.: f. 23r). Rudio wrote in a treatise on the pulse (*De pulsibus libri duo*, 1602):

> I just want you to know that in our age an instrument, which it is possible to call a *pulsilogium*, has been invented in order to discern the quickness and slowness of the pulse. Its author is Sanctorius Sanctorius, a physician, a philosopher, and a man provided with all kinds of erudition (Rudio 1602: 23v).[21]

---

[21] "Sed pro crebritate & raritate dignoscenda unum volo vos admonere, hac scilicet nostra tempestate quoddam instrumentum, quod pulsilogium vocari potest, fuisse excogitatum à Sanctorio Sanctorio Medico & Philosopho, & omni eruditionis genere praestantissimo, …." See: Rudio 1602: 23v. For the English translation, see: Bigotti and Taylor 2017: 58.

Hence, Sanctorius must have already shown the *pulsilogium* to his friends and colleagues in Venice around 1600. In a collection of opinions on medical and philosophical problems published in 1611, the Venetian physician Antonio Fabri (life dates unknown) stated that he had had the opportunity to participate in a demonstration of the *pulsilogium* by Sanctorius. Johannes Ravius (1578–1621), a physician from the German town Rinteln (today, in Lower Saxony), reported from a visit to Padua in 1618 that Sanctorius's instruments were especially remarkable, among them, a *pulsilogium*. Three years later, in 1621, another German physician, from Rostock, Peter Lauremberg (1589–1635), claimed to have replicated and successfully applied the *pulsilogium* to examine the usually imperceptible differences in the pulse rate. Lauremberg's account is interesting, since at this time Sanctorius had not yet published his *Commentary on Avicenna*, which contained the illustrations of his *pulsilogia*. Consequently, Lauremberg could not rely on any printed depiction of the *pulsilogium* for the design of his replica. It seems that he was not in direct contact with Sanctorius either, as he explained that he had heard from others that Sanctorius was the inventor of such an instrument (*"qualia à Sanctorio excogitata accepimus"*). This implies that he had to rely on oral accounts or manuscript sheets describing the instrument, and also supports Sanctorius's complaint of 1625, that his instruments were known to, and copied by his disciples across Europe. However, given that Lauremberg neither published an illustration of his version of the *pulsilogium* nor gave any details of its design or use, it remains uncertain whether he really did devise and deploy such a device. A few years later, according to his own testimony, Isaac Beeckman took inspiration from Sanctorius's *pulsilogia* for his observations on vibrating chords (Bartholin 1611: Exercitatio Nona, Problema VIII; Johannes Ravius to Ernst Schaumburg-Holstein 1618; Lauremberg 1621: 28 f.; Beeckman and de Waard 1945: 174 f.).

The list of references to Sanctorius's *pulsilogia* in the first half of the seventeenth century and beyond could be extended much further.[22] However, the few names cited should suffice to show that Sanctorius's *pulsilogia* were well known among physicians and scholars in Europe and, probably, also copied. As stated above, Marek Marci put forward his own *pulsilogium* based on the properties of the pendulum (Sect. 7.2.1). The same is true of Athanasius Kircher (1602–1680), but neither scholar alluded to Sanctorius. Whether Marci and Kircher had direct knowledge of Sanctorius's devices or not, their instruments further attest the spread of *pulsilogia* in the seventeenth century. It seems therefore that Sanctorius, in inventing the *pulsilogium*, had put a finger on the pulse of his era—if you will excuse the pun. The contemporary interest in, and preoccupation with the pendulum phenomenon, combined with the concern for timekeeping methods, including the counting

---

[22] Further examples for references to Sanctorius's *pulsilogia* are Malvicini 1682: 213, Schwenter 1636: 415 f. While Giulio Malvicini was a student of Sanctorius in Padua and therefore probably saw the instruments in Sanctorius's university courses or private lessons, the German scholar Daniel Schwenter heard about the device from a physician (*doctore medicinae*) and erroneously assumed that Sanctorius lived and practiced in Paris ("Santes Sanctorius ein sehr berühmter Medicus zu Paris hat ein Instrumentum von ihme Sphigmaticum genennet erfunden: ....").

of the pulse, can certainly explain the immediate and enthusiastic reception and broad dissemination of Sanctorius's *pulsilogia*. What is more, the utmost relevance of pulse lore as one of the physician's main diagnostic tools probably further fueled interest in a device that heralded a marked improvement in his daily practice of "taking the pulse" of his patients (Marci 1639: Propositio XXXXI, Problema II; Kircher 1665: 51 f.).

Establishing the extent to which *pulsilogia* instruments actually entered into daily medical practice would require further research, however. Around 1714, Giovanni Battista Morgagni mentioned Sanctorius's *pulsilogium* in his university lectures in Padua on Galenic pulse lore, but it is clear from his words that it had not yet become a standard tool for physicians and that even if Morgagni himself used such a device, it played no major role in his medical practice.[23] Interestingly, it is not clear even how much Sanctorius used his *pulsilogia* in his daily practice. He did repeatedly use the plural when referring in his published works to the subjects of his measurements (*sani/aegri homines*), and stated that he was able to distinguish 133 variations in the frequency of the pulse. This implies that he made considerable use of the *pulsilogia*. However, Sanctorius did not further specify the 133 distinctions; how exactly he determined them as well as how they relate to the scales on the *pulsilogia* remains unclear. Furthermore, the ratio that he put forward regarding respiration and pulse is the only numerical outcome of his procedures with the *pulsilogia* that he mentioned. In addition, the technical interpretation of the dial type of *pulsilogia* casts some doubt on their practicability and usability (Sect. 7.2.1). Still, it should not be overlooked, here, that physicians at the time were very experienced in the practical challenges of determining a patient's pulse, and this assured at least some of them a considerable sensitivity also to minute variations in pulse. Thus, handling Sanctorius's *pulsilogia* might have been much easier for contemporary practitioners than one can imagine today. In fact, the experiments made by Bigotti and Taylor with their replica of the second, more advanced, beam type of *pulsilogium*, showed that at least this type of instrument can be conveniently used even by non-physicians and provides very reliable measurements. Last but not least, Sanctorius often directed his readers to his planned but never published book *De instrumentis medicis* for more information on his *pulsilogia*, which might explain the lack of data on the quantitative outcomes of his measurements in his other works. In any case, the French physician François Boissier de Sauvages de Lacroix (1706–1767) explained as late as 1752 that he worked with a Sanctorian *pulsilogium* in order to measure the pulse of his patients, and he recommended its use (Sanctorius 1603: 109r–109v; 1612b: 229 f.; 1625: 77 f., 364 f.; Boissier de Sauvages de Lacroix 1752: 30 f.; Morgagni et al. 1961: 64, 70).

From a preliminary perspective, it does not seem that the measurement of pulse frequency became a major component of pulse theory or of the practice of taking the

---

[23] Morgagni stated with regard to Sanctorius's *pulsilogium*: "Qua in differentia ut omnes mutationes quae possunt contingere certò diagnoscamus, excogitavit Sanctorius Instrumentum Pulsilogium ipsi vocatum, cuius descriptionem, et imaginem dedit in Comment. Fen p. 29 et 310." See: Morgagni et al. 1961: 64.

pulse subsequent to Sanctorius's *pulsilogia*. This notwithstanding, his instruments immediately attracted considerable attention among physicians and were well known across Europe in the seventeenth century. Remarkably, François Boissier de Sauvages de Lacroix still considered a Sanctorian *pulsilogium* a useful device for measuring a patient's pulse more than a century after its invention. Hence, the question of the practical medical application and usability of Sanctorius's own *pulsilogia* as well as of the replicas made by others remains certainly a difficult one. Without attempting to answer it, here, a closer examination of the purposes that Sanctorius ascribed to the instruments will shed more light on his use of them, and provide some possible explanations for their immediate and long-lasting popularity.

7.2.3   The Purpose of the Instruments: What Did the Pulsilogia Measure?

As explained above, Sanctorius applied various scales to his *pulsilogia* which differ in terms not only of form, i.e., beam, dial, or semicircle, but also of division. They range from eighty, seventy, twenty-four, or twelve to seven degrees, each degree of which, in the latter case, is further subdivided into seven degrees.[24] Still, they have one thing in common: they are all linear. Consequently, Sanctorius's *pulsilogia* did not permit a direct reading of pulse frequency and provided only a relative measurement. In fact, Sanctorius intended them exactly for this purpose. According to him, the *pulsilogia* should be used as comparators. In the *Methodi vitandorum errorum*, he highlighted that:

> we should know ... how exactly the pulse of the previous attack [of disease] compares with the present pulse. For only from this comparison can we obtain a certain and infallible judgement on whether the patient is in a better or worse condition. ... I invented "a device that measures the pulse" [*pulsilogium*], by means of which everyone can precisely measure the movement and the rest of arteries, observe and firmly remember, and subsequently make a comparison with the pulses of the previous days (Sanctorius 1603: 109r).[25]

Thus, the *pulsilogia* enabled the physician to take repeated measurements of the pulse of his patients, in health and in disease, which he could remember and directly compare with each other. Comparison was central to this process whereas the measurement of absolute values of the pulse rate was irrelevant. The instruments showed

---

[24]The various scales that Sanctorius applied to his *pulsilogia* are difficult to interpret and most probably represented different measurement resolutions. For more information, see: Bigotti and Taylor 2017: 72–6.

[25]"... sciamus exactè conferre pulsus praeteritarum accessionum cum pulsu praesentis; quoniam solum ex hac collatione certum & infallibile iudicium colligemus, an aeger sit in meliori, vel deteriori statu; ... instrumentum pulsilogium invenimus in quo motus, & quietes arteriae quisque poterit exactissime dimetiri, observare, & firma memoria tenere, & inde collationem facere cum pulsibus praeteritarum dierum; ...." See: Sanctorius 1603: 109r. The English translation is based on: Bigotti and Taylor 2017: 87 f.

the variation of the pulse through time and allowed health trends in patients to be monitored. In this context, the focus was on the small increases or decreases in the pulse frequency, so Sanctorius, since, without a *pulsilogium*, these were very difficult, if not impossible, to perceive, even by well-trained physicians. Accordingly, his instrument was meant to overcome the limits of the physician's senses and to allow him to determine what would otherwise remain obscure and unknown (Sanctorius 1625: 21 f.; 1629: 135 f.).

It is pertinent here to again mention the similarities between Sanctorius's use of his *pulsilogia* and Cusanus's proposed measurement of the pulse by means of a water-clock. Like Sanctorius, Cusanus suggested a comparative measurement of the pulse that would result in relative values for the pulse frequency. While Sanctorius used the length of a pendulum cord as the reference parameter, defining it by means of different scales, Cusanus suggested measuring the weight of the water which fell from a water-clock during the time of one hundred pulse beats (Sect. 5.3.2). He probably thought that this would make the measurements easier to communicate and to compare, given the lack of standard units of measurement. However, as standardized water-clocks didn't exist either, at the time, the reliability of his measurements would still have depended on using absolutely identical devices. Be this as it may, the fact that Sanctorius used his *pulsilogia* as comparators, a method that had been already suggested by Cusanus in his *De staticis experimentis*, further implies that Sanctorius was inspired by the latter work (Kümmel 1974: 3).[26]

In keeping with Sanctorius's effort to determine the quantity of diseases by means of his four measuring instruments, Sanctorius's *pulsilogia* was meant to record, as it were, the "latitude of the pulse" and thus help determine the deviation of a body from its natural, healthy state (Sect. 6.1.1). But the frequency of the pulse was only one of several parameters that indicated whether a pulse was healthy or not. In fact, according to Sanctorius, frequency *per se* was equivocal, since the frequencies of a healthy pulse might include frequencies that could also be found in morbid states. Therefore, it was necessary first to assess the general condition of the patient, so that changes in frequency could be associated, for example, with diseases such as fevers. What is more, the physician had to determine how the pulse of his patient fitted within the Galenic classification of pulses. As mentioned earlier, this was a complicated body of theory, according to which variations of the pulse were broken down into several different components, which included, besides frequency, the dimensions of the arteries and the strength of pulse beats. A good illustration of the way in which Sanctorius used his *pulsilogia* within this context of standard Galenic medicine is his repeated emphasis on the possibility of distinguishing, with his *pulsilogia*, between the *pulsus invalidus* and the *pulsus humilis*, two types of pulses specified by Galenic pulse lore. He explains:

---

[26] It is interesting to note in this context that the French physicist Guillaume Amontons (1663–1705) tried to substitute Sanctorius's *pulsilogium* with a water-clock in 1695. Most probably he was ignorant of Cusanus's work *De staticis experimentis*, as he made no reference to it. See: Amontons 1695: Avertissement, Bigotti and Taylor 2017: 69.

> if the pulse that was previously strong and frequent decreases in strength and frequency it
> will be called *humilis*, whereas it will be called *invalidus* when it mostly does not present
> such a condition, that is, of becoming quieter: if the difference between the major or minor
> frequency is very small, physicians cannot distinguish it without the *pulsilogium* (Sanctorius
> 1629: 137).[27]

Thus, it was important for the physician to detect not only the frequency of the pulse, but also its strength, in order to decide whether the patient had a *pulsus humilis* or *invalidus*. Only with regard to frequency could the *pulsilogium* provide reliable measurements, due to its ability to detect also minute variations. This illustrates again that Sanctorius's *pulsilogia* were strongly integrated into Galenic medicine. Against this backdrop, they served to ascertain one of several variables that occurred in the pulse over time: its frequency. For the other variables, the physician still had to rely on traditional qualitative assessments of the pulse. Given the persistence of Galenic pulse theory—Galenic observations of the pulse being still included in standard sphygmology textbooks in the late nineteenth century (Nutton 2019: 472 f.)—I consider it plausible that it was precisely its strong roots in Galenic pulse lore that guaranteed the enduring popularity of Sanctorius's *pulsilogia*—a hypothesis that does, however, need further investigation.[28]

To conclude, Sanctorius's *pulsilogia* allowed physicians to collect, record, and compare the frequency of the pulse of their patients. By this means, not only health trends in individual patients could be detected, but general ranges of healthy or morbid pulses could be determined, provided that enough measurements were taken in healthy and sick people. In this connection, comparisons would be made between the data measured not only of one patient, but of several. However, before using the instrument, the general condition of the patient had to be assessed and the measurements always needed to be related to Galenic pulse lore with its classification of different pulse species and its qualitative methods of determining the pulse. Still, within the intricacies of pulse lore, the measurements with the *pulsilogia* could provide reliable reference points, permitting the "latitude of the pulse" to be shown, i.e., its variation in healthy and unhealthy conditions, expressed comparatively in degrees. Since Sanctorius published neither records nor results of his measurements, it remains unclear how often he used the device and on how many different people. Notably, the *pulsilogium* was the first and only measuring instrument whose description Sanctorius published as early as 1603. From his written records it can be assumed that he had already been engaged for several years in his observations of insensible perspiration with the weighing chair, by this time, and it is also probable that he conducted the studies on his two steelyards to measure climatic conditions before 1603 (Sects. 2.2 and 6.1.2). This notwithstanding, it was only the *pulsilogium* that Sanctorius mentioned in the *Methodi vitandorum errorum*, without

---

[27] "… si pulsus, qui antea fuit vehemens & frequens remittat vehementiam & frequentiam, dicitur humilis: invalidus verò ut plurimum caret hac conditione, quod scilicet fiat quietior: haec maior vel minor frequentia si perexigua sit, à medicis sine pulsilogio dignosci non potest." See: Sanctorius 1629: 137. The English translation is taken from: Bigotti and Taylor 2017: 96 f.

[28] On the long-lasting relevance of the Galenic doctrine of the pulse, see: Tassinari 2019: 514 f.

reference to any other devices. Moreover, nowhere did he relate the use of the *pul-silogia* to the six non-natural things or to insensible perspiration, despite their evident strong connection, as illustrated, for example, by the fact that certain emotions or forms of exercise accelerate the pulse. It seems, therefore, that Sanctorius did not use the *pulsilogia* in relation to his static observations, but nonetheless ascribed great importance to these instruments, probably also with regard to his own medical practice, given their early description in his published work. The strong interest in the device among other scholars, such as his colleague Rudio, might have confirmed him in this view.

7.3 The Thermoscopes

The historical development of the thermometer has long attracted intensive scholarly attention and Sanctorius's thermoscopes are no exception in this regard.[29] In the following, I will briefly outline the basic features of these instruments and Sanctorius's use of them, and consider the historical context in which they emerged. Adding to the existing literature, I will present some further reflections on Sanctorius's application of the thermoscopes to medical practice within the framework of Galenic fever theory and the doctrine of the six non-natural things. Moreover, I will examine his exceptional use of the device to demonstrate the falsity of an astrological argument.

7.3.1 Design and Basic Functioning of the Thermoscopes

Sanctorius first referred to his thermoscopes in 1612, in the second volume of his *Commentary on Galen*. Two years later, he mentioned them again in an aphorism of the *De statica medicina*, but, as with all of his instruments, he published descriptions and illustrations of the thermoscopes only in the *Commentary on Avicenna,* in 1625. He put forward six different versions of the device, most of them equipped with a scale and thus, already representing thermometers (Figs. 7.13, 7.14, 7.15, 7.16, 7.17, and 7.18).[30]

The basic design and functioning of these thermoscopes is the same and can be summarized as follows. Each consists of a tiny vessel full of water at its base, from which a thin pipe vertically emerges, the upper part of which mostly leads to a bowl.

---

[29] Recent studies on the historical development of the thermometer are Borrelli 2008, Valleriani 2010: 155–90 and Bigotti 2018. The following account of Sanctorius's thermoscopes draws largely on these studies.

[30] I refer to "thermoscopes" rather than "thermometers," when writing generally about the devices that Sanctorius suggested to measure degrees of heat and cold, since not all of them are thermometers, i.e., equipped with a scale.

The bowl and the upper part of the pipe are empty, i.e., filled only with air. By, for example, touching the bowl, the air is heated and expands, pushing the water downwards. In the absence of touch, the air cools and contracts, which tends to create a vacuum and hence pull the water upward. Since the device is not hermetically sealed, expansion and contraction of the air occur, owing to changes not only in temperature, but also in pressure. However, the influence of atmospheric pressure was still unknown in Sanctorius's day (Sanctorius 1612b: 62, 105, 229, 375; 1614: 20v–21r; 1625: 7, 22 ff., 76 f., 144, 215, 219–22, 304 ff., 346, 357, 360).[31]

Without analyzing the design of Sanctorius's different versions of thermoscopes in detail, I just want to point out a few aspects which highlight the particular, medical application that Sanctorius foresaw for these instruments. In the *Commentary on Avicenna*, he explained that he had adapted the device "so that it serves to discern the cold and hot temperature of the air, and of all parts of the body, and to learn the degree of hotness of those who have fever" (Sanctorius 1625: 23).[32] Following the basic design described above, in order to determine the "temperature" of his patients, Sanctorius had them touch the bowl at the top of the device, which enabled him to measure the heat of the skin of their hand (Figs. 7.13 and 7.18). Furthermore, Sanctorius put forward thermoscopes in which the bowl was exchanged either for a small ball that the patient could take into his mouth (Fig. 7.15), or for a semicircular top piece that could be attached to different body parts (Fig. 7.16) or, in a slightly different version, could be breathed into (Fig. 7.17). As stated above, most of these devices have a graded scale (Figs. 7.14, 7.15, 7.16, 7.17, and 7.18) and only in the initial versions of the thermoscope were measurements recorded, as Sanctorius himself explained, by means of a compass.[33] In the first illustration of a thermoscope in the *Commentary on Avicenna*, one can see two threads round the tube, which presumably could be moved to mark the level of liquid in it (Fig. 7.13).[34] Even though Sanctorius depicted most of his thermoscopes with a scale already in the *Commentary on Avicenna*, he explained only five years later, in a second revised edition of the *Commentary on Galen*, how he obtained such a scale. He determined terms of comparison for its extremities, i.e., for the hottest and coldest temperature, so that he could divide the scale as he wished. He found those terms in snow and the fire (or flame) of a candle (Sanctorius 1612b: 62, 229; 1625: 23, 219–22; 1630: 762; Middleton 1969: 45 f.).

---

[31] For an entirely different view, see: Bigotti 2018. In conjunction with his assertion that Sanctorius adopted a corpuscular theory, Fabrizio Bigotti has argued that Sanctorius was aware of the influence of atmospheric pressure and that some of his thermoscopes worked as sealed instruments.

[32] "Nos verò illud accomodavimus, & pro dignoscenda temperatura calida, & frigida aeris, & omnium partium corporis, & pro dignoscendo gradu caloris febricitantium, …." See: Sanctorius 1625: 23. The English translation is taken from: Borrelli 2008: 109 f.

[33] The marks along the tube of the thermoscope, which Sanctorius designed to measure the heat in the patient's mouth, most probably represent tick marks (Fig. 7.15).

[34] Interestingly, Sanctorius replaced this illustration with a less elaborate sketch of a thermoscope in the second edition of his *Commentary on Avicenna*, published only one year after the first, in 1626. See: Sanctorius 1626: 22, Bigotti 2018: 80, 92.

Fig. 7.13 Sanctorius's first
illustration of a
thermoscope (Sanctorius
1625: 22). (© British
Library
Board 542.h.11, 22)

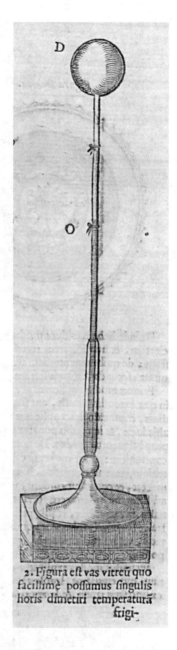

2. Figura eft vas vitreū quo
facillimē poffumus fingulis
horis dimetiri temperaturā
frigi-

7.3.2   What Did the Thermoscopes Measure?

The above quote shows that Sanctorius applied his thermoscopes to the outside air
and to the his patients' body parts. With regard to the latter, the theoretical context
was, of course, again the Galenic concept of latitudes (Sects. 5.2.1, 5.2.2, and 6.2.1).

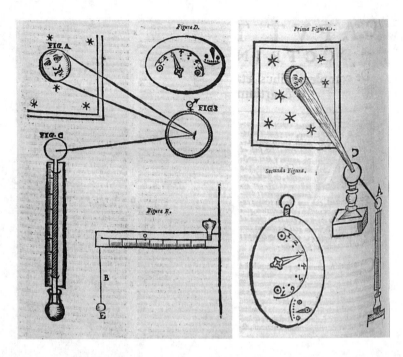

Fig. 7.14 Illustrations of a procedure to measure the heat of the moon by means of thermometers (left: Fig. C, right: letter A) (Sanctorius 1625: 77, 346). (© British Library Board 542.h.11, 77, 346)

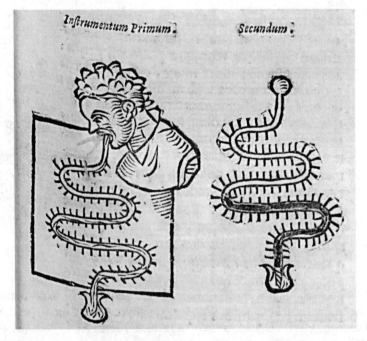

Fig. 7.15 Thermometers to measure the heat in the mouth of the patient (Sanctorius 1625: 219). (© British Library Board 542.h.11, 219)

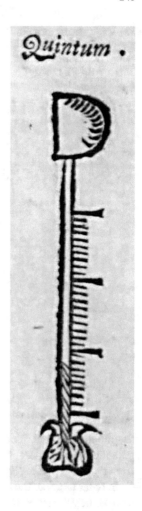

By using a thermoscope to determine the degrees of hot and cold in the complexion
of a patient, the physician no longer needed to rely on his sense of touch, as
Sanctorius remarked, but had an instrument at his disposal that allowed measurements to be taken repeatedly, and to be compared, even more accurately so when the
instrument was equipped with a scale. The direct connection to Galen's teachings is
particularly evident in the way in which Sanctorius developed the scales for his
instruments. As mentioned earlier, Galen already had suggested measuring the temperate complexion as the medium against the extremes found in reality—ice and
boiling water, or fire (Sect. 5.2.2). Thus, Sanctorius replaced Galen's subjective
appreciation of the primary qualities of hot and cold by means of the hand of the
physician with his thermoscopes. But, he used the same method as proposed by
Galen in order to define a point of reference for comparing degrees of hot and cold.
This point was a body's natural, healthy state, the balanced complexion. Accordingly,

Fig. 7.17 Thermometer to measure the breath of the patient (Sanctorius 1625: 221). (© British Library Board 542.h.11, 221)

Sextum Inſtrumentum.

despite the fact that Sanctorius used the same vocabulary as is used today for temperature measurement, his notion of "temperature" was, to be sure, very different from the modern one (Sanctorius 1625: 357, 360; 1630: 262 f., 762).

In order to know the quantity of diseases, that is, the deviation of a body from its natural, healthy state, it was important, per Sanctorius, to measure the temperature not only of the skin of the hand, but also of other body parts. As was seen, he adapted the design of the thermoscopes, so that they could measure the temperature of the breath, or of inside the mouth, or of other body parts. In this context, Sanctorius especially referred to the ability of the instruments to measure the "hot or cold temperature of the heart." Measuring the temperature of the heart was of particular importance since, according to Galenic medicine, the body's innate heat originated in the heart and was distributed from there throughout the whole body. It was specifically relevant for fever patients, so Sanctorius. Following contemporary Galenic views of fever, Sanctorius conceived of the disease as a qualitative change in the innate heat. Hence, the organ which was first affected by fever was the heart. From here, Sanctorius explained, the febrile heat arrived at the other organs, affecting their innate heat, too. It seems that Sanctorius understood the measurement of the temperature of the heart with his thermoscopes in the literal as well as the figurative sense. He explicitly stated that the thermometer with the semicircular top piece

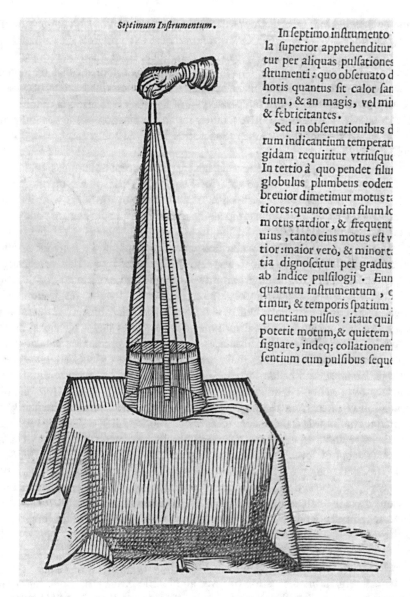

Septimum Inſtrumentum.

In ſeptimo inſtrumento
la ſuperior apprehenditur
tur per aliquas pulſationeſ
ſtrumenti : quo obſeruato d
horis quantus ſit calor ſar
tium , & an magis, vel mii
& febricitantes.

Sed in obſeruationibus d
rum indicantium temperatı
gidam requiritur vtriuſque
In tertio à quo pendet filuı
globulus plumbeus eodeɱ
breuior dimetimur motus tɜ
tiores:quanto enim filum lc
motus tardior , & frequent
uius , tanto eius motus eſt v
tior:maior verò, & minor tɜ
tia dignoſcitur per gradus
ab indice pulſilogij . Eun
quartum inſtrumentum , ҫ
timur, & temporis ſpatium ,
quentiam pulſus : itaut quiſ
poterit motum,& quietem
ſignare, indeq; collationeɱ
ſentium cum pulſibus ſequҽ

Fig. 7.18 Thermometer to measure the heat of the palm (Sanctorius 1625: 221). (© British Library
Board 542.h.11, 221)

(Fig. 7.16) should be applied to the region of the heart to detect whether the heat of
the heart increases, decreases, or remains constant. Given that the principal product
of respiration was thought to be the heat which was generated and distributed by the
heart and the arteries, Sanctorius could determine the heat of the heart also by measuring the breath of his patients (Fig. 7.17). The thermometer whose upper part

could be inserted into the mouth was probably also intended by him to measure the breath of the patient (Fig. 7.15). Or, he referred here to an indirect measurement of the heart's heat because all body parts received heat from this organ. Sanctorius most certainly had this in mind with regard to the thermometer that the patient should touch with the palm of his hand (Fig. 7.18). As described above, according to Galenic medicine, the most temperate part of the body was the skin of the hand, which was why it was able to give a rather reliable indication of the general complexion of the patient (Sect. 5.2.2) (Sanctorius 1625: 219–22; 1629: 312, 355 f.; Lonie 1981: 20–8).

As the preceding paragraph suggests, besides the concept of latitudes, traditional fever theory provided the framework for Sanctorius's use of his thermoscopes. Since fever was a central preoccupation of early modern medicine with its main diagnostic sign being heat, an instrument that could measure changes in bodily heat must have seemed especially promising to practicing physicians.[35] Yet, the "quantity" of heat was only one parameter that served to differentiate between normal, healthy heat and febrile heat, which was also characterized by qualitative adjectives such as "sharp" (*acris*) or "biting" (*mordax*)—by differences of kind rather than just of degree. Moreover, besides hotness, a patient's pulse, tongue, respiration, skin, visage, eyes, bowels, or urine indicated a fever's character. Indeed, Galenic fever theory specified many different fever types and sub-types that a physician had to distinguish. On the basis of the substances involved in the production of heat—vital spirits, humors, and flesh—three main genera of fever were identified: *ephemeral*, *putrid* or *humoral*, and *hectic* fevers. Another distinction of fevers referred to its frequency of presentation, an example being the intermittent fevers. Here, the differentiation resulted from the observation that there was a specific regularity or intermittency that was independent of age, constitution, diet, and other variables. These brief remarks on traditional fever lore should be enough to delineate that the Galenic concept of fever was certainly very different from today's, with hotness, albeit important, being only one of several symptoms of the disease. Within this framework, the repeated and comparative measurement of heat by means of a thermoscope certainly helped the physician to observe health trends in his patients and to diagnose fevers. However, since there was no single measure to diagnose fever, the degrees of hot and cold that he determined always needed to be related to the Galenic teachings on fever with their qualitative methods and classification of various fevers (Siraisi 2012: 504; Hamlin 2014: 4–64; George 2017: 31 f.).

Interestingly, in his discussion of fever and febrile heat in the *Commentary on Hippocrates*, Sanctorius did not even mention his thermoscopes. Nor did he refer to the instruments in the *De statica medicina*, when dealing with the topic of fevers.

---

[35] For more information on Renaissance fever pathology, see: Lonie 1981. This is, however, the only relevant study that I could find which specifically focuses on Renaissance concepts of fever. With regard to Renaissance *practices* of diagnosing and treating fevers, I was unable to find any study.

Instead, he emphasized the role of blocked perspiration as a cause of the disease.[36] In fact, as mentioned in Sect. 3.2.9, already in traditional medicine, hindered or blocked perspiration was identified as a major cause of diseases, also of fevers. Despite the connection between febrile heat and an accelerated pulse in Galenic medicine, Sanctorius did not explicitly relate his *pulsilogia* to fever diagnosis; and when he used the thermoscopes and *pulsilogia* as complementary instruments, the latter served only as timekeepers.[37] According to Christopher Hamlin, this was, however, very much in line with early modern Galenic medicine, which neglected the pulse as a mode of assessment for fevers, even though Galen had fixated on the pulse as the chief indicator of the disease.[38] While these observations do not allow certain conclusions to be drawn, they do call into question the importance of Sanctorius's thermoscopes in the diagnosis and treatment of fevers. They show that traditional qualitative fever lore, and likewise the monitoring of insensible perspiration by means of the weighing chair, were of great relevance for Sanctorius in this context, too. They caution us to not let our modern concept of fever and its very familiar measurement with a fever thermometer distort our view of Sanctorius's thermoscopes (Sanctorius 1629: 170 f., 222–5, 300–4).

Sanctorius's application of the thermoscopes to the outside air, for its part, must be considered against the backdrop of traditional dietetic medicine, according to which air was one of the six non-natural things and as such influenced the health and disease of a body (Sect. 3.3.1). The temperature of air, just like the temperature of human bodies, was conceived in complexional terms and, thus, a healthy human body and a temperate climate would have the same temperature, namely a balanced one. Notably, Sanctorius noted the connection between his measurements with the thermoscopes and the doctrine of the six non-natural things only once in his books. In the *Commentary on Galen*, he addressed the requirement that the non-natural factors be optimally temperate and asked how the physician could know this in terms of degrees (*quo ad gradum*). With respect to air, Sanctorius explained that he was able to detect the medium between too hot and too cold with his thermoscope. In an earlier passage of the same work, he referred to his thermoscopes in a discussion of the most temperate climate or region, stating that each climate had its own temperate climate depending on the complexion of its inhabitants. This shows that the temperature of ambient air was intrinsically tied to the temperature of human bodies and that the physician needed to measure both, the temperature of the ambient air as well as the temperature of the human body, in order to make a diagnosis. However, as mentioned earlier, Sanctorius alluded to the thermoscope only in one

---

[36] For Sanctorius's references to fever in the *De statica medicina*, see: Sanctorius 1614: 3v, 10v–11r, 27r, 28r–28v, 30r–30v, 53r, 54r–54v, 76v–77r, 1634: 14r–14v, 16r–17r, 18v, 40v.

[37] Only in one passage of the *Commentary on Avicenna* did Sanctorius refer to fever in his description of *pulsilogia*. He explained that the *pulsus humilis* decreased in frequency during fever, while the *pulsus invalidus* did not; and that this could be detected only by means of his *pulsilogium*. See: Sanctorius 1625: 22.

[38] Unfortunately, Christopher Hamlin has not explained *why* early modern Galenic physicians neglected the pulse in the diagnosis of fevers. See: Hamlin 2014: 72.

of the many aphorisms in the *De statica medicina* and did not relate his measurements with the instrument to the observation and quantification of insensible perspiration. Thus, in view of the strong connection between the ambient air and the complexion of human bodies, Sanctorius was surprisingly silent about the use of his thermoscopes within the medical framework of the doctrine of the six non-natural things (Sanctorius 1612b: 62, 104 f.; 1614: 20v–21r).

7.3.3   Measuring the Heat of the Moon

Besides determining the temperature of the ambient air and the body parts of his patients, Sanctorius used his thermoscopes also for another purpose: to measure the heat of the moon. An opponent of astrology, he hoped to show that the moon did not emit cold rays, as some astrologers claimed. Since this use of the thermoscope—to physically demonstrate the falsity of an astrological argument—was quite extraordinary, I will consider it in some detail. In the *Commentary on Avicenna*, Sanctorius explained and illustrated how he measured lunar heat (Fig. 7.14) (Sanctorius 1625: 76 f., 346).

On the night of a full moon, he took a large concave mirror made of glass (Fig. B in Fig. 7.14, left) and used it to "collect" moonbeams (Fig. A in Fig. 7.14, left).[39] He positioned the mirror at such an angle that the reflected moonbeams would touch on the upper part of a thermometer (Fig. C in Fig. 7.14, left), which would then reveal their temperature. The next day, around noon, he repeated the same procedure, this time measuring the heat of sunbeams and comparing their temperature to that of the moon. According to his measurements, in a time period of ten pulse beats, the lunar heat was ten degrees, while the solar heat reached 120° after only one pulse beat. Hence, Sanctorius not only measured lunar heat, but also solar heat, in order to have a point of comparison. Given the lack of standard units of temperature, his contrasting juxtaposition of the effects on the earth of the moon and the sun, respectively, served to emphasize that the physical effects on the earth of any heavenly body other than the sun were extremely small. He was able thus to refute astrologers' claims, by demonstrating that the moon neither influenced the body through any supposed heat, nor emitted cold rays (Sanctorius 1625: 76 f.).

It is interesting to note that Sanctorius compared here not only the measurement results that he obtained with the thermometer, but also the duration of his observations. To this end, he used his pocket watch type of *pulsilogium* (Fig. 7.14, left: *Figura D*, right: *Secunda Figura*). In fact, with regard to measurement of the temperature of the ambient air and the body parts, Sanctorius, too, noted the need to register time intervals by means of his *pulsilogium*. However, while he usually recommended that measurements be taken at equal intervals of time, he referred to

---

[39] In a later passage of the *Commentary on Avicenna*, Sanctorius proposed to replace the glass mirror with a crystal sphere, or with a water-filled drinking cup (*phiala*) (Fig. 7.14, right: letter C, printed inversely). See: Sanctorius 1625: 346.

different time periods for the measurement of lunar and solar heat. Most probably, this was to highlight the great difference between them; or because he simply could not register higher degrees of solar heat on his thermometer's scale and therefore decided to limit the measurement to the duration of one pulse beat only. It is in any case doubtful whether he determined the sun's heat to be 120° using a thermometer whose scale spans only eighty degrees (Fig. 7.14, on the left). What is more, it is now known that the illumination of the full moon on the earth is 440,000 times weaker than that of the sun (each in zenith). The moon's illumination level is 0.27 Lux which corresponds to a 100-watt bulb at a distance of 19 m. Consequently, the degrees of heat that Sanctorius allegedly determined for the moonbeams must have been due to other factors, such as still air. Remarkably, Sanctorius's measurement of lunar and solar heat is one of the rare occasions on which he expressed the outcome in numerical values. It is the only instance in which he specified a quantity for the procedures undertaken with his thermoscopes.[40] That he conducted the measurement as a public event therefore seems likely, also given his statement, that he showed it to a large crowd of students (*magna scholarium frequentia*). Moreover, he noted that the measuring results would vary depending on the time intervals and different instruments used. Hence, Sanctorius gave ample cause to assume that he actually took his students outside the university classroom, at night and in the daytime, to demonstrate the insignificance of any supposed lunar heat compared with that of the sun and to do away, once and for all, with the claim that the moon emitted cold rays (Sanctorius 1625: 23 f., 76 f., 219 ff.; Siraisi 1987: 289; Kuphal 2013: 103).

Sanctorius included the description of his measurement of lunar and solar heat in a lengthy and virulent assault on astrology, which was at least partially inspired by his greatest critic, the astrologizing physician of Ferrara, Ippolito Obizzi (Sect. 5.3.2). Without going into the details of Sanctorius's diatribe, which have been described elsewhere, I want to point out a few aspects that elucidate the probable reasons and context of Sanctorius's open and explicit attack on astrology, and his use of the thermometer in this regard.[41] As a prominent feature of Renaissance culture, astrology was closely related to medicine, with astrological knowledge being integrated into the university curriculum of medicine.[42] Yet, as Nancy Siraisi has shown, by the latter part of the sixteenth century there were good reasons for Galenic teachers of medicine to be skeptical about the idea of astral and/or occult causes. They disapproved of the attempts made by certain neoteric physicians to overcome the limitations of physiological and therapeutic complexion-based explanations by

---

[40] For Sanctorius's references to the thermoscopes in his published works, see: Sanctorius 1612b: 62, 105, 229, 375 f., 1614: 20v–21r, 1625: 7, 22 ff., 76 f., 144, 215, 219–22, 304 f., 346, 357, 360, 1629: 24, 137, 326, 1630: 262 f., 762. I do not refer here to the 1626 edition of the *Commentary on Avicenna*, as the pagination is identical to that of the 1625 edition.

[41] Nancy Siraisi has analyzed Sanctorius's critical discussion of astrology and occult celestial influences in the *Commentary on Avicenna*. See: Siraisi 1987: 284–9.

[42] For a general account of Renaissance astrology, see: Dooley 2014; and for a study on astrology in medieval medical practice, see: French 1994. With regard to the role of occult qualities in the Renaissance, see: Hutchison 1982.

drawing on occult qualities, celestial influences, and insensible characteristics specific to individual diseases, remedies, or patients. According to the more orthodox Galenists, this was detrimental to reason, system, and *scientia*. Furthermore, the growing emphasis on the collection and precise recording of data derived from sense experience, as exemplified by anatomy, together with the attack by the Counter-Reformation church on most astrology and all magic most probably made these physicians less sympathetic to medical explanations that propounded and amplified the role of celestial influences and occult forces. It is within this framework that Sanctorius's attack on astrology has to be seen (Sanctorius 1625: 72–83; Siraisi 1987: 284; Hübner 2014: 26).

Thus, Sanctorius was not exceptional in refuting astrology and, as Siraisi has argued, his criticism of astrology might well reflect a common professional concern in his day about a new type of unlicensed medical practitioner: exorcists who administered their own medications.[43] What was extraordinary and certainly did set him apart from his colleagues, however, was his recourse to the thermometer—to measurements and observations made by means of his senses—in order to refute an astrological argument. It is indicative of the importance that he ascribed to quantitative methods not only for a strictly medical use, but also regarding meteorological-astronomical observations, as was already seen with regard to his steelyards to measure climatic conditions (Sect. 7.1). In this context, it is important to keep in mind that Sanctorius did not completely reject the concept that the earth was affected by celestial influences. He held that the celestial bodies influenced things on earth only through their motion, light, and heat. According to his understanding, the measurement of lunar and solar heat therefore did not only serve him to refute an astrological idea, but also to show that the rays of these celestial bodies affected the earth through their heat, the moon to a very small, the sun to a very high degree (Sanctorius 1625: 73).

7.3.4   The Thermoscopes in Context

As mentioned above (Sect. 7.3.3), Sanctorius highlighted in the *Commentary on Avicenna* that many students attended his demonstration of measuring lunar and solar heat by means of his thermometer. Already thirteen years earlier, in 1612, in the first reference to the device in the *Commentary on Galen*, Sanctorius had explained that he showed it "very freely to everybody" at his house in Padua (Sanctorius 1612b: 62; 1625: 76). In June of the same year, the Venetian nobleman

---

[43] For a brief discussion of Renaissance physicians who condemned astrology, see: Wear 1981: 245–50. Although Sanctorius's more general arguments against astrology did not include much that was new, Nancy Siraisi has emphasized that his views about witchcraft, sorcery, and magic "show considerable independence of spirit, since they were written at a time when the Venetian Inquisition was much preoccupied [with these subjects]" (Siraisi 1987: 287 f.).

Giovan Francesco Sagredo (1571–1620) sent a letter to his good friend Galileo Galilei, reporting:

> The Lord Mula was at the patron fair and told me he has seen an instrument by Lord Sanctorius with which one measures the cold and the heat with the divider. Finally he communicated to me that it is a large bowl of glass with a long neck and I immediately applied myself to producing some of them very exquisitely and beautifully (Sagredo 2010: 229).

Evidently, Sanctorius also presented his thermoscope at the fair of the patron saint of Padua, where Agostino da Mula saw it. Da Mula's report to Sagredo and Sagredo's subsequent letter to Galileo show that the thermoscope was among the subjects investigated and discussed by Sanctorius's network of friends in Venice, as all of these men belonged to the *Ridotto Morosini*. Matteo Valleriani has pointed out that it is, however, quite impossible to determine who "invented" the device first. This is because the appearance of the thermoscope was not really a new invention, but rather the result of a gradual process—the transformation of an old pneumatic device into an instrument to measure temperature—which took place in the early seventeenth century. Involved in this process were Galileo, Sagredo, Sanctorius, and various other scholars scattered far and wide, geographically. Yet, current historical research supports the idea that it was Sanctorius, who first applied the thermoscope to medicine. In any case, Sanctorius's development and use of thermoscopes illustrates once again that he was part of a vibrant intellectual and social milieu: fertile ground in which to develop and test his new ideas related to quantification and instrumentation (Sect. 7.2.2, fn. 20) (Valleriani 2010: 156 f.; Siraisi 2012: 505).

Sanctorius first mentioned and presented his thermoscope at the very moment it was about to become a very common instrument. Already by 1624, thermometers were being produced and sold for profit in many workshops and markets. Sanctorius's instruments, too, quite soon gained in popularity. It is striking how repeatedly he stressed that the thermoscopes especially should be integrated into his Paduan lectures, owing to the avid interest of his students, who, as he wrote, "did observe this novelty not without great admiration" (Sanctorius 1612b: 105).[44] Ironically, there is evidence to suggest that even Ippolito Obizzi, the "Great Astrologer" (*Astrologus Magnus*) against whom Sanctorius directed the diatribe, including the measurement of lunar heat in the *Commentary on Avicenna*, had the pleasure to attend a demonstration by Sanctorius of an early type of his thermoscopes. Despite the fact that Sanctorius's thermoscopes were known to many contemporary physicians, it does not seem that they found considerable application in daily medical practice.[45] Only in the nineteenth century did the thermometer come into general clinical use—a

---

[44] "… quod Patavi ostendimus auditoribus nostris, eiusque usus docuimus: quam novitatem non sine magna ipsorum admiratione intellexerunt." See: Sanctorius 1612b: 105.

[45] In 1633, the Bohemian philosopher and physician Johan Caspar Horn (life dates unknown) donated a so-called *Hydrolabium Sanctorii* (*Sanctorius's water-catcher*) to the German Nation of the University of Padua and, according to Fabrizio Bigotti, Sanctorius's texts and devices continued to be copied and studied by many medical students, as attested by the notes in the Marmi Collection in the Wellcome Library, London. See: Rossetti 1967: 341 f., Bigotti 2018: 84, 92 f.

development which deserves extended consideration in its own right, although that is, unfortunately, not possible here (Grmek 1952: 43).[46]

To sum up, the development and use of thermoscopes show once again that Sanctorius was receptive to the intellectual and technological trends under examination in his day by different scholars, not only in Italy, but also elsewhere. Having presumably discussed the devices with his Venetian circle of friends, Sanctorius recognized their potential for medical practice and was the first to apply them to medical diagnosis. In this context, he adapted their design and developed a scale in order to better use their most important feature: the recording and comparison of temperatures. Like all of his measuring instruments, the thermoscopes were intended to enhance the physician's perception, in this case, his sense of touch, and to note even minor variations in temperature. As modern as his descriptions of temperature measurements may sound, they can be understood only against the backdrop of Galenic medicine. As we have seen, Sanctorius's use of thermoscopes was related not only to the concept of latitudes, but also to traditional fever lore. Within this framework, fever was not reducible to the "quantity" of heat and thus, a physician needed more than a thermoscope to diagnose the disease. Accordingly, besides highlighting the usefulness of the thermoscopes for fever patients, Sanctorius emphasized in the *De statica medicina* the importance of impeded perspiration as a sign and cause of fevers. In the *Commentary on Hippocrates*, he discussed fevers and febrile heat without alluding either to the thermoscopes or to the weighing chair. Hence, it remains unclear which role the thermoscopes played, along with traditional qualitative methods and the weighing chair, in Sanctorius's diagnosis and treatment of fevers. Indeed, it is generally difficult to assess even the extent to which Sanctorius actually used the thermoscopes in his medical practice. He sometimes used the plural when writing about the subjects whose temperatures he measured (*sani, febricitantes*), and he described how he tested with his thermometer whether the heat in children and young men was the same. But when explaining the procedure, he referred to *a child* and *a young man* and this use of the singular implies that he did not scale-up the procedure to more subjects. At the same time, it also suggests that he thought it possible to move from individual bodies to general groups of people—yet a form of generalization according to age groups was already present in traditional Galenic medicine (Sect. 3.1.3). From the evidence at hand, it therefore seems that Sanctorius used his thermoscopes to observe health trends and fevers in several individual patients without, however, making generalizations about healthy or unhealthy temperatures and their applicability to many individual cases (Sanctorius 1625: 23 f., 159 f., 219, 222, 357; 1629: 170 f., 222–5, 300–4).

---

[46] Volker Hess has shown that the lack of interest in the quantitative registration of body heat on the part of physicians can be explained by the fever concepts of the seventeenth and eighteenth centuries, according to which the measurement of body heat was simply irrelevant (Hess 2000: 19 ff.). In the same vein, Christopher Hamlin has remarked: "Periodically fever writers had published temperature data, but usually temperatures were facts without signification" (Hamlin 2014: 252).

7.4   The Hygrometers

In the *Commentary on Galen* and in the *De statica medicina* Sanctorius referred to several methods of determining the humidity of air. Thirteen, respectively eleven years later, in 1625, in the *Commentary on Avicenna,* he described and illustrated two instruments for this same purpose. Contrary to the *pulsilogia* and the thermoscopes, Sanctorius's hygrometers have not been dealt with specifically in recent secondary literature.[47] Since an in-depth study of the devices is not possible here, I will limit myself in the following to describing their basic features and to briefly outlining Sanctorius's use of them. Furthermore, I shall summarize the broader historical context of their emergence and consider their relation to traditional dietetic medicine, according to which the environment, including the climate, had an important influence on the health and disease of a body. Against this backdrop, hygrometric measurements could provide physicians with helpful information regarding the diagnosis and treatment of patients. The connection of the hygrometers to Sanctorius's other quantitative observations will be examined, too (Sanctorius 1612b: 105, 229 f.; 1614: 20v–21r; 1625: 23 f., 144, 215, 305).

7.4.1   Four Methods to Measure the Humidity of Air

Sanctorius's earliest mention of a method to measure the humidity of air dates back to 1612, when he wrote in his *Commentary on Galen*:

> … we have found a very certain way of diagnosing the humidity of the air, that is to say how much of it there may be each day. This is to take salt of tartar, commonly called alum of the lees; it is exposed to the air, but first it is weighed very exactly.[48] Then in the morning it is weighed again. Now it always weighs more after exposure to the air, but considering the different weights we say that the greater the weight, the greater the humidity, and the less the weight, the less the humidity that reigns in the air (Sanctorius 1612b: 105).[49]

---

[47] According to my research, the most recent study on the historical development of hygrometers to have considered Sanctorius's instruments in some detail is: Middleton 1969: 81–132. Furthermore, Mirko Grmek has dealt with Sanctorius's hygrometers in his monograph on Sanctorius and his instruments, see: Grmek 1952: 45 ff. Another, still older account of the devices can be found in: Miessen 1940: 22–6, but it contains several flaws and inaccuracies.

[48] W.E. Knowles Middleton, whose English translation I follow here, has interpreted the term *alumen faecis* ("alum of the lees") as referring to the tartar that builds up in wine barrels (Middleton 1969: 86, fn. 37). In my translation of the aphorism of the *De statica medicina* in which Sanctorius put forward different methods to measure the humidity of air, I translate *aluminis faecum* with "sediment of alum" (Sect. 5.3.2). In his translation of the same aphorism, Fabrizio Bigotti has simply written of "several types of salt" (Bigotti 2018: 88). While it is impossible to verify exactly what Sanctorius meant, when he wrote of *alumen faecis*, he certainly had some kind of salts in mind.

[49] "… nos invenimus modum certissimum pro dignoscenda aeris humiditate, quanta videlicet quotidie sit: & talis est, sumimus tartarum combustum, quod à vulgo dicitur alumen faecis: hoc expo-

Thus, Sanctorius tried to determine the humidity of air by detecting the change in weight of a hygroscopic substance, a substance that absorbs water vapor from the air. This recalls the passage in the *De statica medicina*, quoted above, in which Sanctorius suggested the same method in order to measure the "weight of air" which was, according to him, directly related to its humidity (Sect. 5.3.2). In this aphorism, he specified that the salt had first been dried in the sun before being then exposed to night air. Consequently, Sanctorius does not seem to have measured the humidity at a given moment, but determined it rather over a longer period, namely one day or one night. This was probably because he could detect notable differences in air's humidity only after longer time spans. In fact, it is known today that relative humidity is often considerably higher at night than in the daytime.[50] Hence, it is likely that Sanctorius managed to observe major differences in the humidity of air only by using his salt-weighing method. However, this is purely speculation, as he neither mentioned any measuring results nor gave information on the balance he used, let alone its precision (Deutscher Wetterdienst 2015).

In addition to his method of measuring the change in weight of a hygroscopic substance, Sanctorius described in the *De statica medicina* aphorism three other ways to determine the humidity of air. These were, firstly, a greater feeling of cold than what was measured with the thermoscope, since, so Sanctorius, the humidity of air sharpened the sensation of cold (*lima frigiditatis*). Secondly, "the greater or lesser warping of very thin boards, especially of pearwood" and thirdly, "the contraction of lute strings, or hemp cords" (Sanctorius 1614: 20v–21r).[51] With regard to the first, it is interesting that Sanctorius referred to his thermoscope for the measurement of humidity, but the semi-subjective procedure he outlined is somewhat puzzling. According to him, it was the *feeling* of cold that ultimately indicated the humidity of the air, while the measurement of the thermoscope served only as a point of comparison for the subjective perception of cold. Even though Sanctorius recognized the need to measure the humidity and likewise the air temperature in quantitative terms, by means of instruments, he nowhere described the interrelated measurements of both parameters. He did not connect the humidity of air to its temperature, but only to its *felt* temperature. This notion most probably derived from common experience, maybe from walks through foggy Venice. Without further details, however, it is difficult to conclude anything definite from Sanctorius's brief remarks in the *De statica medicina*. Still, it should be noted that Sanctorius differentiated here between "perceived temperature" and "measured temperature,"

---

nitur aeri, sed antequam exponatur exactissimè perpenditur; & deinde mane iterum perpenditur: semper enim expositum aeri magis ponderat: nos enim pro varietate ponderis dicimus maius pondus maiorem humiditatem, & minus minorem in aere dominari: ...." See: Sanctorius 1612b: 105. The English translation is taken from: Middleton 1969: 86.

[50] Relative humidity of the air is the amount of water vapor which is actually present in the air compared to the greatest amount it would be possible for the air to hold at the same temperature. Relative humidity is usually expressed in percent (Cambridge Dictionary 2014).

[51] "... ex maiori, vel minori incurvatione tabulae subtilissimae praecipuè ex piro. ... ex contractione cordarum testudinum, vel ex cannabe." See: Sanctorius 1614: 21r. The English translation is based on: Middleton 1969: 86, Bigotti 2018: 88.

on the basis of the procedures with his thermoscopes. In doing so, he emphasized the discrepancy between what a person feels and what is measured by means of instruments. Yet, despite his claims as to the certainty assured by his quantitative observations and instruments,, Sanctorius proposed using the subjective sense of cold and not measured degrees of cold as a means to determine the humidity of air (Miessen 1940: 24; Bigotti 2018: 88).

The other two methods that Sanctorius described in his aphorism of the *De statica medicina* depended on the measurement of some change in the shape or size of certain substances—pearwood boards, lute strings, or hemp cords. Here again, it can be assumed that Sanctorius drew on common experience. While it is completely unclear how he measured the warping of wooden boards, he revealed eleven years later in the *Commentary on Avicenna* that the contraction of strings, or cords served as the basis for his two hygrometers depicted in that book. This implies that Sanctorius developed these instruments in the time between the publication of the *De statica medicina* in 1614 and the publication of the *Commentary on Avicenna* in 1625. Yet, it is also possible that Sanctorius had already come up with the instruments by the time he released the *De statica medicina*, even though he did not explicitly mention them—because he generally did not offer many details in his concise aphorisms, and did not even describe in this book the weighing chair with which he undertook his static procedures. But the fact that Sanctorius had not referred to the contraction of cords two years earlier, in the *Commentary on Galen*, leads one to suppose that he developed his two hygrometers sometime between 1614 and 1625 (Sanctorius 1614: 20v–21r; 1625: 23 f., 305).

In this context, it is interesting to note that Sanctorius, when dealing with the measurement of humidity in the *Commentary on Avicenna*, did not speak always of his two hygrometers, but rather repeatedly directed his readers to the four methods outlined in the *De statica medicina*. What is more, four years later, in the *Commentary on Hippocrates*, he again referred to "the four ways of measuring dryness and humidity …, which we proposed in the fourth aphorism of the second section of our statics" (Sanctorius 1629: 24).[52] Hence, it seems that Sanctorius still considered these methods valid, even after he had developed his two hygrometers; he evidently did not regard the latter as more advanced or superior. This is especially perplexing with respect to Sanctorius's suggested use of the subjective perception of cold in determining air's humidity, which, in this light, can hardly be interpreted as an idea that Sanctorius entertained only during the search for a useful instrument to observe humidity, and later abandoned. From today's perspective, this shows again that Sanctorius was in a phase of transition: on the one hand, he was beginning to use thermoscopes in order to rule out the uncertainties entailed by a physician's subjective sense of touch and, on the other hand, he still adhered to the idea that an individual perception of cold could be used to determine air's humidity (Sanctorius 1625: 7, 144; 1629: 24).

---

[52] "Tertium consistit in quatuor modis dimetiendi siccitatem, & humiditatem …, quos proposuimus in 4. aphorismo 2. sect. staticae nostrae." See: Sanctorius 1629: 24.

7.4.2   Two Hygrometers

As mentioned above, Sanctorius's two hygrometers were based on the contraction of cords.[53] The first version (Fig. 7.19) consisted of a cord, or a thick lute string that was stretched between two pegs in a wall. To the middle of the cord there was attached a lead ball which moved, depending on humidity, along a graduated scale drawn on the wall behind it. Contrary to what one might intuitively suppose, when the air was moist, the cord would contract and the weight would be elevated, indicating a high degree of humidity on the scale. Conversely, when the air was dry, the cord would loosen and the weight would be dropped, pointing to a low degree of humidity. However, the illustration shows that, according to Sanctorius's graduation of the scale, the highest degree of humidity would be 1, while the lowest degree would be 10. Therefore, the scale might be also interpreted as referring to the dryness of the air rather than its humidity. In fact, Sanctorius wrote of "degree[s] of moisture or dryness." Judged from this technical understanding of the instrument, it is conceivable that this version of the hygrometer enabled Sanctorius to measure what today is called relative humidity (Sect. 7.4.1, fn. 50). What is more, measurements did not need to be recorded over lengthy time spans, as was the case with the salt-weighing method. This device could most probably also determine immediate changes in the humidity of air (Sanctorius 1625: 23; Miessen 1940: 24; Hodgson 2008: 50).

In the second version of a hygrometer that Sanctorius presented in the *Commentary on Avicenna*, a thick and long flaxen cord was wound up around a clock-like disc (Figs. 7.20 and 7.21). He explained that the cord was attached to a peg at the back of the disc, which was, in turn, connected to the sun-shaped hand on the front of the device. Depending on the humidity or dryness of the air, the cord would contract or loosen, thereby moving the hand so as to indicate the measured degrees on the dial. Contrary to the first hygrometer, the scale ranged not from 1° to

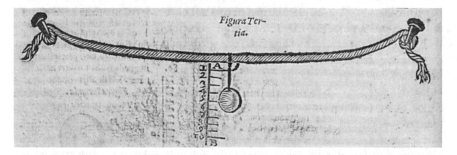

Fig. 7.19 First version of a hygrometer (Sanctorius 1625: 23 f.). (© British Library Board 542.h.11, 23 f.)

---

[53] I use here the term "hygrometer" instead of "hygroscope," because the instruments that Sanctorius depicted in the *Commentary on Avicenna* are both equipped with a scale (Figs. 7.19 and 7.20).

Fig. 7.20 Second version
of a hygrometer, modeled
on a clock (Sanctorius
1625: 23, 215). (© British
Library
Board 542.h.11, 23)

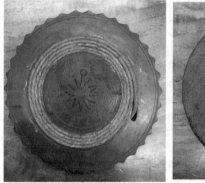 

Fig. 7.21 Replica of Sanctorius's clock type of hygrometer. (The replica was made by Loris Premuda for an exhibition held in 1961 at the University of Padua, where it can still be found today. The replica is imperfect, however, since the cord is affixed to the disc by glue and nails, and is thereby prevented from contracting or loosening (Biblioteca medica 'Vincenzo Pinali antica' dell'Università degli Studi di Padova, © Philip Scupin))

10°, but from 1° to 12°. This might imply that the device had a higher measurement resolution. Or Sanctorius simply chose this range because the hygrometer was modeled on a clock, whose dial is usually divided into twelve sections (Sanctorius 1625: 23 f.; Del Gaizo 1936: 15).

From the written and pictorial evidence in the *Commentary on Avicenna*, it is difficult to assess whether the instrument actually worked and could be used in the way Sanctorius described. On the one hand, he specified materials to be used (a flaxen cord) and gave instructions regarding the dimensions of the cord, which to "better serve the purpose" should be as thick and long as possible (Sanctorius 1625: 23 f.).[54] This suggests that he really built and worked with the device. On the other hand, questions arise as to how the winding of the cord affects its contraction or loosening and whether the long length of the cord changes so much due to humidity that a differentiation into twelve degrees is feasible. In any case, Sanctorius did not present this version of hygrometer as an advancement on the first and there is nothing to suggest a stated preference for one or the other device. Notably, however, in his study of the historical development of hygrometers, W.E. Knowles Middleton has argued that Sanctorius's clock-hygrometer "has had no offspring," while the first hygrometer "was the parent of a large family of hygrometers" (Middleton 1969: 87). The fact that Sanctorius seems to have been the only one who proposed a clock type hygrometer can be seen as an indication that it was difficult, or maybe even impossible, to realize measurement of the air's humidity with this design (Sanctorius 1625: 23 f.).

7.4.3   What Did the Hygrometers Measure?

As outlined above, the hygrometer was one of the instruments that Sanctorius put forward in order to measure the quantity of diseases, i.e., the deviation of a body from its natural healthy state (Sect. 6.1).[55] Surprisingly though, Sanctorius gave no hint that he applied the instrument to the body of his patients, but only referred to the measurement of ambient air. Since he thus did not determine degrees of dryness or humidity in the complexion of bodies, but rather was concerned with determining these degrees in the complexion of air, the doctrine of the six non-natural things rather than the concept of latitudes must be seen as the theoretical context in which Sanctorius employed his hygrometers (Sect. 3.3.1).

In order for the air to be healthy, its complexion, or temperature did not need to be balanced only with respect to the primary qualities of hot and cold, but also concerning the other two primary qualities of wet and dry. Hence, the thermoscope and the hygrometer allowed Sanctorius to measure all of the four primary qualities contained in the complexion of air. However, as pointed out earlier (Sect. 7.3.2), the complexion of ambient air was intrinsically tied to the complexion of human bodies. With regard to his hygrometer, Sanctorius therefore stated that the healthiest degree of humidity and dryness for each person "varies according to the variety of complexions, seasons, and regions" (Sanctorius 1625: 215).[56] Accordingly, the optimal

---

[54] "… sumitur corda ex lino satis crassa, & longa: quia quo crassior, & longior eo melius inservit huic officio." See: Sanctorius 1625: 23 f.

[55] For the sake of simplicity, in what follows I subsume under the term "hygrometer" both, the methods of measuring air humidity as well as the two instruments Sanctorius developed for this purpose, unless otherwise indicated.

[56] "… variatur pro varietate temperaturae, temporis, & regionis: …." See: Sanctorius 1625: 215.

degree of humidity differed for Venetians and Paduans, so Sanctorius. In fact, the hand of his clock type hygrometer, presented in the *Commentary on Avicenna*, points to two degrees of humidity, a value that was, based on Sanctorius's experience, "more beneficial to Venetians than to Paduans" (Sanctorius 1625: 215).[57] But how could he know this, if he was not able to measure the degrees of humidity or dryness in the complexion of his patients? Since Sanctorius gave no further information on this point, the only conclusion that I can draw is that he based his diagnosis of a patient's degree of humidity and dryness on traditional, qualitative methods relating to sign theory and the collection of a syndrome of signs.

Sanctorius's specification of a healthy degree of humidity for Venetians is interesting also for another reason. It is the only instance in which he mentioned a numerical outcome of his procedures with the hygrometers. Remarkably, he referred here, to the clock type of hygrometer, which implies that, despite the aforementioned doubts regarding its functioning, he actually was able to measure differences in air's humidity with this version of the hygrometer. What is more, it suggests that Sanctorius used the instrument in different locations, Padua and Venice, and compared his measurements. As a result, he was able to make regional generalizations, as when defining a healthy value of humidity for the inhabitants of Venice. However, at the same time, he highlighted that the measurements needed always to be related to individuals, as a healthy degree of humidity also varied with bodily complexions. This reflects the tension between patients' broad-ranging individual differences, on the one hand, and the need to generalize, on the other: a balancing act which physicians faced then and still face to this day, in their daily practice (Sanctorius 1625: 215).

In the description of the first hygrometer, in the *Commentary on Avicenna*, Sanctorius indicated how he determined the scale of the instrument (Fig. 7.19). Similar to the procedure that he followed in developing the scale for his thermometers, he searched for terms of comparison for the extremities of air's humidity and dryness. These were, according to him, air from the south (*aer austrinus*) and north winds (*venti septentrionales*). He explained that the "air from the south moistens and shortens the cord so much that the ball rises to the letter A" and that "while the north winds blow, they dry it [the cord] until the ball reaches B" (Sanctorius 1625: 23).[58] But in contrast to snow and fire—used by Sanctorius as the extremities to determine the scale of thermometers—southern air and north winds were not always, but only "sometimes" (*aliquando*) or "often" (*saepe*) extremely humid or dry. Thus, Sanctorius had to measure these winds at least a few times to determine their extremes. Developing a scale for his hygrometer was therefore more difficult than for the thermoscope and required repeated measurements and careful comparison of the results (Sanctorius 1625: 23, 306).

---

[57] "… magis proficuus est Venetijs, quam Patavij, sicuti experti fuimus." See: ibid.: 215.

[58] "… aliquando nam aer austrinus ita humectat, & contrahit cordam, ut attollatur usque ad litteram A. dum verò spirant venti septentrionales ita exsiccatur, ut pila perveniat ad ipsum B." See: ibid.: 23. The English translation is taken from: Middleton 1969: 87.

In defining southern air, or winds as extremely humid and northern winds as exceedingly dry, Sanctorius followed the Hippocratic-Galenic teachings. In the *Commentary on Galen*, he stated that, according to Galen, the complexion of south wind was warm and moist and the complexion of north wind cold and dry. Therefore, north wind cooled and dried, while south wind heated and moistened. In fact, in more recent times, Volker Langholf has shown that the authors of the Hippocratic treatises *De aere, aquis et locis* (On Airs, Waters, and Places) and *De morbo sacro* (On the Sacred Disease) already associated north wind with dry, and south wind with rainy air. This notion most probably derived from an even older ancient folk tradition, per Langholf, since in the ancient Greek poem, the Iliad, attributed to Homer (ca. ninth or eighth century BCE), the north wind was described as dry, while the south wind was referred to as covering the mountaintops with mist. Furthermore, the Greek word for south wind, *nótos*, originally meant "wet wind." Hence, just as with the thermometers, Sanctorius used traditional medical concepts rather than experience as the starting point for the development of a scale for his hygrometers (Sanctorius 1612a: 383 f.; Langholf 1992: 170–4).

In this context, it is interesting to note that Sanctorius measured the humidity of winds, and, more generally, air, with a focus on their impact on health, and did not consider his hygrometers in connection with weather forecasting, as he did with his anemometer. As stated above, he claimed that the latter device could be used to predict sea storms and so mitigate the dangers of flooding (Sect. 7.1.2). Remarkably, despite the strong relation between the hygrometers and the two steelyards to measure climatic conditions, especially the anemometer, Sanctorius did not associate these devices with each other in his works. Moreover, notwithstanding that Sanctorius measured air's humidity and not bodily humidity, he mostly presented his hygrometers in the context of determining the quantity of diseases; only in one passage of his works did he explicitly relate the instruments to the doctrine of the six non-natural things (Sanctorius 1612b: 104 f.). As previously mentioned, he did not include the hygrometers in his observation and quantification of insensible perspiration. The aphorism in the *De statica medicina* in which he put forward the four methods to measure humidity deals with the determination of the "weight of the air." Given that Sanctorius attributed to humidity the cause of air's weight, he regarded his hygrometers as tools to quantify the element and non-natural factor of air (Sect. 5.3.2, fn. 26) (Sanctorius 1614: 20v–21r).

7.4.4   The Hygrometers in Context

In Sect. 5.3.2, I already referred to Nicolaus Cusanus's proposal to measure the weight of the air by means of a method very similar to the one described by Sanctorius, based on the weighing of a hygroscopic substance. Interestingly, contemporary to Cusanus, there was another writer who explained that he "determine[d] the heaviness or dryness of the air and winds" by putting a sponge on a balance (Alberti 1986: 214). This writer was Leon Battista Alberti, the aforementioned inventor of a swinging-plate anemometer (Sect. 7.1.2), who referred to this sponge method in the tenth book of his work *De re aedificatoria* (On Architecture, completed in 1452). According to W.E. Knowles Middleton, it is unclear whether Cusanus and Alberti came independently to such notions, or through concerted efforts. But what their works clearly show is that the measurement of air's humidity by weighing a hygroscopic substance—Alberti suggested a sponge, Cusanus, wool—was known as early as the mid-fifteenth century. A few years later, this method was revived by Leonardo da Vinci, who made a drawing in his notebooks that showed a sponge counterbalanced by a weight, and to which he added the following note: "Mode of weighing the air and of knowing when the weather will change" (Da Vinci and Richter 1970b: 220, fn. 999).[59] Historical research suggests that Leonardo was familiar with Cusanus's work and thus most probably also knew of, and took inspiration from, the latter's proposed method of measuring air's humidity (Middleton 1969: 85–90; Alberti 1986: publisher's note, 214).

Hence, several scholars before Sanctorius examined the possibility of measuring the humidity of air, which they, like Sanctorius, assumed was related to its weight. From the evidence at hand, it can be quite safely assumed that Sanctorius was inspired by Cusanus's book *De staticis experimentis*, in his effort to measure air's humidity (Sect. 5.3.2). Most probably, he also read Alberti's famous and influential work *De re aedificatoria* and, therefore, might well have been familiar with the hygrometric procedure presented in the book. However, Sanctorius went further than these earlier writers and investigated different methods to measure humidity. Most importantly, his newly invented hygrometers were based not on the weighing of a hygroscopic substance, but on the contraction and loosening of cords. Another aspect that distinguishes Sanctorius's approach from earlier ones is that he considered the measurement of humidity in a medical context, in an attempt to improve the daily work of physicians. In keeping with this, traditional dietetic medicine provided the framework in which he developed the scales of his hygrometers.

In the following, I will make some general remarks on the reception of Sanctorius's hygrometers. Sanctorius's first cord hygrometer, displayed in the *Commentary on Avicenna*, was further developed and improved on in Italy in the 1660s. The physician Francesco Folli (1624–1685) and the mathematician Vincenzo Viviani (1622–1703), for example, made similar yet superior instruments (Fig. 7.22).

Both recognized that their hygrometers would have a nonuniform scale, contrary to the devices illustrated by Sanctorius. In fact, already in 1636, Marin Mersenne discussed in his work *Harmonicorum libri* (Books on Universal Harmony), problems regarding the interpretation of the scales of Sanctorius's hygrometers. While Mersenne explicitly referred to Sanctorius, the relation of Folli's and Viviani's hygrometers to Sanctorius's devices seems to be unclear, and requires further investigation. From a preliminary perspective, the practical medical use of hygrometers

---

[59] "Modi di pesare l'arie eddi sapere quando s'à arrompere il tempo" See: Da Vinci and Richter 1970b: 220, fn. 999. The English translation is taken from: ibid. The drawing of the hygroscope can be found in: Da Vinci and Richter 1970a: 297. Another drawing of a similar hygroscope made by Leonardo da Vinci is preserved in the Codex Atlanticus, see: Da Vinci and Pedretti 2000: 30v.

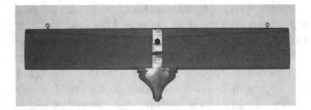

Fig. 7.22 Vincenzo Viviani's rope hygrometer (*Museo Galileo*—Istituto e Museo di Storia della Scienza, Florence. Inv. 799). (Museo Galileo, Firenze. Photo Franca Principe)

emphasized by Sanctorius did not have much resonance among contemporary and later scholars or physicians, who usually employed the instruments for meteorological studies (Mersenne 1636: 43; Grmek 1952: 46; Robens et al. 2014: 337; Bigotti and Taylor 2017: 108, fn. 18).

In conclusion, the hygrometers, just like the *pulsilogia* and the thermoscopes, allowed Sanctorius to determine, record, and compare degrees, in this case, of the humidity of air. Yet, in contrast to his other two measuring instruments, Sanctorius did not apply his hygrometers to the patient's body, but to ambient air alone. In my view, it is somewhat surprising that he did not try, or at the least, did not mention that he tried, to adapt the design of his hygrometers to measuring also bodily humidity, for example through his patients' breath, in a similar way to how he did this with some of his thermoscopes. It adds to this puzzlement to recall that he referred in the *De statica medicina* to a means of determining the amount of daily respiration, and even specified a quantity thereof. This was based on weighing the water drops that collect on a mirror placed before the patient's mouth (Sect. 3.2.4). Thus, according to Sanctorius, the humidity of breath was related to its weight, exactly as was the humidity of air. Why then, did he not mention his hygrometers in this context? It is a question that must remain unanswered here. In any case, in the *Commentary on Avicenna,* Sanctorius claimed to have successfully treated patients who suffered from moist or dry diseases with the help of his hygrometers. This implies that he frequently used the instruments in his medical practice and related the humidity of air to an imbalance in the moist and dry qualities in his patients' complexion, which he diagnosed by other, most probably qualitative means. Still, since Sanctorius left no clues as to how he knew, for example, that it was his hygrometers, ultimately, that contributed to healing his patients, it cannot be ruled out that this was merely a rhetorical statement (Sanctorius 1614: 2r; 1625: 24).

7.5 The Sanctorian Chair

In the preceding chapters, I have written much about the *De statica medicina*, but little about its actual protagonist: the steelyard that Sanctorius designed in order to conduct his weighing procedures. In a way, this reflects Sanctorius's own silence on the instrument—he published an illustration and a short description of it only in his

Commentary on Avicenna, eleven years after the *De statica medicina* had been released (Sect. 6.1.2). Notably, although historical accounts ascribe an important role to Sanctorius's static medicine, supporting the identification of Sanctorius as the founder of a new medical science, up until now the design of his weighing chair and the method of measurement have not been closely analyzed. The aim of this chapter is to close that gap. Through a collaboration between the Max Planck Institute for the History of Science and the workshops of the Technical University (TU) of Berlin (Institute of Vocational Education and Work Studies), Sanctorius's weighing chair was reconstructed and used to conduct certain experiments. This was partly undertaken in the framework of a seminar in the History of Science Department at the TU Berlin.[60] The project opened up new perspectives on Sanctorius's works and his doctrine of static medicine, and led to a review of the function and purpose of his weighing chair.[61]

7.5.1 Sanctorius's Presentation and Use of the Weighing Chair

In keeping with Pamela Smith's apt description of the historian's use of reconstruction, I was obliged to approach Sanctorius's original method in reverse order.[62] Thus, while the early modern physician tackled the difficult task of translating his making and doing into (preferably published) images and words, I toiled to retranslate his codified output into processes and products. Such "reverse engineering" requires both textual and pictorial research, as well as hands-on research involving reconstruction.[63] The starting point for my investigation was the illustration and attendant description of the weighing chair provided by Sanctorius in the *Commentary on Avicenna*. As these are the only known primary sources on the original instrument, I quote Sanctorius's description at length:

---

[60] For a detailed visual documentation of the reconstruction project, see: https://www.mpiwg-berlin.mpg.de/research/projects/reconstruction-sanctorian-chair. The website was created with the kind support of Stephanie Hood.

[61] The following chapter is largely based on my article "The Weighing Chair of Sanctorius Sanctorius: A Replica," published in 2018. See: Hollerbach 2018.

[62] With regard to authorship and terminology, and depending on the context, the "I" in the following refers to Teresa Hollerbach, the author of this book, and to the various participants in the reconstruction project. These include Katharina Wegener, Volker Klohe, Matteo Valleriani, and Jochen Büttner as well as the participants in a seminar at the History of Science Department at the Technical University Berlin, during which some parts of the reconstruction and experimentation were undertaken.

[63] According to Eilam 2011, reverse engineering describes "the process of extracting the knowledge or design blueprints from anything man-made." The concept probably dates back to the time of the Industrial Revolution and is usually practiced to obtain missing knowledge, ideas, and design philosophy, when such information is unavailable, either because it is owned by someone who is not willing to share it, or because it has been lost or destroyed (ibid.: 3). Pamela Smith, for example, uses the term in her Making and Knowing Project, to describe the process of reconstructing techniques from a Renaissance manuscript (Smith 2016: 217).

The proposed aphorisms and those that are contained in our book of statics … are proven true by the use of this chair, from which we draw two advantages. First, how much *perspiratio insensibilis* of our bodies occurs daily: which, if not rightly weighed, renders medicine altogether vain. For nearly all bad illnesses usually originate from a smaller or larger perspiration than is proper. Secondly, sitting in this chair and easily eating in between, we observe when we reach the due quantity of food and drink, in excess of which or in shortage of which, we are injured. The chair is arranged as it appears in the figure [Fig. 7.23], in which the steelyard is suspended from the beams above the dining room, in a hidden place because of the nobles, as it renders the room less appealing, and because of the ignoramuses, to whom all unusual things appear ridiculous. The chair remains lifted from the floor at a finger's height, stable in such a way that it cannot be easily moved; when, due to the ingested food, one reaches the expected weight and the measure previously set, then the outermost part of the balance ascends a little and contemporaneously the chair descends a little. This descent immediately indicates to the sitter that he has arrived at the stabilized quantity of food; which quantity, or weight, of salutary food is advisable for somebody, and how high the insensible transpiration in the individual bodies should be, one weighs comfortably with the chair. This is easily understandable for everyone who reads our book *De statica medicina* (Sanctorius 1625: 557 f.).[64]

Hence, the Sanctorian chair consisted of a chair suspended from one of the beams of a large steelyard and was designed to monitor bodily losses by means of systematic weighing procedures. These losses indicated the quantities of sensible and insensible excretions and allowed Sanctorius to define a healthy quantity of the *perspiratio insensibilis*. Interestingly, the weighing chair also had another purpose, which was to determine the optimal healthy consumption of food for each person using the chair. Before a meal, one had to set a measure corresponding to the quantity of food one intended to ingest. During the weighing procedure, the weighing chair would drop. As soon as one had reached the set measure, the meal would end.

In Sect. 3.3.2, the close connection between insensible perspiration and the non-natural pair of food and drink was already outlined and it was shown how, according

---

[64] "Propositi aphorismi, & illi, qui continentur in libro staticae nostrae aliquot iam per annos in lucem edito, veritate comprobantur ex usu istius sellae: ex qua duo beneficia colligimus. Primum quanta quotidiè fiat corporis nostri perspiratio insensibilis: qua non rectè perpensa, vana fermè redditur medicina: namq; ob iusto pauciorem, vel largiorem perspirationem omnes ferè malae valetudines fieri solent. Secundum, in hac sella sedendo facilè intercomedendum animadvertimus, quando pervenimus ad debitam cibi & potus quantitatem, ultra vel citra quam, laedimur. Sella accommodatur, sicuti in hac figura apparet, in qua statera ad tigna supra caenaculum in loco ab. dito est appensa propter proceres, quia cubiculi gratiam tollit, ac propter indoctos, quibus omnia insolita videntur ridicula: Sella verò digiti interstitio à pavimento elevata manet, stabilis, ne facilè quassari possit: dum igitur ob cibum ingestum ad debitum pondus, & mensuram antea praescriptam devenimus: tunc staterae extrema pars paululum attollitur, ac una sella illicò paululum descendit: Hic descensus est ille, qui statim admonet sedentem ad debitam ciborum quantitatem pervenisse: quaenam verò ciborum salubrium. quantitas seu pondus unicuique conveniat: & quanta in singulis corporibus debeat esse perspiratio insensibilis quae per sellam commodè perpenditur, ex lib. nostro de statica medicina quisque facilè intelliget." See: Sanctorius 1625: 557 f. The English translation is made with the help of the Italian translation according to Sanctorius and Ongaro 2001: 33.

Fig. 7.23 The original
illustration of the
Sanctorian chair
(Sanctorius 1625: 557).
(© British Library Board
542.h.11, 557)

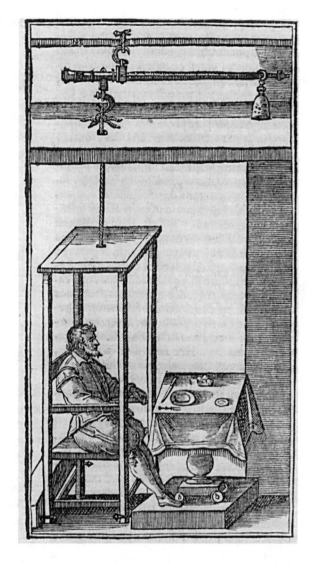

to Sanctorius, the healthy amount of food was directly related to the quantity of
perspiratio insensibilis. The fact that Sanctorius specified the monitoring of food
intake as one of the two functions of his weighing chair, in the *Commentary on
Avicenna*, shows that food and drink were particularly important with regard to the
use of the instrument. This is very much in line with the prominent place that these
non-natural factors had in the *De statica medicina* (Sect. 4.1.2). It seems thus that
the function and use of the weighing chair, just like the content of the *De statica
medicina*, responded to the great contemporary demand for food guidance.

One can only speculate why Sanctorius did not add an illustration and a description of the instrument to the original editions of the *De statica medicina*, even though the book's content is so closely connected with it. He simply might not have felt the need to do so. However, once later publishers or printers added the illustration of the *Commentary on Avicenna* to their editions of the *De statica medicina*, released after Sanctorius had died in 1636, it contributed much to the success of the work, as Giuseppe Ongaro has argued in the introduction to his edition of the *De statica medicina* (Sanctorius and Ongaro 2001: 34). Similarly, Lucia Dacome has identified the illustration of the weighing chair in her article on Sanctorius's doctrine of static medicine as "an integral, non-verbal and crucial component of static medicine's rhetorical apparatus" (Dacome 2001: 475). I shall demonstrate that the development of this illustration is not only indicative of the reception of the *De statica medicina*, but also reveals something about the material dimensions of Sanctorius's weighing procedures themselves.

The Original Use of the Sanctorian Chair No detailed records of Sanctorius's static experiments have been found. It is therefore commonly assumed that he did not leave any. Nevertheless, we know that he conducted them over a long period of time. According to his own claim, Sanctorius observed more than ten thousand subjects over the course of around thirty years (Sect. 2.2, fn. 9). To believe Sanctorius himself, he must have conducted the experiments constantly, as he wrote in the preface to the *De statica medicina*: "… the same experiments, in which I was daily engaged through continued studies for many years, …."[65]

Perusal of this work shows how carefully Sanctorius carried out his experiments. In one of the aphorisms, he specified the quantities of excrement expelled in one night: sixteen ounces of urine and four ounces of stool. This number, together with knowledge of the quantity of the food previously ingested, enabled Sanctorius to determine the quantity of the *perspiratio insensibilis* that was expelled in one night. According to his measurements, it amounted to forty ounces or more (Sanctorius 1614: 13v). In addition to the evacuation of feces, urine, and *perspiratio insensibilis*, Sanctorius also referred to sweat, although in these cases he did not specify exact quantities, but remained vague.[66] Moreover, Sanctorius did not only weigh

---

[65] "… quandoquidem ipsa experimenta, quibus quotidie assiduis multorum annorum studijs incumbebam, …." See: Sanctorius 1614: Ad lectorem.

[66] Sanctorius was most probably unable to differentiate between sweat and insensible perspiration in his weighing procedures (see below Sect. 7.5.3), which might explain why he did not give any numerical value for the amount of sweat. For his references to sweat in the *De statica medicina*, see ibid.: e.g., 4r, 5v, 10r, 14r–14v. For an analysis of Sanctorius's concept of sweat and its relation to insensible perspiration, see Sect. 3.2.7.

people before and after meals, but at regular intervals during the day and night.[67] Following the list of the six non-naturals, he tried to include parameters like climate, sleep, exercise, age, and even affections of the mind in his weighing experiments. Besides monitoring variations in the *perspiratio insensibilis*, Sanctorius also tried to regulate these variations in order to establish the parameters of an ideal balance between ingestion and excretion:

> How much perspiration is necessary for everyone, in order to preserve a state of perfect health, you will thus know. Observe in the morning, after a more abundant supper, the greatest perspiration which can occur in yourself in the space of twelve hours: suppose it be fifty ounces; some other morning, observe the same, but after having fasted and provided that there was no excess in the previous day's lunch: suppose it be twenty. With this established, choose such a temperance in eating and in the other non-natural causes, which can bring you every day to a mean between fifty and twenty, which is thirty-five ounces. In this way, you will lead a most healthy and long life, lasting to a hundred years (Sanctorius 1614: 14v–15v).[68]

A few aphorisms show that Sanctorius also observed the absolute weight of individuals using the chair. In this context he put forward an example of a healthy weight range between 200 *libbre* and 205 *libbre*. It can be assumed that this weight range referred to adults, perhaps even to Sanctorius himself, since the unit of *libbra* was equivalent to approximately one-third of a kilogram. Given that Sanctorius suggested here a *supposed* ideal weight range, he certainly allowed for other healthy weight ranges, too, dependent on the individual constitutions of people (Sanctorius 1614: 18v, 25r, 47v; Sanctorius and Ongaro 2001: 46).

In view of the scant information Sanctorius left us regarding his experimental setup and the experiments themselves, one might imagine that his brief description of the weighing chair together with the illustration would have given rise to many different interpretations. Indeed, some authors (among them Giuseppe Ongaro in his study of 2001) have felt the need to highlight that there was only a chair—and not a table or a bed, as others claim—hanging from the steelyard (Ettari and Procopio 1968: 64; Sanctorius and Ongaro 2001: 34). However, there seems to be a general consensus on the overall functioning of the weighing chair, and there is little or no discussion at all with regard to the exact design or the measuring method

---

[67] In the original Latin description of the weighing chair, Sanctorius wrote "… in hac sella sedendo facilè intercomedendum …." See: Sanctorius 1625: 557 f. See also the English translation above. The Latin preposition *inter* can be translated as either "between" or "during." In connection with the verb *comedere*, I consider the translation "between" to be more accurate.

[68] "Quanta conveniat perspiratio cuilibet, ut conservetur in statu saluberrimo, sic dignosces. Observa manè post aliquam pleniorem caenam illam maiorem perspirationem, quae in teipso duodecim horarum spatio fieri possit: esto esse quinquaginta uncias; alio mane; sed post ieiunium, hac tamen lege, ne in prandio praeteritae diei excesseris, idem observa; ponamus esse viginti; hoc praecognito, eligas illam cibi, & aliarum causarum non naturalium moderationem, quae te ad medium inter quinquaginta & viginti quotidie ducere poterit; medium erit triginta quinque unciarum; hoc modo sanissimam, & diutissimam seù centum annorum vitam duces." See: Sanctorius 1614: 14v–15v.

Sanctorius used.[69] Against this backdrop, I set out to reconstruct the Sanctorian chair.[70] Things soon began to look different, as I will show.

7.5.2   The Reconstruction of the Sanctorian Chair

I used the replication method to develop a deeper understanding of the mechanical knowledge involved in the *De statica medicina*. This approach can be summarized in three stages: reconstruction of the apparatus, replication of the experiments, and contextualization of the experience gained in the first two stages.[71]

Without discussing this methodology in detail, some aspects of how I applied it to Sanctorius's experiments must be mentioned to explain its potential to elucidate the practical aspects of the weighing experiments. My aim in the reconstruction was not a full *historical* replication, but rather what Hasok Chang would describe as a *physical* replication (Chang 2011: 320). First and foremost, I wanted to build a functional instrument with which I could reproduce the mechanical phenomena that formed the basis of Sanctorius's physiological observations. By using the technical potential of modern tools, my motivation was not to check the historical results, but to develop an understanding of historical practice. Given the anachronism inherent in the project, it was essential to proceed with a keen eye on both research methodology and modern

---

[69] As Lucia Dacome has pointed out in her article (Dacome 2001), many scholars performed Sanctorius's weighing experiments well into the eighteenth century, in France, Britain, Ireland, Colonial America (South Carolina), and the Netherlands. However, these imitators did not prioritize the historical accuracy of Sanctorius's experiments, but were interested rather in his novel idea and method of quantification. To them, the output was more important than the design and measuring method Sanctorius used. Thus, they left detailed static tables that indicate their commitment to drawing general conclusions concerning the relationship between intake, weight, and health, based on minute calculations of bodily excretions. Most of them did not even describe their experimental setup. Hence, it is not known which balances they used for their re-trials, whether they tried to reconstruct the original Sanctorian chair, or invented novel constructions. There are, however, two exceptions. In his French translation of the *De statica medicina*, the French scholar Louis-Augustin Alemand (1653–1728) pointed out some inconveniences that occurred when using the design of the weighing chair as proposed by Sanctorius in the *Commentary on Avicenna*. To overcome these problems, Alemand proposed another design based on an equal-armed balance. But from his illustration and short description of this device, it seems that he discussed and tried to improve Sanctorius's design of a weighing chair only in thought and not in deed (Sanctorius and Alemand 1695: Explication des Figures). The other exception is Jacob Leupold (1674–1727), who described in detail his own design of a weighing chair and also criticized the design illustrated in the *De statica medicina*. He even stated that this design cannot have been used in the way Sanctorius described it in his *Commentary on Avicenna*. See: Leupold 1726: 63.

[70] Examples for the common discussion in the secondary literature of Sanctorius's weighing chair and, more generally, the *De statica medicina* are: Miessen 1940, Ettari and Procopio 1968, Dacome 2001, Sanctorius and Ongaro 2001, Guidone and Zurlini 2002.

[71] The replication method is an attempt to analyze historical experimental practice, as applied systematically by members of the Oldenburg Group. This group was established in Oldenburg in 1987 under the direction of Falk Rieß (Heering 2008: 350, fn. 15). For an extensive study of the replication method and what the authors call an "experimental history of science," see: Breidbach et al. 2010.

assumptions. The focus of the project was not the historical details of how the balance was produced and used, but rather how it might possibly have been used. Thus, when I staged the experiments on the basis of the information provided in the source material, I tried to develop a deeper understanding of the experimental procedures and the skills involved in conducting them. Simultaneously, I reflected on my own practices with the instrument and how these practices developed over the course of the project (Heering 2008: 350, fn. 15; 2010: 796).

On the basis of the original source material, I developed a design proposal for the weighing chair. The illustration of the weighing chair (Fig. 7.23) indicates that Sanctorius used a Roman steelyard. As mentioned above (Sect. 7.1.1), scales of this type were widely in use at the time, and steelyards the size of the Sanctorian chair were used to weigh sacks of flour or other commodities. Therefore, it can be assumed that Sanctorius used an instrument already in circulation for his weighing chair, as in the case of his balances to measure climatic conditions. In contrast to Sanctorius, who suspended his weighing chair from the ceiling, I had to construct a stable framework in order to make my replica mobile, as I planned to exhibit it in different locales (Fig. 7.24). Moreover, the limited space in the TU workshops did not allow for a permanent installation of the instrument. Consequently, I had to calculate measurements that guaranteed a manageable size. At the same time, I had to make sure that the chair could be used by people of varying weights. I used a beam with a length of 1.5 m and defined a maximum load of 100 kg, including the weight of the chair. To keep the counterweight as light as possible, I decided to work with a ratio of 1:5, which corresponds to a counterweight of 20 kg for a load of 100 kg. This resulted in the following lengths for the arms that flank the pivot: a short arm of 25 cm and a long arm of 1.25 m. With regard to the materials, I chose structural steel for the beam and the pivot, and timber for the chair and the framework. The simple reason for this was that these materials were convenient, economical, and readily available through the stock of the TU workshops. After many hours of work in the wood and metal workshops, I finished a prototype with which I could begin experimenting (Fig. 7.24).

But this is only half the story.

The Measuring Method   At first, I assumed that Sanctorius used his model of a Roman steelyard in the traditional way described above (Sect. 7.1.1). But in the course of discussions, I recognized two difficulties. Firstly, Sanctorius wrote very clearly in his description of the weighing chair that a certain measure, which is set before the weighing starts, can be determined from the descent of the chair, that is, the chair's distance from the floor. This indicates that the weight of the load is not read from the position of the counterweight hanging from the beam of the steelyard. Secondly, the actual steelyard was hidden behind the ceiling above the dining room. Thus, the arms of the beam and the counterweight were very difficult to access.[72] Given the fact that Sanctorius used the weighing chair to monitor metabolic changes

---

[72] Interestingly, Louis-Augustin Alemand already referred to the inconvenience of reaching the counterweight above the false ceiling. See: Sanctorius and Alemand 1695: Explication des Figures. See also above, Sect. 7.5.1, fn. 69.

Fig. 7.24 The first prototype of the Sanctorian chair. (© Philip Scupin)

in many individuals of varying weights, he would have to balance the arms of the steelyard by moving the counterweight for every individual sitting on the chair anew—if he used the steelyard in the common way.

With these considerations in mind, I took another look at the original illustration of the Sanctorian chair. This time I specifically examined the lower part of the weighing chair. I could clearly identify little pointers or pegs at the bottom of the chair, attached to each leg. What were they intended for? Did they point to a scale that indicated to the sitter when he had reached the proper weight? Were they used to add weights to the sitter? Or did they serve to stabilize the suspended chair and prevent it from swinging? It is striking that this detail varies in later reproductions of the original illustration, and that the variation has never been discussed. In the following, I shall briefly refer to one of the reproductions: the frontispiece of a Dutch edition of the *De statica medicina*, written by the physician Heidentryk Overkamp (1651–1693) and published posthumously as part of his *Opera Omnia* (Overkamp 1694).[73]

The frontispiece shows a version of the Sanctorian chair (Fig. 7.25), in which one can clearly identify one little pointer or peg at the rear end of the chair's right-hand stretcher. Moreover, in contrast to the original illustration, it shows not only the person sitting on the weighing chair, but three other people, too. The two on the right appear to be discussing the beam of the steelyard, the part of the weighing chair that is hidden behind the ceiling, in the original. On the left, another person seems to be bending to reach the lower part of the chair, close to the point where the pointer or peg is placed. From this illustration alone, one cannot deduce with any certainty the purpose of the pointer or peg. Nor can it be known whether the person leaning forward is a craftsman, a servant, or a spectator interested in the weighing process. It is known, however, that this person was not there to remove feces from the chair, as the weighing chair was not designed to be used as a lavatory. Sanctorius

Fig. 7.25 An illustration of the Sanctorian chair (Overkamp 1694: frontispiece). (Courtesy of Niedersächsische Staats- und Universitätsbibliothek Göttingen (SUB Göttingen))

---

[73] My thanks to Ruben Verwaal, who drew my attention to this illustration of the Sanctorian chair.

stated as much in his own defense, in response to his detractor Ippolito Obizzi's harsh allegation, that the weighing chair was used for the inappropriate practice of weighing fecal excreta (Obizzi 1615: 3, 38 ff.; Sanctorius 1634: 69v).[74]

Therefore, even though many questions remain open, it is evident that the lower part of the chair and its descent are of significance with regard to the weighing procedures, and most probably for the measuring method as well. These were interpreted in the reception of Sanctorius's *De statica medicina* in different ways, but had never yet been included in a historical reconstruction. It was with this in mind that I started the experiments with my replica.

7.5.3   Experimenting with the Reconstruction

The experimentation process can be divided into four phases. In the first phase I used the prototype mentioned above (Fig. 7.24), with two people of differing weights. In a second phase I experimented with an adapted and improved version of the prototype, which I constructed in the light of the experience gained in the first phase (Fig. 7.26). In this second series of experiments, seven different individuals used the chair. Subsequently, I again set out to further adapt and improve my prototype. I planned to conduct my next experiments with many different people and had to prepare my reconstruction accordingly.[75] In the fourth and final phase of experimentation, in order to more closely approximate Sanctorius's experimental practice, I took the reconstruction home with me.

The First Two Phases In the first two phases, I conducted the experiments over several hours on one day. The aims were to test the functioning and precision of my reconstruction, to analyze different possible measuring methods, and to define potential scales. I thereby hoped to better understand the mechanical knowledge involved in the weighing procedures Moreover, in performing Sanctorius' experiments myself, I aimed to develop a better understanding of the methodology underlying them. As the purpose and use of the weighing chair are closely connected to its design, these objectives could not be analyzed separately but had to be considered as complementary factors. In the following, I will give a brief overview of the two series of experiments and present the conclusions that I drew from them.

Before any actual weighing can begin, a starting point must be defined. This point guarantees the universal validity of the measuring process, with universal

---

[74] "Staticus scit pondus faecum, licet eas nec videat, nec perpendat. Corpus ante perpendit & iterum post omnem excretionem: quod deficit est earum pondus: Sic non est indignum perpendere faeces, ut ait trico." See: Sanctorius 1634: 69v. In the secondary literature, it is also sometimes erroneously stated that Sanctorius weighed feces by means of a balance. See e.g., Major 1938: 374, Poma 2012: 215.

[75] I found an opportunity to do so in the framework of the Long Night of Sciences in Berlin, an established public science fair regularly held in Germany (see below).

Fig. 7.26 The adapted prototype of the Sanctorian chair used by Matteo Valleriani and Teresa Hollerbach. (© Paul Weisflog)

means valid for everyone, regardless of each individual's constitution. Sanctorius weighed many people of different weight, so always had to ensure that the beam of his weighing chair was optimally weighted for the person in question, before he could begin his experiments. The beam could be in any position, as long as it was the same for everyone using the chair, but there is good reason to suppose that the preferred starting point was the balanced, horizontal position of the beam about the pivot—the right-angle being a common reference point, most easily measured with the eye. Most probably there was a marking somewhere at the bottom of the chair that indicated when the chair had arrived at the starting point. There are various ways to define this point. The most obvious is to use the steelyard in the classic way, by moving the counterweight attached to the beam above the ceiling. This would have had to be done for every person anew. Another method, not so obvious but far more convenient, is to add weights to the person sitting on the weighing chair, to compensate for the differences in weight. Thus, the beam of the steelyard is balanced once for a rather heavy weight of test person, and further weights are added, as necessary, but this time to the load (that is, to the chair). With this method, the counterweight does not have to be moved for each individual to reach the starting point. In Sanctorius's case, this would have made it possible to balance the weighing chair without always having to climb up to the ceiling above the dining room.[76]

In addition to the starting point, there least one other marking at the bottom of the chair. As mentioned above, Sanctorius referred in his description of the weighing chair to a certain measure that was set before the weighing started and that indicated to the sitter when he had ingested the sufficient amount of food and drink. Sanctorius explicitly stated that the quantity of ingested food and drink was indicated by the *descent* of the chair. Thus, he used the weighing chair not only to observe weight loss, but weight gain, too. Where exactly this second mark would have had to be made—i.e., the mark for the quantity of food and drink Sanctorius would advise an individual to ingest—remains vague for the modern reader. Given the character of the *De statica medicina* as a dietetic handbook and its orientation to the six non-natural things, Sanctorius most probably connected it not only with the amount of excreted insensible perspiration, but also with the six non-natural factors, thereby including a variety of parameters that influenced the quantity of food and drink that an individual person should ingest. This leads one to conclude that Sanctorius based the position of the second mark on contemporary dietetic knowledge and the experience he gained during the weighing procedures.

So, when I tried to define this second mark in my experiments with the replica, I was not dealing with an exact quantity but rather attempting to determine how a certain value (the position of this second mark) could be universally determined for the various individuals using the chair. Here again, there are various options, depending on the method used to determine the starting point. If one balances the beam by moving the counterweight suspended from the steelyard above the ceiling—with regard to the weight of each individual, as explained above—the descent of the chair is proportional to the weight of the load. Therefore, whatever the weight of the occupant of the chair, a single mark would suffice to show when she or he had consumed the amount of food and drink necessary to lower the chair to this preset

---

[76] At the beginning of the eighteenth century, Jacob Leupold designed a portable weighing chair, the *machina antropometrica*, to which he applied a similar measuring method. In his work *Theatrum Staticum Universale* (1726), Leupold described how the person using the *machina* should determine the counterweight on the basis of an estimate of his own body weight. According to Leupold, this estimate did not have to be accurate, as there were additional weights that the sitter put on the arms of the chair to compensate for inaccuracies and to balance the two arms in a horizontal position about the pivot. See: Leupold 1726: 64–6; table XVIII.

measure. Another option would be to use the steelyard in the common way to determine the weight of a person as well as to set the desired weight of food and drink
that this person should ingest, and then also move the counterweight to a position
such that the chair not only descends toward, but actually rests on the ground, once
the preset measure has been reached. Using the ground floor in this way as an indicator of when the desired amount of food and drink has been ingested is of course
also possible for the method mentioned before, instead of setting a mark close to the
bottom of the chair. If one uses the other method, namely adding weights to the
person sitting on the chair to balance the beam of the steelyard, the amount of food
ingested can be indicated by using a graduated scale. Here, the initial weight of the
load is the same for everyone using the chair. Thus, the descent of the chair after a
meal is not proportional to the weight of the individual person. The addition or
removal of weights to or from the chair might have enabled Sanctorius to identify
exact quantities not only by looking at the position of the counterweight on the
beam of the steelyard installed behind the false ceiling (presumably, a difficult task),
but also by noting the chair's distance from the floor.

On the basis of my practical experience of the different measuring methods
paired with the examination of the source materials, it appears most plausible that
Sanctorius used the steelyard in the classic way, in order to define the starting point
for the measurements. Most probably, an assistant climbed up above the false ceiling to move the counterweight until the beam reached the balanced position. The
height of the chair was then noted. A second mark was made at the bottom of the
chair to indicate when a person had ingested the required amount of food and drink,
which was measured exclusively in terms of how far the chair had descended toward
the floor. As outlined above, there are good reasons to assume that Sanctorius did
not work much with the counterweight and that his daily weighing practice centered
rather on measuring the chair's distance from the floor. Yet, although the method of
adding weights to the person sitting on the chair necessitates the least use of the
counterweight and relies wholly on measuring the descent of the chair, it turned out
that it also easily leads to errors. The added weights must be distributed equally over
the chair, in an identical position for each measurement, so that the chair does not
descend more on one side than on the other. This is extremely difficult, unless a
special storage place for the weights is integrated into the chair. Perhaps this is the
reason there is no evidence in Sanctorius's illustration and description of the weighing chair either of this solution being used, or, more generally, of weights being
added to, or removed from the chair.

Precision of the Sanctorian Chair and of the Replica As soon as I started to
include the descent of the chair in my procedures and to test its possible function as
an indicator of changes in weight, flaws in the reconstruction came to light. It turned
out that the chair was very unstable and sensitive to any kind of movement. Thus,
the various persons using the chair not only had to keep still during the measuring
process, but also had to adopt an identical seating position. To prevent the chair
from rotating to one side, I replaced the rope that suspended the chair in my first
version of the prototype with a steel chain that I attached to the chair with the help

Fig. 7.27 The suspension
of the chair. (© Paul
Weisflog)

of a U-bolt (Fig. 7.27). Additionally, on the basis of the original illustration of the
Sanctorian chair, I placed a wood panel behind the chair, attached to the framework
(Figs. 7.24 and 7.26). This also helped me prevent major oscillations, even though I
had to be careful to keep the friction between the wood panel and the chair legs to a
minimum, so as not to falsify the measurements. Even a minor disequilibrium
caused perceptible differences in the descent of the chair. As stated above, this
became even more obvious when I started to add weights to the person sitting on the
chair. The added weights had to be distributed equally over the chair to prevent it
from descending more on one side than on the other. A spirit level attached to the
top of the chair helped me to monitor its horizontal position.

My experiences showed that suspending the chair from the beam, as in the original illustration, makes the chair prone to rotation, its descent uneven, and hence the
measurements hard to read accurately. However, Sanctorius was well aware of this
difficulty, as he stated in the description of the weighing chair: "the chair remains …
stable in such a way that it cannot be easily moved; …" (Sanctorius 1625: 558).[77]
Unfortunately, he did not reveal to the reader how he achieved stability. Thus, I can
only speculate that he might have used the pegs near each chair leg for stabilization.
Arranged between the wood panel behind the chair and the platform beneath the dining table, the pegs might have served to guide the chair's descent and make it as
steady as possible. Perhaps the pegs were actually iron nails, whose shanks against
the chair's uprights were meant to limit its rotation and prevent it swinging from side
to side, while their heads would prevent it from swinging forwards. Furthermore,
another detail in Sanctorius's illustration is interesting with regard to the stabilization
of the chair. The feet of the man seated on the chair rest on the dais on which the table
is placed (Fig. 7.23).[78] Since this makes no sense with regard to the weighing

---

[77] "Sella verò … manet, stabilis, ne facilè quassari possit: …." See: Sanctorius 1625: 558.

[78] It is interesting to note that this detail does not vary in later reproductions of Sanctorius's original
illustration of the weighing chair except for the frontispiece to Heidentryk Overkamp's edition of
the *De statica medicina* (Fig. 7.25), in which the feet of the person, sitting in the chair, do not rest
on the platform but on the chair's bar.

procedure and would even falsify the measurements, it is conceivable that the man is in fact steadying himself *before* the measurements are made. By this interpretation, the illustration does not represent a snapshot view of the weighing procedure, but rather combines discrete operations. This reading of the illustration is very much in line with the assumption that Sanctorius weighed people only *before* and *after* and not *during* meals, even though the illustration shows a laid table (Sect. 7.5.1, fn. 67).[79]

When I tried to determine where to make the second mark, I realized that the chair's descent was not proportional to the weight of the load and, moreover, was affected only by large differences in weight. As the figure shows (Fig. 7.24), in the first version of my reconstruction the pivot is located between two steel rings that are welded together and form the fulcrum, which is attached to the stable frame-work, my substitute ceiling. In order to make my weighing chair more precise, I had to minimize the distance between the pivot and the lever. However, I had to be care-ful to find the right distance, as minimizing the distance between the pivot and the lever not only makes the steelyard more precise but simultaneously causes smaller inclinations of the beam, which makes it more difficult to determine minor differ-ences in weight. Hence, I had to find a solution that on the one hand, guaranteed the necessary precision of the weighing chair and on the other hand, still allowed me to read the measurements at the bottom. My modern solution to this problem was a ball bearing (Fig. 7.28). Sanctorius, of course, had to find another method. The original illustration of the weighing chair shows that he connected the lever directly to the hook on the ceiling with some kind of box or rectangular guide, which made the distance between the pivot and the lever relatively small.

The precision of a steelyard also depends on the length of the beam. To adapt this parameter to my needs in relation to the different persons using the chair and to guarantee maximal precision, I replaced the initial suspension hook with three hooks at different positions on the beam of my prototype (Fig. 7.28). This resulted

Fig. 7.28 The ball bearing to minimize the distance between the pivot and the lever. (© Paul Weisflog)

---

[79] I am grateful to Roger Gaskell for pointing out to me the interpretation of the pegs as iron nails and for suggesting I read the original illustration of the weighing chair not so much as a snapshot but as a stop-motion image, in which the man in the chair might have placed his feet on the dais in order to stabilize himself *before* the measurements were made.

in the following lengths for the arms flanking the pivot. First hook, short arm: 17.5 cm; long arm: 1.325 m. Second hook, short arm: 23.5 cm; long arm: 1.265 m. Third hook, short arm: 29.5 cm; long arm: 1.205 m. My experiments with the different hooks showed that the third, foremost hook (the one closest to the beam's front end), was ideal for my load weight range of 66–75 kg when working with a movable counterweight of 20 kg.

On the basis of the original illustration of the Sanctorian chair, it can be assumed that Sanctorius used a beam with a length of around 3 m—twice as long as the beam in my reconstruction. This enabled him to achieve great precision in his measurements and to reduce the counterweight. Sanctorius might well have equipped his weighing chair with different hooks, too, even though the illustration does not clearly indicate this. Steelyards with up to three suspension hooks had been in use for weighing objects of varying weights since the Roman Empire (Robens et al. 2014: 169).

The adapted and improved version of my prototype with regard to the oscillation of the chair, the distance between the pivot and the lever, and the length of the beam allowed me to measure differences in weight by means of the descent of the chair with a precision of up to 100 g, in the second series of experiments. This comes close to the precision that Sanctorius claimed to have measured in the *De statica medicina*. The minimum quantity to which Sanctorius referred in his aphorisms is four ounces, which, if calculated with the Venetian *oncia sottile*—corresponds to around 100 g (Sect. 5.4.2, fn. 39).[80] In the aphorism mentioned above (Sect. 7.5.1), Sanctorius stated that up to sixteen ounces of urine were usually expelled in one night. In several other aphorisms, especially of the third section, *Food and Drink*, he gave quantities of six, twelve, fourteen, eighteen, and twenty-two ounces. He wrote for example: "Very nourishing foodstuffs, except for mutton, usually do not perspire more than eighteen ounces in the time between supper and lunch" (Sanctorius 1614: 32v).[81] This indicates that he worked with a steelyard that had a precision of one ounce. This in itself is nothing out of the ordinary: at the time, steelyards were used to weigh loads ranging from ounces to tons. But merchants and traders who had to weigh small, ounce-sized merchandise usually used small, portable steelyards of only some ten centimeters in length (Robens et al. 2014: 169). In contrast, steelyards of the size of Sanctorius's weighing chair were commonly used to weigh sacks or barrels of commodities in which precision to the ounce was hardly needed. Thus, the mechanical challenge of the Sanctorian chair is to develop a design that, on the one hand, allows the weighing of heavy loads up to around 80 or 90 kg, and on the other, guarantees precision enough to be able to note even minor variations in weight.

---

[80] For Sanctorius's references to four ounces in the *De statica medicina*, see: Sanctorius 1614: 13v, 33r, 40r–40v.

[81] "Cibi multum nutrientes, excepta carne vervecina, à caena ad prandium non solent perspirare ultrà octodecim uncias." See: ibid.: 32v. For further examples, see: ibid.: 32r, 39r–39v, 40r–40v.

Reading of Measurements I developed and tested various methods for reading the measurements. I made marks on the beam of the weighing chair to indicate the respective starting point for each person using the chair. This was relatively easy and became difficult only when I tried to discern differences in weight. Calibration of the longer arm requires skill and great accuracy. Since I worked with a counter-weight of 20 kg, it was extremely difficult to record minor weight differences, which corresponded to only very short lengths of the beam. Sanctorius probably did not face these problems, as we can assume that he worked with a calibrated steelyard, a type widely in use at the time.

To monitor the descent of the chair, I developed various solutions that I tested in my experiments. Figure 7.29 shows that I attached to one leg of the weighing chair a wooden arrow, whose height could be marked and then measured on the wood panel behind the chair. Inspired by another reproduction of the illustration of the Sanctorian chair, I attached to a different leg a wooden duct that served to hold upright a steel bar resting on the ground. The steel bar helped me ascertain the chair's distance from the floor (Fig. 7.30).[82] My experiments showed that the use of the arrow to indicate the chair's descent was problematic. Although the arrow's

Fig. 7.29 Wooden arrow as indicator of the descent of the weighing chair. (© Philip Scupin)

[82] For the illustration, see: Beugo n.d.

Fig. 7.30 Steel bar as indicator of the descent of the weighing chair. (© Philip Scupin)

height could be marked on the wood panel, reading the measurements in this way was very difficult. Even though I had already enhanced the stability of the chair in my second prototype, the person seated on the chair still had to remain in a very stable, balanced position to prevent the chair from descending more on one side than on the other. For every measurement, the distribution of the load on the chair had to be identical. The steel bar proved far easier to handle and permitted a highly accurate reading of the measurements. As the figure indicates, a graded scale was still missing at this point. In the next version of the reconstruction, however, I attached a ruler to the steel bar.

Given the depiction of pointers or pegs inserted into each chair leg in the original illustration of the Sanctorian chair, it can be assumed that these might have served as indicators of the chair's descent, similar to the arrow that I used in my experiments. However, this cannot be deduced with certainty. As mentioned above, they might also have served as stabilization. Further, they possibly had a dual function. The two pointers or pegs at the rear end might have served as fixed guides to ensure stability, and the ones at the front end as indicators of the descent of the chair, pointing to the platform on which the table is placed. There is no evidence that Sanctorius used a steel bar as an indicator, since one appears only in a later reproduction of the Sanctorian chair; it also differs slightly from the one I used in my experiments. I applied the steel bar to my reconstruction to investigate different possibilities for measuring the descent of the chair.

The Exhibition of the Sanctorian Chair In a next step, I wanted to test my reconstruction on many different individuals. The Long Night of Sciences in Berlin (Fig. 7.31) seemed a perfect opportunity both to do so and, at the same time, to present my research project to a wider audience. During this annual event, science and research institutions that are usually closed to the public open their doors to visitors. In different formats, such as lectures, demonstrations, or exhibitions, the institutions present themselves to the general public and give an overview of their research topics.[83] In preparation for this third phase of experimentation, I further adapted and improved the prototype. The original balance beam was fitted with an

Fig. 7.31 The exhibition of the Sanctorian chair at the Long Night of Sciences in the Max Delbrück Center in Berlin 2018. (© Stephanie Hood)

---

[83] For more information on the Long Night of Sciences in Berlin and Potsdam, see: https://www.langenachtderwissenschaften.de. Three years later, I exhibited my reconstruction of the Sanctorian Chair again, on the occasion of the City of Science Berlin 2021, a project to showcase Berlin as one of the most exciting locations for science and research in Europe. For more information and images, see: https://www.mpiwg-berlin.mpg.de/news/mpiwg-exhibits-wissensstadt-berlin-2021-review.

extra length of structural steel, extensible up to 50 cm, as required. This served to enhance precision and extend the weight range of suitable candidates for testing the chair. Moreover, I equipped the chair with a more stable suspension, made of wood and a ball bearing, to prevent any lateral movement. I planned to use the measuring method that I had identified as the most viable one, in the light of the first two phases of my experiments with the reconstruction. However, instead of making marks on the wood panel behind the chair to indicate the the chair's distance from the floor, I would use the steel bar (to which I had meanwhile attached a ruler) to do so, and then record the values on a sheet of paper. Only the starting point for each individual was marked on the instrument itself—on the beam, namely, to show the position of the counterweight when the test person was seated in the chair and the beam was perfectly horizontal. Besides requirements concerning the design and functioning of the reconstruction, the special setting also posed other challenges. While I had previously worked in a closed environment, I was now engaging an audience that was completely unfamiliar with the subject and had no background in historical research. Furthermore, members of this audience became not only "guinea pigs" (test objects) by using the chair, but also factors integral to my ongoing research. This became especially obvious as, despite my extensive planning and preparation, the new experimental setup produced different results than expected.

My initial idea for the public exhibition of the Sanctorian chair was to offer visitors bananas and water between their "weigh-ins," so as both to make weight changes visible by means of the chair's descent and to illustrate the concept behind Sanctorius's weighing procedures. But as it was a very hot day, and visitors were not very eager to eat bananas, I altered the test while sticking to the measuring method. Once an individual was seated, I marked the position of the counterweight on the beam as soon as the balanced position was reached. Simultaneously an assistant noted the chair's distance from the floor, using the steel bar. I then asked our volunteers to neither eat, drink, nor use the toilet prior to their second weigh-in, which was to reveal how much they had perspired. If they cheated, the instrument would betray them. A surprisingly large number of individuals accepted the assignment, and returned at different time intervals to learn more about their *perspiratio insensibilis*. Of course, it was not only insensible perspiration that my instrument measured, but also, and probably to a large extent, sweat.[84] This, however, disclosed a fundamental problem that Sanctorius must have encountered as well. How did he differentiate between sweat and insensible perspiration? Did he do so at all? My experience at the Long Night of Sciences helped me grasp what it must have meant for Sanctorius to indirectly measure invisible bodily losses by means of a balance and changes in weight. It gave good reason to assume that Sanctorius did not give any numerical values for the amount of sweat because he simply was not able to.

---

[84] In the following, I do not differentiate between sweat and invisible losses, when referring to the experiments with the reconstruction, for the simple reason that we cannot distinguish between the two.

The alternation in the experiment, from measuring gains resulting from the bananas and water consumed, to measuring invisible losses, also had an important consequence for my methodology and conclusions. While the previous tests in the TU workshop mainly served to assess the function and precision of my reconstruction of the chair, as well as to examine different possible measuring methods and the definition of potential scales, they did not concern Sanctorius's experiments in the stricter sense. Before I could tackle Sanctorius's actual research interest, the measurement of *perspiratio insensibilis*, I had to make sure that my replica was working as it should. But in the public setting, encouraged by the active engagement of the visitors, with the knowledge that the steelyard could detect weight changes with a precision of up to 100 g, I endeavored to further the experiment—and I did so with success. The measuring results showed that it was possible to detect invisible losses by means of my weighing chair and with the measuring method that I employed.

Yet, the experiments at the Long Night of Sciences also disclosed some problems arising from using the weighing chair for numerous people. On a general level, my measurements worked and the measuring method that I had chosen proved relatively easy to implement. But the fact that I measured, in quick succession, many different people of a different weight made the weighing process feel laborious. I constantly had to work with the counterweight in order to first determine the starting point for each individual and then to return the counterweight to that same customized position for the second measurement. During the phases when the test persons entered and exited the chair, I had to exercise caution to prevent the counterweight from rising or dropping down in an uncontrolled manner. This required attention on the part of the test persons, too. For their first measurements, an assistant was on hand to help them into and out of the chair, although with a little practice this can easily be done alone. Furthermore, despite having fitted the beam with an extensible component, I was unable to cover the entire weight range of the children and heavy adults among the test persons. By contrast, reading the chair's distance from the floor was unproblematic.

These observations allow some further conclusions to be drawn regarding Sanctorius's weighing procedures. If his claim to have weighed more than ten thousand people is true, he must have encountered problems similar to my own during the Long Night of Sciences. Although there can be little doubt that he had an assistant who was much better trained and more familiar with the handling of a steelyard than I, the frequent moving of the counterweight still must have been exhausting and time-consuming; and all the more so, given that the mechanism was hidden behind a false ceiling. But at the same time, this detail might have been useful. When the weighing chair was not in use, the counterweight could simply sit on the ceiling; and when a person entered or left the chair, it would move only a little; thus, the danger of its uncontrolled movement was greatly limited. Still, Sanctorius needed to instruct every single test person on how to properly enter, leave, and sit on the chair ideally, in an always identical manner. Their level of cooperation and skill would thus influence the measurements and affect the comparability of the measurements gained from the various individuals. To cover a broad range of weights, Sanctorius probably used different counterweights and might have also worked with various suspension hooks. But if several people with differing weights used the chair consecutively, the individual adjustment of the weighing chair would be quite complex and time-consuming. In this regard, the inclusion of the descent of the chair in the measurements might not have been so practical, since one had to work a lot with the counterweight and the hidden steelyard anyway. Contrary to me, Sanctorius most certainly worked with a calibrated steelyard and it is therefore conceivable that he used the steelyard exclusively in the classic way, when observing differences in weight in many people. However, the weighing procedures that I undertook during the exhibition were still far removed from the observations that Sanctorius described in his work *De statica medicina*. To further approximate his experimental practice, I took the replica home with me.

Reenacting the Weighing Procedures When reenacting the weighing procedures of Sanctorius, I had to consider the different parameters that the Venetian physician allegedly included in his measurement of insensible perspiration. Following his reinterpretation of the doctrine of the six non-natural things, he tried to examine the effect of climate, sleep, exercise, coitus, and even states of mind on the excretion of the *perspiratio insensibilis* (Sect. 3.3). In an attempt to find out if it is truly possible to take into account all of these parameters in the weighing procedure, I decided to commence a test series in which I myself would be the guinea pig. This required that I meticulously record my food intake, my tangible and intangible excretions, my sleep patterns, the weather, and my mood in the intervals between the measurements. I weighed myself before and after eating and drinking, before and after going to the toilet, before and after exercise, before and after sleeping, and whenever something occurred that might potentially influence my physiology.[85]

As this suggests, my imitation of Sanctorius's procedures demanded a high level of self-discipline and a regular and uniform lifestyle, always within reach of the weighing chair. I needed to develop an intimacy with the balance akin to that which some people share with their smartphones or fitness trackers. The big difference, however, was that while "wearable technology" can easily be transported, I had to stay close to the weighing chair, to make sure that no change took place unnoticed. I could not simply go out and meet friends, but had to invite them to my flat. When I did so, they became direct witnesses of my weighing procedures, which provoked mixed reactions: sometimes interest, always astonishment, and occasionally perplexity or even amusement. My regular sports activities had to be adapted, too. No longer could I go to the gym for longer periods of time, since I was not supposed to drink or go to the toilet without monitoring any changes in my weight before and afterwards. Moreover, I had to work from home, without the constant exchange with colleagues, or the technical facilities of my usual working environment. In short: experimenting with the weighing chair entailed inflexibility, isolation, and a complete orientation of daily life toward the requirements of the weighing procedures.

---

[85] For a brief description of a pilot phase of the reenactment of Sanctorius's weighing procedures, see: Hollerbach 2018: 141 f.

Due to these constraints, I only stayed the course for two days. For another four days, I confined myself to measuring how much I perspired when asleep at night.

My experience over the six-day test period unveiled an important dimension of the Sanctorian weighing procedures: the problematic status of living beings in an experiment. In her study on nutritional physiological experiments in the nineteenth century, Elizabeth Neswald pointed out some important characteristics when experimenting with living beings, as opposed to with inorganic objects (Neswald 2011). There is, for example, a large variability not only among different individuals, but also in a single individual at different times. Hence, even if I recorded intangible bodily losses, it was difficult to determine whether these were caused by my mood or the weather. How did I know which parameter caused which effect, or whether they influenced my body simultaneously? Without the help of statistical methods, one would need to considerably scale up the weighing procedures in order to at least detect some tendencies. Contrary to inorganic objects, living beings can be standardized and manipulated only to a limited extent. They have individual needs, interests, preferences, and boundaries. Hence, the test persons actively participate in experiments and thereby add a new dimension to the interaction already at work between the experimenter and her instruments. As Neswald suggests, the success of any physiological experiment depends on the level of cooperation between the different actors, human/animal and material (Neswald 2011: 61 f.). In my case, the situation was unique because I conducted experiments on myself.

Once a helper brought the steelyard into a balanced position for my weight, I could use the weighing chair all by myself. I just needed a ladder on which the counterweight could sit when I was not using the instrument, a small stool to help me enter the chair, and my smartphone to film the ruler attached to a chair leg to indicate the descent of the chair. Knowing the distance from the chair to the floor for my initial weight, I was able to detect weight changes by measuring the descent of the chair: 1 mm on the ruler corresponded to 100 g.[86] As this implies, the experimental situation was quite different from the previous ones. The replica entered into my private life and I interacted with the instrument on an immediate level—without any spectators or the assistance of fellow researchers. However, the struggles to discipline my behavior and to adapt my daily routine to the requirements of the weighing procedures affected my body and therefore also the outcome of the experiments. I experienced firsthand what Neswald meant when she wrote that the needs and constraints of the test persons in nutritional physiological experiments forced the researchers to modify their experiments, to shorten their planned duration, to prepare for new variables, and to accept the imprecision that resulted from these changes (Neswald 2011: 69). Although I was both the experimenter and test person in one, and was thus highly motivated to conclude the experiments successfully, my body signaled resistance. The isolation and

---

[86] Before I started the experiments, I again tested the proportionality and precision of the instrument by simulating weight changes through adding weights to the person seated in the chair. This showed that the chair descending by 1 mm corresponded to 100 g of weight change. However, the measurements were still prone to inaccuracies when working with very small weight changes, due to the difficulty of, for example, always adopting the exact same posture.

loss of freedom that the weighing procedures entailed were difficult for me to cope with. I had never before experienced such constraints in my daily life and quickly reached the point where I found them unbearable.

My "resistance" was more psychological than physical. Given that the weighing procedures structured every aspect of my life, I thought about them nonstop. Knowing that I had to weigh myself whenever I did something that possibly influenced my physiology, I had to train myself to recognize the situations requiring me to sit on the weighing chair. But this resulted in a certain bias that impacted my behavior. Even though the weighing itself was easy to conduct, I found myself trying to limit the weighing procedures as much as possible. Usually, I drink small amounts of water very often throughout the day. During the experiments, I tried to switch to drinking larger amounts of water only a few times a day. Similarly, I stopped eating snacks throughout the day and ate only three larger meals daily instead. Thus, also here, my behavior actively shaped the experimental practice and the outcome of the weighing procedures. My emotions and my mind influenced the way I dealt with the artificial experimental situation and made me deviate from my "normal" routines. As Neswald aptly put it with regard to nutritional physiological experiments in the nineteenth century: "normality, the normal metabolism, could only be studied under normal conditions, which, however, ran counter to the conditions of the experiment" (Neswald 2011: 73). Already during the whole reconstruction process, I had become fluent in handling the instrument in order to realize my research agenda. And yet, it turned out that I was not prepared for the dictates that the instrument imposed on me once it was installed next to my bedroom. This was, indeed, a very instructive experience.

As previously mentioned, the aim of my experiments was not to verify Sanctorius's exact results, but to develop an understanding of his method. The calculations for my *perspiratio insensibilis* were intended to give me a general idea of how Sanctorius's weighing practice might have looked; they did not provide reliable data to verify Sanctorius's measurements. In order to reach a certain comparability between the present-day procedures and those undertaken by Sanctorius, one needs far more than a functional replica. In order to conduct the experiments in an identical climate to that of the historical setting, one would have to feed the test persons a Renaissance diet and move the weighing chair to Venice. But even if such measures were taken, problems like the different physiologies of early-modern and present-day individuals would remain. Here again, the fact that the experiments were physiological and undertaken with living beings complicates matters. Yet, despite these limitations, the observations made during my reenactment of Sanctorius's weighing procedures may be reinterpreted in order to gain a better understanding of his work.

7.5.4 The Weighing Procedures of Sanctorius

Just like the experiments that I conducted with my replica, Sanctorius's weighing procedures and their outcome were actively shaped by his test persons. Hence, in order to be successful he had to find cooperative and suitable research subjects. My experience during the experiments on myself revealed that the use of the Sanctorian chair must have been very demanding, requiring participants to put themselves wholly at the service of the experiment. Moreover, Sanctorius had to find people who were ready to closely follow his instructions and thus to interact not only with the instrument, but with the physician, too. Since the weighing chair most probably stood in Sanctorius's house, they had to stay there at the least for the duration of the weighing procedures. This shows that there must have been a high level of intimacy between the experimenter Sanctorius and his research subjects. He would monitor every visit to the toilet, every bite they ingested, and even any sexual activity. At the Long Night of Sciences, I could easily tell if a person did not comply with my request and went to the bathroom between measurements. But, of course, I was very hesitant to address their cheating or carelessness, since most people are not too eager to talk openly about their excretions.

So, who might the persons have been, willing not only to adapt their whole lifestyle to the weighing procedures, but also to let Sanctorius control and monitor their excretions? Based on my research, I think that Sanctorius could have conducted long-term measurements only with people from his immediate environment, probably with colleagues or friends. As Neswald has pointed out, for a willingness to subject oneself to the constraints of physiological experiments, it is very helpful to have an interest in, and an understanding of the research involved (Neswald 2011: 70 f.). Another possibility would be that Sanctorius paid people to sit on his chair. But given the intimacy and diligence required of the test person, I do not consider this very likely. Another, in my opinion, far more plausible scenario, is that Sanctorius used no one but himself to make long-term measurements; yet, he nonetheless issued an open invitation to sit on the chair to all and sundry who visited him at home. He accordingly was faced with the great variability of his test persons and all the challenges this involved, such as the need to frequently adjust the steelyard (Sect. 7.5.3). At any rate, the results that Sanctorius presented in the *De statica medicina* imply that it was mostly middle-aged men who helped him test the weighing chair, since he scarcely made a reference to age or gender (Sects. 3.3.5 and 4.1.2). It is also conceivable that Sanctorius avoided the problems connected with weighing many different people by conducting more experiments on himself than he cared to admit. As my experience with the replica showed, taking measurements is much easier when only one person uses the chair. The counterweight needs then be put in position once only, after which it is possible to work solely with the descent of the chair, without further need of assistance. What is more, in the course of his research, Sanctorius certainly developed great skill in using the chair properly, skill he could not expect of other test persons. In addition to this, he must have been highly motivated to conduct the measurements successfully and diligently. Yet, even though his willpower and stamina were perhaps greater than mine, it is doubtful whether he strictly observed his insensible perspiration in connection with the various measuring parameters, the six non-natural things, over a long period of time. Furthermore, if it is true that he conducted a lot of experiments on himself, then he must have faced the difficulty of translating his very individual measurements into more generally valid statements.

Along with the issues relating to Sanctorius's test persons came the problem of including the many different parameters in his measurements. In Sect. 3.3, I have analyzed how deeply embedded in traditional Galenic medicine was Sanctorius's

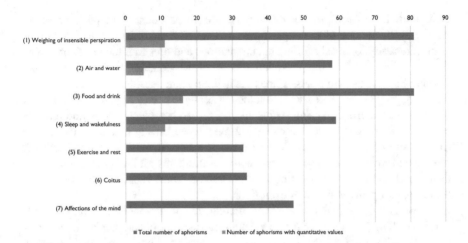

Fig. 7.32 The mention of quantitative values in the first edition of the *De statica medicina* (Sanctorius 1614)

reinterpretation of the doctrine of the six non-natural things. Perusal of the *De statica medicina* shows that Sanctorius often complemented his measurements of insensible perspiration with qualitative conclusions and general observations that he made during the weighing experiments, or that he knew of from the medical literature. Interestingly, he only specified precise quantities in the first four sections, whereas he confined himself to more general and rather qualitative statements in sections V to VII. As the diagram illustrates (Fig. 7.32), even in those sections in which Sanctorius mentioned quantitative values, he did so in only a small proportion of the aphorisms overall. This is especially remarkable with regard to the first section which, as its title says, deals with the weighing of insensible perspiration. Contrary to what one might expect, Sanctorius mentioned quantitative values in fewer aphorisms in this section than in the section on food and drink.

Looking at the 1634 edition of the *De statica medicina*, to which Sanctorius added 108 aphorisms, it is striking that none of them contains any quantitative value, except for one aphorism in the section on food and drink; yet, this refers to an assumed quantity of ingested food rather than to a measurement of insensible perspiration (Sanctorius 1634: 40r).

That being said, against the backdrop of my experiences during the reenactment of Sanctorius's weighing procedures, these observations no longer seem so surprising. In fact, they must be taken as an indication of Sanctorius's ability (or inability) to measure certain parameters. My experiments with the reconstruction have shown that it is not complicated to apply the weighing procedures to food and drink, as their quantities can be controlled and monitored relatively easy. Hence, this was most certainly the case for Sanctorius, too, and he therefore was able to specify in this section the most quantitative values. Consequently, the prominent place of food and drink in the *De statica medicina* cannot be explained solely by the great contemporary demand for dietary guidance, but also by Sanctorius's ability to quite easily measure this non-natural factor. But how did his weighing chair precisely indicate weight changes that were directly connected to his test person's mood? How could Sanctorius possibly have included all the six non-natural factors in his measurements and considered them in conjunction with each other? Since I failed to achieve this in my experiments on myself, it is most probable that Sanctorius actually also had difficulties in doing so.

In view of the lack of numerical values in the section on the influence of mood on insensible perspiration, for example, it can be assumed that Sanctorius was unable to determine any. When reenacting the experiments, I often found it hard to tell what emotions I had or what mood I was in. Furthermore, these were also influenced by the many constraints that the experimental situation imposed on my daily life. For the chair to measure my "affections of the mind," I would have had to use the weighing chair as soon as I recognized a mood change, to determine how this was affecting my weight and the excretion of the *perspiratio insensibilis*. While it was already very difficult for me to somehow detect a mood change in myself, especially under the artificial circumstances of the experiment, it was nigh on impossible to isolate its impact from that of the other parameters simultaneously influencing my physiology. For example, if I rested for a longer period of time and then noticed a mood change, was it the mood change, or the long rest, or a combination of both factors that was responsible for the weight change I measured with the replica chair? Adding to these difficulties, if Sanctorius did experiment not on himself, but on test persons, he would have had to completely rely on their own assessment of their mood and emotions, and on their diligence in using the weighing chair in relation to them. Thus, in all likelihood, it was issues such as these that made Sanctorius confine himself in this section of the *De statica medicina* to outlining general tendencies, for example that some emotions provoke weight loss, whereas others provoke weight gain (Sect. 3.3.6). Regarding the section on coitus, it might also have been issues of privacy and shame that prevented him from arriving at quantitative measuring results. On a more general level, the scarce references to precise quantities in the *De statica medicina* could imply that Sanctorius did not conduct as many experiments as he claimed. My experience with the replica revealed that both, the weighing of many different people over a short period of time, and long-term measurements with one person only, each bring difficulties of their own.

All things considered, it is most probable that Sanctorius tinkered with different factors until he found the most practical combination of design, measuring method, test duration, and test person. The research with the reconstruction strongly implies that he, just like me, varied the number of research subjects, the duration of the weighing procedures, the counterweights, the length of the balance beam, and his measuring methods. Whenever possible, he tried to include the six non-natural factors in his measurements, but certainly struggled, as I did, to consider all of them simultaneously. His mention or omission of quantitative values in the different sections of the *De statica medicina* reflect these struggles. Moreover, given the problems that I faced in the reenactment of the weighing procedures, Sanctorius's claim that he conducted the experiments with more than ten thousand people and over a time span of around thirty years seems highly exaggerated.

7.5.5 *Measuring Respiration*

In the following, I shall take up a specific aspect which does not directly relate to the
preceding paragraphs, but is still important to consider in the context of Sanctorius's
measurements of insensible perspiration. As explained in Sect. 3.2.4, according to
the Venetian physician, *perspiratio insensibilis* resulted not only from the digestive
activities of the body, but also from respiration. In the *De statica medicina*, he specified a quantity of daily respiration, which suggests that he differentiated between
the two different forms of insensible perspiration in his weighing procedures.
Furthermore, Sanctorius also described a way in which he arrived at this quantity,
which can be interpreted as a measuring method rather than only as a simple quantitative reference, because he included it in one of his static aphorisms. However, the
method which Sanctorius allegedly used to measure breathing is far from clear. He
simply stated that "the drops on a mirror placed in front of the mouth" indicated that
the daily respiration usually amounted to about half a pound (Sanctorius 1614: 2r).[87]
It seems thus that Sanctorius placed a mirror on a balance in order to weigh the
water drops on its surface caused by breathing. Given that it would be impossible to
conduct such a measurement over a period of one whole day, Sanctorius most likely
determined the amount of respiration for a shorter period and projected the result for
the whole day. For this purpose, he might have used his pocket watch type of *pulsilogium*, which he also employed to register the duration of his observations with the
thermoscopes. Or he used one of his dial type *pulsilogia*, with which he claimed to
be able to measure the respiration cycle. Yet, since the water drops on the mirror
would quickly evaporate, Sanctorius must have worked with very brief periods of
time. This, in turn, would result in exceedingly small quantities measured, since the
value that he determined for the daily amount of respiration was only half a pound.[88]
Hence, it is quite questionable how Sanctorius actually conducted his measurements
of breathing and how he arrived at a quantity for daily respiration. What is more, as
mentioned above, it is unclear why he did not refer to his hygrometers in this context (Sect. 7.4.4). Still, some valuable clues to Sanctorius's dealing with respiration
as an origin of insensible perspiration can be found in the medical tradition.

Sanctorius upheld the Galenic conception that insensible perspiration resulted
from the respiratory and digestive activities of the body. In her analysis of Galen's
notions of perspiration, Armelle Debru has argued that the function of respiration,
oral as well as cutaneous, was, according to Galen, only qualitative, namely to balance body heat.[89] Contrary to this, *perspiratio insensibilis*, which resulted from the
digestive process, fulfilled a quantitative function. Being a bodily evacuation, just

---

[87] "Perspiratio insensibilis … fit per respirationem per os factam, quae unica die ad selibram circiter ascendere solet; hoc enim indicant guttae in speculo, si ori apponatur." See: Sanctorius
1614: 2r.

[88] If calculated on the basis of Venetian *oncia sottile*, half a pound corresponds to 150 g (Sect. 5.4.2,
fn. 39).

[89] According to Armelle Debru, Galen did not include oral respiration, but only cutaneous respiration in his concept of perspiration (Debru 1996: 183–7).

like urine or feces, it entailed a material loss. However, as Debru has further out-lined, Galen did not strictly differentiate between the two forms of insensible per-spiration, but sometimes confounded them in his works (Debru 1996: 153–91). This might explain why Sanctorius did not explicitly refer to cutaneous respiration in the context of insensible perspiration, and why respiration, more generally, played no major part in his weighing procedures, since he referred only in one aphorism to oral respiration and to a dubious method of weighing it. At the same time, however, the fact that Sanctorius included this aphorism in the *De statica medicina* implies that he departed from the Galenic teachings according to which the measurement of the quantity of inhaled air and exhaled matter was not only impossible, but also unimportant, owing to the exclusively qualitative function of respiration. It seems then that Sanctorius, unlike Galen, considered important the quantity of respiration, as a form of insensible perspiration, but struggled to measure it.

Interestingly, in the second half of the sixteenth century, Girolamo Cardano had already tried to quantify "inspired" air. In his commentary on the Hippocratic trea-tise *Nutriment*, the same work in which he examined the quantitative relation between pulse and respiration (Sect. 7.2.2), he stated that "we inspire daily eight hundred amphoras" (Cardano 1574: dedication).[90] But here, too, Cardano gave no information on how he determined this amount and whether he used an instrument to do so. It is intriguing that he indicated the quantity of respiration in amphoras, an ancient Roman unit of capacity, especially used for liquid products. Since one amphora is equivalent to about 27.84 liters, the amount that Cardano mentioned is extremely high. While his measurement of inhaled air thus raises more questions than answers and shall not be discussed here in any detail, it is still worth mention-ing that Cardano had dealt with the quantity of respiration in a medical-dietetic context, before Sanctorius did (Encyclopaedia Britannica 2018).

7.5.6 The Sanctorian Chair: A Multifunctional Instrument?

With his weighing chair, Sanctorius repurposed a long-established instrument. Although the balance is one of the oldest measuring instruments, Sanctorius's seventeenth-century scale was the first to be applied to humans.[91] My reassessment of the original source materials in the light of the experience gained through recon-structing the Sanctorian chair and replicating the weighing experiments taught me how this novel application of the steelyard raises challenges for the instrument's mechanical design. It also widened my perspective on the great variety of its

---

[90] "… singulis diebus haurimus mensura mensa DCCC. Amphoras aeris Italicas: …." See: Cardano 1574: dedication.

[91] According to Robens, et al., weighing people was a practice during the witch trials held in Europe between the fifteenth and eighteenth centuries (2014: 470). This was not related to medical consid-erations, however, but to the identification of witches. Since witches were supposed to fly on brooms, they were expected to be light. A person who weighed less than circa 50 kg was thought to be able to fly. A witch trial of this sort took place in the Netherlands (near Oudewater) in 1545.

potential applications. Different measuring methods can be applied that directly affect the design, the functioning, and the precision of the weighing chair. Although my research does not allow me to unambiguously define the measuring method Sanctorius used, it has shown that this method is not as self-evident as has commonly been assumed.

On the basis of my research, it can be assumed that Sanctorius most likely used some variation on the measuring methods mentioned above. He used both the steelyard concealed behind the ceiling and at least two reference points made on the bottom of the chair. Even though the original illustration of the weighing chair gives no clear indication of a scale at the base of the chair or on the wood panel behind it, scaling would have been necessary at these two reference points. In short, Sanctorius had to translate weight into a distance. He thus worked with proportions as well as with exact quantities. Whether the pointers, nails, or pegs at the base of the chair served to indicate these reference points to stabilize the chair, or both, cannot be ascertained using the available sources.

The aphorisms of the *De statica medicina* and the description of the Sanctorian chair imply that the instrument had two functions. On the one hand, it was used as a research tool to monitor variations in the production of *perspiratio insensibilis*; on the other, it helped to determine and maintain an ideal body weight. The measuring methods might have varied in correspondence with these two functions. Based on my experiences with the reconstruction, it seems likely that Sanctorius used the steelyard in the traditional way, especially in the initial phase of his experiments, when he tried to define the healthy quantity of insensible perspiration. In this connection, he most probably observed weight changes in many different people over shorter periods of time. As soon as he managed to stabilize this quantity, he could determine the ideal body weight for individual persons and determine the healthy amount of food and drink that they should ingest. To this end, he might have used the descent of the chair as an indication of changes in weight, as described in the *Commentary on Avicenna*. My own experiments have shown that this would have enabled individuals, even laymen, to use the chair on their own, without any need of an assistant to move the counterweight along the longer arm of the weighing chair. In this regard, the weighing chair would not have been meant for use by multiple individuals, but only by one person; the beam of the steelyard would therefore be balanced only once, for that person's respective weight. Due to the rather easy measuring method and the narrow focus on keeping an ideal weight, it is indeed conceivable that Sanctorius tested this second type of use of his steelyard over a longer time span, most certainly on himself. This fits with his suggestion that the beam of the steelyard be hidden above the ceiling to obviate the astonishment of guests, to whom the weighing device might have looked ridiculous. It implies—as did the longer quote in Sect. 7.5.1—that Sanctorius may have conceived of the chair for use by a larger public, to regulate their eating habits.[92]

---

[92] Lucia Dacome has also pointed out the possibility that Sanctorius's proposal to hide the beam of the weighing chair above the ceiling implies that he may have conceived the chair for a larger public, beyond the community of physicians. See: Dacome 2001: 476.

In this context, it is important to keep in mind that Sanctorius published the description and illustration of the Sanctorian chair eleven years after the *De statica medicina*. Based on the insights I gained during my research, I have come to imagine a possible chronological use of the instrument, which might reflect the development of Sanctorius's research during these years. After beginning with the aim of determining the quantity of the *perspiratio insensibilis* within the frame of contemporary dietetic medicine, he might have realized that the chair not only helped the physician to monitor changes in weight and, on this basis, to issue rules of health, but also offered an opportunity to find and maintain an ideal weight. Of importance here, certainly, is the fact that the weighing procedures could be applied with relative ease to food and drink, as my own experience with the reconstruction has shown. In order to make the chair accessible to laymen, Sanctorius might have adapted the design and measuring method with regard to this newly discovered function and published both in his *Commentary on Avicenna*.[93] The great contemporary demand for health handbooks, especially food guides, and the general awareness of the importance of regulating food intake in quantitative terms (Sect. 5.1) most certainly played their part, too. With his weighing chair, Sanctorius was able to offer dietary guidance not only in the form of written advice, as in the *De statica medicina*, but also in the form of an instrument. He enabled his audience to conduct by themselves weighing procedures that allowed them to monitor their weight—without the help of a physician. As mentioned before (Sect. 5.1), dietetics in the Renaissance became a field in which laypeople—and not only physicians—might gain a certain level of authority and this propelled their efforts to regulate personal hygiene. In all likelihood, Sanctorius's weighing chair was a response to this trend.

However, it should not be forgotten that Sanctorius presented the illustration and description of the weighing chair in a lengthy medical commentary addressed, in Latin, to an audience within the university realm. Outside of this context, the work was reserved to learned physicians, scholars, or other well-educated persons fluent in Latin. Furthermore, in order to copy the Sanctorian chair, prospective weight watchers would have needed money, materials, equipment, and technical support.

7.5.7   The Reception of the Sanctorian Chair—A Few Thoughts

Without aiming to provide a detailed history of the reception of the Sanctorian chair, I will focus rather in the following on those aspects that I consider relevant to the present study. Despite the great success of the *De statica medicina* and the popularity of the weighing chair, Sanctorius repeatedly stated that he anticipated criticism

---

[93] In their paper (Valleriani and Pearl 2017), Matteo Valleriani and Yifat-Sara Pearl highlight the use of images as low-threshold educational tools, particularly in scientific texts, since this makes knowledge accessible to wider audiences.

of his novel quantitative approach. In the dedication to the *De statica medicina*, he wrote that he had long reflected on whether or not to publish the treatise. He was worried about its reception by "ignorant and malevolent people, who either disapprove of the novelty, or do not understand the subtleties" of static art (Sanctorius 1614: dedication).[94] In the preface to the *De statica medicina*, he similarly warned that people usually tried to suppress novelties because of envy, instead of advancing them through studies. He further explained that he expected that "many, not only among the vulgar, but also among the learned, … will rise up against this new art and will heavily inveigh against it" (Sanctorius 1614: Ad lectorem).[95] Moreover, in the dedication in the *Commentary on Avicenna*, he emphasized that many people did not accept his "new and extraordinary way of dealing with medical theory" and that he was therefore in need of a most learned and most celebrated patron—whom he found namely in Ferdinando Gonzaga (1587–1626), Duke of Mantua and of Montferrat (Sanctorius 1625: dedication).[96] In the dedication in the *Commentary on Hippocrates*, he referred to his static medicine, explaining that he hoped to promote longevity with it. According to him, matters as important as longevity depended solely on the "patronage of truth." But given that truth was in itself troublesome and the origin of hatred, he required the support of the "greatest man," who was, in this case, the Duke of Urbino, Francesco Maria II della Rovere (1549–1631) (Sanctorius 1629: dedication).

Of course, issues of authority, legitimation, and credibility were a common concern of scholars at the time, as they are still today, and it is anything but unusual that Sanctorius glowingly praised his patrons. Furthermore, citations similar to those by the Venetian physician can be found, for example, in the works of William Gilbert (1544–1603), Francis Bacon (1561–1626), and Galileo Galilei. They reflect a general attitude among the scholars of the sixteenth and seventeenth century, their sense of the dawning of a new era in which anyone who did not approve of their innovations could rightly be attacked as a backward ignoramus. Sanctorius's recurrent mention and anticipation of criticism is therefore remarkable and even more so considering that, at the time when he published the *De statica medicina*, he already held one of the most prestigious positions at the University of Padua—the chair for medical theory. The other two works in which he referred to others' disapproval, the *Commentary on Avicenna* and the *Commentary on Hippocrates*, were both published after the *De statica medicina*, when Sanctorius had already resigned his professorship. Apparently, his innovative approach to physiology and to the teaching of

---

[94] "… ex una parte erat imperitorum, & malevolorum hominum magna acies, qui vel nova improbantes, vel subtilia non intelligentes, hanc artem, divinam licet, damnaturi essent: …." See: Sanctorius 1614: dedication.

[95] "… scio multos non solum vulgares, sed etiam ex literatorum censu, … contra artem hanc novam insurrecturos, eamque graviter detracturos esse, …." See: ibid. The English translation is taken from: Sanctorius and D. 1676: Sanctorius to the reader.

[96] "… hic novus, & propemodum inusitatus stylus tractandi Theoricam …." See: Sanctorius 1625: dedication.

theoria was controversially received. But being a recognized physician and an (emeritus) medical professor, why was Sanctorius so worried about criticism?

The answer certainly lies in part in his rivalry with Ippolito Obizzi. As mentioned earlier (Sect. 3.1, fn. 2), only one year after the appearance of the *De statica medicina*, Obizzi published a violent attack on the work (Obizzi 1615). In fact, already in a letter (*epistola*) dated July 1613, the physician from Ferrara had criticized Sanctorius's first book *Methodi vitandorum errorum* (Obizzi 1618: 25–32). Hence, Obizzi's objections were not exclusively directed against static medicine and Sanctorius was most probably aware of his critic before he published the *De statica medicina*. It is conceivable that personal motives, unknown to us today, were involved in the dispute, too (Grmek 1952: 10, 37; Sanctorius and Ongaro 2001: 40 f.).

However, in my opinion, Sanctorius's worries about criticism cannot be explained solely by Obizzi's attacks. It seems to me that they equally stemmed from a more general skepticism about his novel quantitative approach to physiology, which Sanctorius claimed to detect in his contemporaries, both educated and uneducated, as the citations above show. Since physiology, as a university subject, was a highly theoretical discipline at the time, Sanctorius's introduction of mechanical procedures into this field of medicine was most likely perceived as particularly radical. Accordingly, Sanctorius feared the mockery of his colleagues, and anticipated his patients' irritation upon being confronted with a huge steelyard, installed in the middle of their physician's living room. It was to mitigate this irritation that he hid the beam of his weighing chair behind a false ceiling. Interestingly, the illustration of Sanctorius's *lectus artificiosus* (Fig. 4.15) shows that the crank mechanism, serving to lift and lower the bed, was likewise concealed by a false ceiling. Even though Sanctorius did not comment on this in his description of the device, it can be assumed that, here again, he wanted to hide this novel and unorthodox feature of the instrument. Hence, Sanctorius's introduction of mechanical devices and procedures known from other contexts into the world of medical practice was not uncontroversial. In order to give a pair of scales a medical identity, Sanctorius had not only to materially adapt the device, but also to build trust in his new medical technology. Hiding the mechanism of the device was his attempt to integrate the Sanctorian chair as smoothly as possible into the domestic sphere and, more generally, into people's lives.

Putting ourselves in Sanctorius's shoes, for a moment, let's consider how he might have sold his weighing chair to a colleague or friend, without needing to rhetorically defend his novel approach. Perhaps he would have explained that, given the relevance to health of maintaining an ideal balance between ingestion and excretion, it was of the utmost importance to observe this balance in quantitative terms; and that the physician could now do so, for the first time ever, thanks to his, Sanctorius's, newly invented weighing chair. He might have uttered his conviction that most diseases resulted from hindered or blocked insensible perspiration—a physiological process which was no longer obscure, but detectable, with his instrument; and the weighing procedures he had devised would allow his colleague or friend to make a better diagnosis, prognosis, and treatment. In addition, Sanctorius would most certainly have pointed out the second purpose of his weighing chair: to define and monitor a person's ideal weight. In this regard, he probably emphasized that the weighing chair would enable everyone to keep track of their own weight. To put this in a nutshell, if Sanctorius were to advertise his instrument on today's market, he might use a slogan like: "The Sanctorian Chair—Creating Healthier and Longer Lives!" Of course, this is only playful speculation, yet it allows us to see the instrument in a new light.

Whatever words Sanctorius used to promote his weighing chair, it is difficult to ascertain how successfully he did so. Testimonies of people who built their own versions of the Sanctorian chair and imitated the weighing procedures date only from the late seventeenth century and especially, the eighteenth century, while little is known to us of the instrument's earlier reception.[97] The available sources show that, over the course of the eighteenth century, Sanctorius's weighing chair drew mixed reactions and that its two functions were hotly debated (Dacome 2001: 475). Who should use the Sanctorian chair? Was it designed for medical or lay practice? With the primary sources at hand, I still cannot unambiguously answer these questions. However, the methodological approach of replication enabled me to find a possible connection between the different functions of the Sanctorian chair and its design and measuring methods. During my research, I developed a new understanding of the mechanical and practical knowledge involved in Sanctorius's weighing procedures—an understanding that I could hardly have developed on the basis of the written sources alone.

In conclusion, while we can be sure that Sanctorius did build his weighing chair, questions still remain regarding how he actually used it. My experience with the reconstruction has shown that it is possible to measure very small quantities with a steelyard the size of the Sanctorian chair. Moreover, my own experimentation revealed that the instrument can easily be used by just one person, when the distance from the chair to the floor is to be measured. Other issues remain open, however. It is, for example, still unclear how Sanctorius dealt with the problem of including all of the six non-natural factors in his quantitative observations, or how he handled the high variability of his test persons and their influence on his weighing procedures. Furthermore, we do not know how he coped with the constraints that the experiments imposed on the test person and how this affected his weighing practice. This notwithstanding, I think there is no reason to doubt that Sanctorius actually measured the quantities to which he referred in the *De statica medicina* with an instrument that was at least similar to the one depicted in the *Commentary on Avicenna*. But his claims regarding the duration, range, and frequency of the weighing procedures are a different matter. As my experience with the reconstruction has demonstrated, it is highly questionable that he conducted his experiments over a period of thirty years. The many travels he undertook in the late sixteenth century—the time

---

[97] Ippolito Obizzi claimed that Sanctorius's friend Hieronymus Thebaldus used the Sanctorian chair and that the weighing procedures made him ill (Obizzi 1615: 24). Given that I was unable to find any other reference to Thebaldus's use of the instrument and that Obizzi was an opponent of both Sanctorius and Thebaldus, this statement must be taken with a grain of salt. On the quarrels between Obizzi and Thebaldus, see: Sanctorius 1625: 82.

when he allegedly started using his weighing chair—reinforce this assumption (Sect. 2.2). Similarly, Sanctorius certainly exaggerated when he wrote in his letter to Galileo Galilei that he had observed the insensible perspiration of more than ten thousand subjects (Sanctorius 1902). Indeed, as Evan Ragland has shown, it was common at the time to invoke a rhetorically large number of trials to substantiate new claims. Galileo himself claimed to have repeated experiments "one hundred times" and the physician and anatomist Gabriele Falloppia reported that he had tested a prophylaxis against the French disease "in a thousand and one hundred men" (Ragland 2017: 515).

In the same letter to Galileo, Sanctorius mentioned that the famous scholar was among the subjects who had sat in his weighing chair. Would Galileo not have protested, had this been untrue? Would Sanctorius's Venetian circle of friends, the *Ridotto Morosini*, not have been suspicious, had Sanctorius never showed them his device? And what about Sanctorius's many pupils at the University of Padua, whom he introduced his static observations to? In view of Sanctorius's renown and his large network of friends, one can imagine that it would hardly have gone unnoticed, had his static medicine been mere rhetoric. Still, it is striking that none of Sanctorius's students, friends, or colleagues seems to have written about the original weighing chair and Sanctorius's presentation of it. While there are such reports on his thermoscope and *pulsilogium*, there is no known evidence of this regarding the Sanctorian chair. From a preliminary perspective, it seems therefore that the instrument sparked enthusiasm only later, toward the end of the seventeenth century. Although Sanctorius's anticipation of criticism was certainly in part rhetorical, it might also reflect his immediate contemporaries' hesitant reception of his static experiments. Apparently, they were not prepared to install a Sanctorian chair in their homes.

7.5.8   Sanctorius's Measuring Instruments in Context

In the foregoing paragraphs, I have examined Sanctorius's measuring instruments from a broad perspective, analyzing their development and use in various contexts—theoretical, social, practical. This has revealed their deep integration into Galenic medicine and made clear that they can only be understood within such framework. Sanctorius's interest in, and receptivity to contemporary technological developments came to the fore, as illustrated by his use of a pendulum for his *pulsilogia*, for example, or his attempt, inspired by the practical hydraulics of his day, to measure the *impetus* of water currents. Moreover, the chapter has shown that his socio-intellectual milieu in Padua and Venice, most importantly, the *Ridotto Morosini*, brought him into contact with distinguished scholars and aristocrats and gave him a platform to discuss the latest technological and intellectual trends. The meetings in the *palazzo* on the Grand Canal certainly spurred him in his use of quantification and measurements—they were fertile ground in which to develop and test new ideas. Although I could often not unambiguously clarify how Sanctorius's measuring instruments were related to earlier similar ideas, such as those of

Cardano, it was hopefully instructive to highlight that these ideas did arise independently of Sanctorius. Ultimately, however, it was he who applied these ideas, concepts, instruments, and techniques to medical practice. And this is not at all trivial. As my reconstruction of his most famous instrument, the weighing chair, has clearly illustrated, the path from the intellectual conception of an instrument to its actual application in research and practice is often long, and surely was, in the case of Sanctorius, since he applied his measuring instruments to human physiology. Therefore, caution is advised, if analyzing Sanctorius's devices solely on the basis of his written and pictorial accounts of them, without further inquiry into his making and doing. In any event, Sanctorius's strong interest in practical technologies, especially mechanics, was anything but ordinary for a Renaissance physician. With his innovation of various measuring instruments, whether he actually used all of them or not, he opened up new perspectives—in medicine and beyond.

References

Alberti, Leon Battista. 1986. *The Ten Books of Architecture: The 1755 Leoni Edition*. New York: Dover Publications.

Amontons, Guillaume. 1695. *Remarques et expériences phisiques sur la construction d'une nouvelle clepsidre sur les baromètres, thermomètres et higomètres*. Paris: Claude Jombert.

AWWA Meter Manual. 1959. Chapter I–Early History of Water Measurement and the Development of Meters. *Journal (American Water Works Association)* 51: 791–799.

Bacalexi, Dina, and Mehrnaz Katouzian-Safadi. 2019. Touching the Patient: Galen's Treatise *On the Pulse for Beginners* and its Reception in the Medieval Latin, the Islamic Oriental and the Renaissance World. *Scientiae* 2019, Belfast. https://halshs.archives-ouvertes.fr/halshs-02356368. Accessed 29 Mar 2020.

Bartholin, Caspar. 1611. *Problematum philosophicorum & medicorum nobiliorum & rariorum miscell. exercitationes*. Wittenberg: Ex Typographia Andrea Rüdingeri. Apud Bechtoldum Raaben, Bibliopolam.

Bedford, D. Evan. 1951. The Ancient Art of Feeling the Pulse. *British Heart Journal* 13: 423–437.

Beeckman, Isaac, and Cornelis de Waard. 1945. *Journal tenu par Isaac Beeckman de 1604 à 1634*, Vol. III. The Hague: Nijhoff.

Beugo, John. n.d. *Sanctorius in His Balance*. https://collections.nlm.nih.gov/catalog/nlm:nlmuid-101428295-img. Accessed 6 Mar 2020.

Bigotti, Fabrizio. 2018. The Weight of the Air: Santorio's Thermometers and the Early History of Medical Quantification Reconsidered. *Journal of Early Modern Studies* 7: 73–103.

Bigotti, Fabrizio, and David Taylor. 2017. The Pulsilogium of Santorio: New Light on Technology and Measurement in Early Modern Medicine. *Society and Politics* 11: 55–114.

Bizzarrini, Giotto. 1947. Curiosità ed attualità. *Minerva medica* 38: 436–450.

BNMVe n.d.: Mss. Ital. VII 2342 (= 9695), Bolis, *Notizie cavate dalli libri di Priori*.

Boissier de Sauvages de Lacroix, François. 1752. *Pulsus et circulationis theoria*. Montpellier: Apud Augustinum-Franciscum Rochard.

Borrelli, Arianna. 2008. The Weatherglass and Its Observers in the Early Seventeenth Century. In *Philosophies of Technology: Francis Bacon and His Contemporaries*, ed. Claus Zittel, Gisela Engel, Nicole C. Karafyllis, et al., 67–130. Leiden: Brill.

Breidbach, Olaf, Peter Heering, Matthias Müller, and Heiko Weber. 2010. *Experimentelle Wissenschaftsgeschichte*. Paderborn: Fink Wilhelm.

Büttner, Jochen. 2008. The Pendulum as a Challenging Object in Early-Modern Mechanics. In *Mechanics and Natural Philosophy Before the Scientific Revolution*, ed. Walter Roy Laird and Sophie Roux, 223–237. Dordrecht: Springer.

———. 2019. *Swinging and Rolling: Unveiling Galileo's Unorthodox Path from a Challenging Problem to a New Science*. Dordrecht: Springer.

Cabeo, Niccolò. 1646. *In quatuor libros meteorologicorum aristotelis commentaria, et quaestiones,* Vol. I. Rome: Haeredum Francisci Corbelletti.

Cambridge Dictionary. 2014. Relative Humidity. In *Cambridge Advanced Learner's Dictionary & Thesaurus*. Cambridge: Cambridge University Press. https://dictionary.cambridge.org/de/worterbuch/englisch/relative-humidity. Accessed 20 Feb 2020.

Cardano, Girolamo. 1550. *De subtilitate libri XXI*. Paris: Ex Officina Michaëlis Fezandat, & Roberti Granion.

———. 1570. *Opus novum de proportionibus numerorum, motuum, ponderum, sonorum, aliarumque rerum mensurandum, …*. Basle: Ex Officina Henricpetrina.

———. 1574. *Commentaria in librum Hippocratis de alimento*. Rome: Apud haeredes Antonij Bladij.

Chang, Hasok. 2011. How Historical Experiments Can Improve Scientific Knowledge and Science Education: The Cases of Boiling Water and Electrochemistry. *Science and Education* 20: 317–341.

Comstock, John Lee. 1836. *A System of Natural Philosophy*. New York: Robinson, Pratt, & Co.

Da Vinci, Leonardo, and Carlo Pedretti. 2000. *Il Codice Atlantico della Biblioteca Ambrosiana di Milano,* Vol. 1. Florence: Giunti.

Da Vinci, Leonardo, and Jean Paul Richter. 1970a. *The Notebooks of Leonardo da Vinci: Compiled and Edited from the Original Manuscripts,* Vol. 1. New York: Dover Publications.

———. 1970b. *The Notebooks of Leonardo da Vinci: Compiled and Edited from the Original Manuscripts,* Vol. 2. New York: Dover Publications.

Da Vinci, Leonardo, Dietrich Lohrmann, and Thomas Kreft. 2018. *Leonardo da Vinci: Codex Madrid I*, Vol 2: *Theorie der Mechanik, Außenblätter*. Vienna: Böhlau Verlag.

Dacome, Lucia. 2001. Living with the Chair: Private Excreta, Collective Health and Medical Authority in the Eighteenth Century. *History of Science* 39: 467–500.

Debru, Armelle. 1996. *Le corps respirant: la pensée physiologique chez Galien*. Leiden: Brill.

Del Gaizo, Modestino. 1936. Santorio Santorio nel terzo centenario della morte. *Il giardino di Esculapio* IX: 4–21.

Deutscher Wetterdienst. 2015. *Feuchte ist nicht gleich Feuchte*. https://www.dwd.de/DE/wetter/thema_des_tages/2015/2/13.html. Accessed 20 Feb 2020.

Di Fidio, Mario, and Claudio Gandolfi. 2011. Flow Velocity Measurement in Italy between Renaissance and Risorgimento. *Journal of Hydraulic Research* 49: 578–585.

Dooley, Brendan, ed. 2014. *A Companion to Astrology in the Renaissance*. Leiden: Brill.

Eilam, Eldad. 2011. *Reversing: Secrets of Reverse Engineering*. New York: Wiley.

Elazar, Michael. 2011. *Honoré Fabri and the Concept of Impetus: A Bridge between Conceptual Frameworks*. Dordrecht: Springer.

Encyclopaedia Britannica. 2018. *Amphora*. https://www.britannica.com/science/amphora-measurement. Accessed 14 Mar 2020.

Ettari, Lieta Stella, and Mario Procopio. 1968. *Santorio Santorio: la vita e le opere*. Rome: Istituto nazionale della nutrizione.

Facciolati, Iacopo. 1757. *Fasti Gymnasii Patavini*. Padua: Apud Joannem Manfrè.

Frazier, Arthur H. 1969. Dr. Santorio's Water Current Meter, Circa 1610. *Journal of the Hydraulics Division* 95: 249–253.

———. 1974. *Water Current Meters in the Smithsonian Collections of the National Museum of History and Technology*. Washington, DC: Smithsonian Institution Press.

French, Roger K. 1994. Astrology in Medical Practice. In *Practical Medicine from Salerno to the Black Death*, ed. Luis García-Ballester, Roger French, Jon Arrizabalaga, et al., 30–59. Cambridge: Cambridge University Press.

George, Mathew. 2017. *Institutionalizing Illness Narratives: Discourses on Fever and Care from Southern India*. Singapore: Springer.

Grmek, Mirko D. 1952. *Santorio Santorio i njegovi aparati i instrumenti*. Zagreb: Jugoslav. akad. znanosti i umjetnosti.

———. 1967. L'énigme des relations entre Galilée et Santorio. In *Symposium internazionale di storia, metodologia, logica e filosofia della scienza: Galileo nella storia e nella filosofia della scienza (1964: Firenze-Pisa) Atti del Symposium internazionale di storia, metodologia, logica e filosofia della scienza: Galileo nella storia e nella filosofia della scienza*, ed. Gruppo italiano di storia delle scienze, 155–162. Florence: Vinci.

Guidone, Mario, and Fabiola Zurlini. 2002. L'introduzione dell'esperienza quantitativa nelle scienze biologiche ed in medicina Santorio Santorio. In *Atti della XXXVI tornata dello Studio firmano per la storia dell'arte medica e della scienza, Fermo, 16–17–18 maggio 2002*, ed. Studio firmano per la storia dell'arte medica e della scienza, 117–137. Fermo: A. Livi.

Hamlin, Christopher. 2014. *More than Hot: A Short History of Fever*. Baltimore: Johns Hopkins University Press.

Heering, Peter. 2008. The Enlightened Microscope: Re-enactment and Analysis of Projections with Eighteenth-century Solar Microscopes. *British Journal for the History of Science* 41: 345–367.

———. 2010. An Experimenter's Gotta Do What an Experimenter's Gotta Do–But How? *Isis* 101: 794–805.

Hess, Volker. 2000. *Der wohltemperierte Mensch: Wissenschaft und Alltag des Fiebermessens (1850–1900)*. Frankfurt/New York: Campus Verlag.

Hodgson, Michael. 2008. *Weather Forecasting*. Guilford: Globe Pequot Press.

Hollerbach, Teresa. 2018. The Weighing Chair of Sanctorius Sanctorius: A Replica. *NTM Zeitschrift für Geschichte der Wissenschaften, Technik und Medizin* 26: 121–149.

Horine, Emmet Field. 1941. An Epitome of Ancient Pulse Lore. *Bulletin of the History of Medicine* 10: 209–249.

Hübner, Wolfgang. 2014. The Culture of Astrology from Ancient to Renaissance. In *A Companion to Astrology in the Renaissance*, ed. Brendan Dooley, 17–58. Leiden: Brill.

Hutchison, Keith. 1982. What Happened to Occult Qualities in the Scientific Revolution? *Isis* 73: 233–253.

Johannes Ravius to Ernst Schaumburg-Holstein, Graf, III. Padua, 20.12.1618. Niedersächsisches Landesarchiv/ Abteilung Bückeburg; Regest [Ulrich Schlegelmilch]. www.aerztebriefe.de/id/00031187. Accessed 20 Jan 2020.

Kepler, Johannes. 1618. *Epitome astronomiae Copernicanae*, Vol. 1. Linz: Johannes Plancus.

Kircher, Athanasius. 1665. *Mundus subterraneus*. Amsterdam: Apud Joannem Janssonium & Elizeum Weyerstraten.

Kümmel, Werner Friedrich. 1974. Der Puls und das Problem der Zeitmessung in der Geschichte der Medizin. *Medizinhistorisches Journal* 9: 1–22.

Kuphal, Eckart. 2013. *Den Mond neu entdecken: Spannende Fakten über Entstehung, Gestalt und Umlaufbahn unseres Erdtrabanten*. Berlin/Heidelberg: Springer Spektrum.

Langholf, Volker. 1992. *Medical Theories in Hippocrates: Early Texts and the "Epidemics"*. Berlin: De Gruyter.

Lauremberg, Peter. 1621. *Laurus delphica, seu consilium, quo describitur methodus perfacilis ad medicinam*. Leiden: Apud Iohannem Maire.

Leupold, Jacob. 1724. *Theatrum machinarum generale. Schau-platz des Grundes mechanischer Wissenschafften*. Leipzig: Christoph Zunkel.

———. 1726. *Pars I. Theatri Statici Universalis, Sive Theatrum Staticum, Schau-Platz der Gewicht-Kunst und Waagen*. Leipzig: Christoph Zunkel.

Lonie, Iain M. 1981. Fever Pathology in the Sixteenth Century: Tradition and Innovation. *Medical History* Supplement No. 1: 19–44.

Maffioli, Cesare S. 1994. *Out of Galileo: The Science of Waters 1628–1718*. Rotterdam: Erasmus Publishing.

References 299

Major, Ralph H. 1938. Santorio Santorio. *Annals of Medical History* 10: 369–381.

Malvicini, Giulio. 1682. *Vtiles colectiones medico-phisicae ad medicinae inscios prolatae a Iulio Maluicino medico-phisico cive veneto sanctorii discipulo & affine*. Venice: Apud Combi, & La Noù.

Marci, Jan Marek. 1639. *De proportione motus seu regula sphygmica ad celeritatem et tarditatem pulsuum ex illius motu ponderibus geometricis librato absque errore metiendam*. Prague: n/a. http://echo.mpiwg-berlin.mpg.de/ECHOdocuView?url=/permanent/archimedes/marci_regul_062_la_1639&viewMode=image&pn=1. Accessed 24 Jan 2020.

Marliani, Giovanni. 1482. *Quaestio de proportione motuum in velocitate*. Pavia: Damiano Confalonieri.

Martin, Craig. 2011. *Renaissance Meteorology: Pomponazzi to Descartes*. Baltimore: John Hopkins University Press.

Mersenne, Marin. 1636. *Harmonicorum libri*. Paris: Guillelmi Baudry.

Mersenne, Marin, Paul Tannery, Cornelis de Waard, and Armand Beaulieu. 1932–1988. *Correspondance du P. Marin Mersenne*, 17 vols. Paris: Éditions du Centre National de la Recherche Scientifique.

Michelotti, Francesco Domenico. 1771. *Sperimenti idraulici principalmente diretti a confermare la teorica, ed a facilitare la pratica del misurare le acque correnti*, Vol. II. Turin: Stamperia Reale.

Middleton, W.E. Knowles. 1969. *Invention of the Meteorological Instruments*. Baltimore: John Hopkins Press.

Miessen, Hermann. 1940. Die Verdienste Sanctorii Sanctorii um die Einführung physikalischer Methoden in die Heilkunde. *Düsseldorfer Arbeiten zur Geschichte der Medizin* 20: 1–40.

Morgagni, G. Battista, Luigi Stroppiana, and Dario Spallone. 1961. *La dottrina galenica dei polsi nell'esposizione didattica di G.B. Morgagni*. Rome: Arti Grafiche E. Cossidente.

Neswald, Elizabeth. 2011. Eigenwillige Objekte und widerspenstige Dinge. Das Experimentieren mit Lebendigem in der Ernährungsphysiologie. In *Affektive Dinge: Objektberührungen in Wissenschaft und Kunst*, ed. Natascha Adamowsky, Robert Felfe, Marco Formisano, et al., 51–79. Göttingen: Wallstein Verlag.

Nutton, Vivian. 2019. Renaissance Galenism, 1540–1640: Flexibility or an Increasing Irrelevance? In *Brill's Companion to the Reception of Galen*, ed. Petros Bouras-Vallianatos and Barbara Zipser, 472–486. Leiden: Brill.

Obizzi, Ippolito. 1615. *Staticomastix sive Staticae Medicinae demolitio*. Ferrara: Apud Victorium Baldinum.

———. 1618. *Iatrastronomicon: Varios tractatus Medicos, & Astronomicos ad rectum medendi usum pernecessarios complectens*. Vicenza: Apud Iacobum Violatum.

Ongaro, Giuseppe. 2009. Santorio e Galilei. *Padova e il suo territorio* 24: 47–51.

Oresme, Nicole, Albert Douglas Menut, and Alexander J. Denomy. 1968. *Le livre du ciel et du monde*. Madison: University of Wisconsin Press.

Overkamp, Heidentryk. 1694. *Nader Verklaringe, over de ontdekte Doorwaasseming, in een dertig-jaarige ondervinding ontdekt op de Weegschaal. Van S. Sanctorius*. Amsterdam: Jan ten Hoorn.

Poma, Roberto. 2012. Santorio Santorio et l'infallibilité médicale. In *Errors and Mistakes. A Cultural History of Fallibility*, ed. M. Gadebusch Bondio and A. Paravicini Bagliani, 213–225. Florence: SISMEL-Edizioni del Galluzzo.

Ragland, Evan R. 2017. "Making Trials" in Sixteenth- and Early Seventeenth-Century European Academic Medicine. *Isis* 108: 503–528.

Renn, Jürgen, and Peter Damerow. 2012. *The Equilibrium Controversy: Guidobaldo del Monte's Critical Notes on the Mechanics of Jordanus and Benedetti and their Historical and Conceptual Background*. Berlin: Edition Open Access.

Robens, Erich, Shanath Amarasiri A. Jayaweera, and Susanne Kiefer. 2014. *Balances: Instruments, Manufacturers, History*. Heidelberg: Springer.

Rossetti, Lucia. 1967. *Acta Nationis Germanicae Artistarum (1616–1636)*. Padua: Editrice Antenore.

Rudio, Eustachio. 1602. *De pulsibus libri duo*. Padua: Apud Paulum Meiettum. Ex Officina Laurentij Pasquati.

Sagredo, Giovan Francesco. 2010 [1612]. G. Sagredo to Galileo in Florence. Venice, June 30, 1612. In *Galileo Engineer*, ed. Matteo Valleriani, 229–230. Dordrecht/London: Springer.

Sanctorius, Sanctorius. 1603. *Methodi vitandorum errorum omnium, qui in arte medica contingunt, libri quindecim*. Venice: Apud Franciscum Barilettum.

———. 1612a. *Commentaria in Artem medicinalem Galeni*, Vol. I. Venice: Apud Franciscum Somascum.

———. 1612b. *Commentaria in Artem medicinalem Galeni*, Vol. II. Venice: Apud Franciscum Somascum.

———. 1614. *Ars Sanctorii Sanctorii Iustinopolitani de statica medicina, aphorismorum sectionibus septem comprehensa*. Venice: Apud Nicolaum Polum.

———. 1625. *Commentaria in primam Fen primi libri Canonis Avicennae*. Venice: Apud Iacobum Sarcinam.

———. 1626. *Commentaria in primam Fen primi libri Canonis Avicennae*. Venice: Apud Iacobum Sarcinam.

———. 1629. *Commentaria in primam sectionem Aphorismorum Hippocratis, &c. ... De remediorum inventione*. Venice: Apud Marcum Antonium Brogiollum.

———. 1630. *Commentaria in Artem medicinalem Galeni*. Venice: Apud Marcum Antonium Brogiollum.

———. 1634. *Ars Sanctorii Sanctorii de statica medicina et de responsione ad Staticomasticem*. Venice: Apud Marcum Antonium Brogiollum.

———. 1902 [1615]. Santorio Santorio a Galilei Galileo, 9 febbraio 1615. In *Le Opere di Galileo Galilei*, ed. Galileo Galilei, 140–142. Florence: Barbera. http://teca.bncf.firenze.sbn.it/ImageViewer/servlet/ImageViewer?idr=BNCF0003605126#page/1/mode/2up. Accessed 5 Nov 2015.

Sanctorius, Sanctorius, and M. Alemand. 1695. *Science de la transpiration ou Medicine statique [par Santorio]*. Lyons: Jaques Lyons.

Sanctorius, Sanctorius, and J. D. 1676. *Medicina statica: or, Rules of Health, in Eight Sections of Aphorisms*. London: John Starkey.

Sanctorius, Sanctorius, and Giuseppe Ongaro. 2001. *La medicina statica*. Florence: Giunti.

Sarpi, Paolo, and Luisa Cozzi. 1996. *Pensieri Naturali, Metafisici e Matematici*. Milan/Naples: Riccardo Ricciardi Editore.

Schemmel, Matthias. 2008. *The English Galileo: Thomas Harriot's Work on Motion as an Example of Preclassical Mechanics*. Dordrecht: Springer.

Schwenter, Daniel. 1636. *Deliciae physico-mathematicae oder mathematische und philosophische Erquickstunden*. Nürnberg: Jeremias Dümler.

Settle, Tom. 1996. *Galileo's Experimental Research*. Berlin: Max Planck Institute for the History of Science, Preprint 52.

Siraisi, Nancy. 1987. *Avicenna in Renaissance Italy: The Canon and Medical Teaching in Italian Universities After 1500*. Princeton: Princeton University Press.

———. 1990. *Medieval & Early Renaissance Medicine: An Introduction to Knowledge and Practice*. Chicago: University of Chicago Press.

———. 2012. Medicine, 1450–1620, and the History of Science. *Isis* 103: 491–514.

Smith, Pamela H. 2017. The Codification of Vernacular Theories of Metallic Generation in Sixteenth-Century European Mining and Metalworking. In *The Structures of Practical Knowledge*, ed. Matteo Valleriani, 371–392. Cham: Springer.

Smith, Pamela H., and Benjamin Schmidt. 2007. Introduction: Knowledge and Its Making in Early Modern Europe. In *Making Knowledge in Early Modern Europe: Practices, Objects, and Texts, 1400–1800*, ed. Pamela H. Smith and Benjamin Schmidt, 1–16. Chicago: University of Chicago Press.

Smith, Pamela H., and The Making and Knowing Project. 2016. Historians in the Laboratory: Reconstruction of Renaissance Art and Technology in the Making and Knowing Project. *Art History* 39: 210–233.

Tassinari, Piero. 2019. Galen into the Modern World: From Kühn to the *Corpus Medicorum Graecorum*. In *Brill's Companion to the Reception of Galen*, ed. Petros Bouras-Vallianatos and Barbara Zipser, 508–534. Leiden: Brill.

Valleriani, Matteo. 2010. *Galileo Engineer*. Dordrecht/London: Springer.

Valleriani, Matteo, and Yifat-Sara Pearl. 2017. Images Don't Lie (?): A Post-Exhibition Reflection. In *Images Don't Lie (?)*, ed. Matteo Valleriani, Yifat-Sara Pearl, and Liron Ben Arzi, 9–12. Berlin: Max Planck Institute for the History of Science, Preprint 489.

Van Dyck, Maarten, and Ivan Malara. 2019. Impetus, Renaissance Concept of. In *Encyclopedia of Renaissance Philosophy*, ed. Marco Sgarbi. Cham: Springer.

Wassell, Stephen R. 2010. Commentary on *Ex ludis rerum mathematicarum*. In *The Mathematical Works of Leon Battista Alberti*, ed. Kim Williams, Lionel March, and Stephen R. Wassel, 75–140. Basle: Springer.

Wear, Andrew. 1981. Galen in the Renaissance. In *Galen: Problems and Prospects*, ed. Vivian Nutton, 229–262. London: The Wellcome Institute for the History of Medicine.

Chapter 8
Sanctorius Revisited

Abstract This chapter reflects on the epistemic processes that made the use of quantification and measurements in medicine conceivable to Sanctorius, and which might explain how these methods made sense to him in ways that they had not before. To this end, I bring into focus the relation between the categories of innovation and tradition as well as the interplay of the realms of theory and practice in Sanctorius's works, unifying the main results of my study. Then, based on my analysis of the measuring instruments in Chap. 7, I reflect on what quantifying health meant to Sanctorius. Finally, I briefly sketch out how his measuring instruments were received. Building upon the historical analyses of the previous chapters, I present a new and revised view of the Venetian physician Sanctorius, which hopefully will contribute not only to a better understanding of his work, but also, more generally, of how knowledge was transformed in the early modern period.

Keywords Historical epistemology · Practical knowledge · Transformation of knowledge

In his first publication, *Methodi vitandorum errorum*, Sanctorius argued that there were forces, or virtues which were not related to the four primary qualities of hot, cold, wet, and dry. In doing so, he referred to the example of the "moving force of the clock" (*potentia motrix horologij*) which was, according to him, "the most apparent example of all" (Sanctorius 1603: 160r).[1] He explained:

> No one of sane mind would say that the force of the clock relates to the temperament [*temperatura*], but rather to the number, position, and figure of the wheels, disks, and springs [*spirae chalibea*]; its inability to move, however, relates to damage to these. ... What prevents us from saying that, using the metaphor of the clock, the moving force [of the body] is not a very simple substance? But that [instead] the force relates to the number, position, and figure of the bodily substance and that in these [bodily substances] is the prime mover

---

[1] "Demum afferri potest exemplum omnium evidentissimum, estque potentia motrix horologij: ...." See: Sanctorius 1603: 160r.

© The Author(s) 2023
T. Hollerbach, *Sanctorius Sanctorius and the Origins of Health Measurement*,
https://doi.org/10.1007/978-3-031-30118-6_8

[*primum mobile*], which moves the others like the spring does.[2] Aristotle discovered the prime matter through an analogy of artifacts; why can we not, in imitating him, in so much simplicity of judgment, philosophize about the hidden forces through the same analogy? (Sanctorius 1603: 160r)[3]

In arguing against Jean Fernel's concept of occult forces which acted on the total substance of the body and could not be attributed to complexion, but were celestial in origin, Sanctorius put forward his own explanation of these hidden, or occult forces.[4] According to him, they were the result of the number, position and figure of the bodily substances, similar to the mechanical parts of a clock. From this citation it is easy to understand how the picture of Sanctorius as the founder of a new medical science based on measurement and quantification evolved. Strikingly, Sanctorius allowed here that some physical effects could be traced back, not to bodily complexions, but to mechanical properties such as movement. In line with his critical attitude toward astrology, he refused the idea of occult, celestial causes. Moreover, in this connection, he used the famous clock metaphor that René Descartes (1596–1650) was to use some 40 years later to defend his mechanistic understanding of the body—the most prominent figure among those who promulgated a machine-like explanation of bodily operations (Siraisi 1987: 280–5; Bertoloni Meli 2016: 91–3).[5]

In the following, I take Sanctorius's use of the clock metaphor as a key example, encapsulating the major findings of my study and demonstrating in what way the prevalent view of Sanctorius as the innovative genius needs to be revised.[6]

In fact, there are several passages in Sanctorius's works in which he put forward the metaphor of the clock. In the *Commentary on Galen*, he compared the

---

[2] The "prime mover" is a concept advanced by Aristotle as a supraphysical entity, a primary cause or "mover" of all the motion in the universe (Bodnar 2018).

[3] "... nemo sanae mentis dicet horologij potentiam à temperatura prodire, sed à numero, situ, & figura rotarum, orbiculorum, & spirae chalibeae; impotentiam vero ab ijs vitiatis; ... Quid prohibet, quin nos quoque hac horologij similitudine dicamus, potentiam movendi non esse substantiam simplicissimam? Sed potentiam ortam à numero, situ, & figura corporeae substantiae; & in ijs esse primum mobile, quod caetera moveat ad spirae chalibeae similitudinem: Aristoteles per analogiam artefactorum invenit primam materiam; cur nos eius imitatione in tanta consilij angustia non poterimus per eandem analogiam de potentijs abditis phylosophari?" See: Sanctorius 1603: 160r. According to the traditional interpretation of Aristotle, "prime matter" refers to the ultimate or first matter underlying the four elements and making elemental change possible. For more information, see e.g., Robinson 1974, Ainsworth 2020.

[4] For more information on Jean Fernel's concept of occult qualities and diseases of the total substance, see: Deer Richardson 1985.

[5] René Descartes used the clock metaphor in different works, such as the *Meditationes de prima philosophia* (Meditations on First Philosophy, 1641) or the *Principia philosophiae* (Principles of Philosophy, 1644), in which he explained that the healthy body "is like a well-made clock" (Hatfield 2003: 273) and that "it is as natural for a clock, composed of wheels of a certain kind, to indicate the hours, as for a tree, grown from a certain kind of seed, to produce the corresponding fruit" (Miller and Miller 1982: 285 f.). In his work *De homine* (A Treatise on Man), published posthumously in 1662, he identified men as nothing but machines (Lokhorst 2018).

[6] Since this last chapter mainly presents the achievements of this research in summarized form, I make no further reference to the sources already discussed, but limit myself to cross-references.

generation of the animal spirits with the movements of the parts of a clock. In doing so, he hoped to illustrate that during the production of the animal spirits, many physiological "movements" occurred at the same time, just as in a clock. Accordingly, the drawing of inhaled air from the lungs to the heart happened, for example, simultaneously to the inhalation of air by the brain through the olfactory tracts (Sect. 3.2.6). In a later discussion of the same work, Sanctorius stated that he used to compare people who trust in physicians with no anatomical experience with people who, when their clock stops working, consult a person who has never seen the inner structure of a clock (Sanctorius 1612a: 267, 738 f.). As mentioned earlier (Sect. 3.3.1), in the *De statica medicina,* Sanctorius compared the course of the plague in the body with the movement of a clock (Sanctorius 1634: 17v–18r). Similar to the first passage in the *Commentary on Galen*, Sanctorius made use of the clock metaphor in the *Commentary on Avicenna*, when trying to explain that the functioning of the body as a whole depended on the satisfactory interaction of many individually functioning parts. The human body resembled a clock, so Sanctorius, in which, if one wheel malfunctioned, the whole clock stopped (Sanctorius 1625: 91).[7]

Hence, these statements seem to imply that Sanctorius had a mechanistic conception of the body, picturing the latter as a machine, in which mechanical processes guaranteed its proper functioning, i.e., good health. In order to understand the human body, he held it necessary to take it apart in anatomical dissections and to observe its inner mechanisms, just as was required for the repair of a machine. Above all, Sanctorius did not refer to any machine, but to the clock, which later became the emblem of the mechanical philosophy of the seventeenth century (Van Lunteren 2016: 767).

Yet, in the preceding chapters I have shown that things are rarely as they seem at first glance. Sanctorius's strong adherence to traditional Galenic medicine has been identified and his innovative, quantitative approach to physiology has been demonstrated to be deeply influenced by the medical tradition, too. In view of this, it does not come much as a surprise that, considered in their broader contexts, Sanctorius's uses of the clock metaphor appear in a different light: Rather than being expressions of a revolutionary, mechanistic conception of the body as a machine, breaking with traditional concepts, Sanctorius presented them as additions or refinements of the ideas of Aristotle and Galen. In the quoted citation from the *Methodi vitandorum errorum*, for example, he explicitly mentioned that he used the analogy of the clock in an Aristotelian spirit. Most probably, Sanctorius had here the artifact-analogies in mind that Aristotle frequently used in describing the workings of living entities,

---

[7] For Sanctorius's other references to the metaphor of the clock, see: Sanctorius 1603: 155r, Sanctorius 1612a: 544. When arguing against Jean Fernel's concept of qualities that acted on the total substance of the body and, more generally, against an excessive reliance on occult causes, Sanctorius repeatedly referred to the example reported by Girolamo Cardano, of a mysterious gem with a mark on it that rotated every 24 h. According to Sanctorius, after Cardano had died, the "gem" in question was found to be a finely wrought clock (Sanctorius 1603: 160r–160v, Sanctorius 1612b: 35, Sanctorius 1625: 90). Interestingly, Sanctorius referred to the same example when discussing the issue of the motion of the earth, for he personally believed the planet to be "at rest in the center of the world" (ibid.: 123 f., Siraisi 1987: 272 f.).

especially regarding discussions of form and matter. Without examining these analogies and Sanctorius's understanding of them, it is important to note that, in using the clock metaphor, Sanctorius saw himself following the tradition rather than making new departures (Grmek 1990: 116; Koslicki 1997: 77).

In the *Commentary on Galen*, Sanctorius referred to the clock when explaining Galenic physiological theory, the generation of the spirits, which ultimately served him to demonstrate that the movement of the brain did not result from the arteries, but from the substance of the brain, as "Galen had taught us" (Sanctorius 1612a: 267).[8] Furthermore, after having emphasized the importance of anatomical knowledge for the physician, which he compared with the knowledge of the inner mechanisms of a clock that one needed in order to repair it, Sanctorius immediately pointed to Galen's statement that "internal diseases can by no means be understood without the cutting of bodies, philosophy, and dialectics" (Sanctorius 1612a: 739).[9] In the *Commentary on Avicenna*, he presented the picture of the body as a machine directly after the conclusion of a *quaestio* in which he had insisted on the qualitative and complexional nature of disease. He simply added his new analogy, without disturbing the underlying base—a process which was, however, very much in line with contemporary commentary tradition on Avicenna's *Canon*, as Nancy Siraisi has shown (Sanctorius 1625: 85–92; Siraisi 1987: 351).

Thus, Sanctorius certainly did not equate the body with a machine. According to him, certain physiological processes could be measured and quantified, but the human body could not be disassembled, let alone reconstructed. It was not a machine, but a humoral body whose functions could be sometimes explained and illustrated by mechanical explanations—in some respects it was *similar* to a machine, but by no means identical with it. As he himself pointed out, even traditional authorities like Aristotle used analogies of artifacts to explain and understand the physical world. The technical developments of his day allowed him to widen the metaphorical field of the ancients and to include with the mechanical clock, a highly innovative instrument of the time, movements and, more generally, complex operations like physiological processes, in the comparison of artifacts and bodies. In doing so, challenging traditional medical theory was not his aim (Farina 1975: 373; Mazzolini 1994: 126).

Sanctorius's use of the clock metaphor impressively illustrates the complex relation between the realms of tradition and innovation in his works, as encountered throughout this study; and it suggests that simply dividing early modern thinking and practice into these two categories is ambiguous and potentially misleading. Similarly, it points to the intricate interplay between the categories of theory and practice, given that Sanctorius brought an innovative technical instrument, the mechanical clock, into the sphere of theoretical medical explanations, mainly within

---

[8] "… quamvis motus dilatationis, & constrictionis cerebri non fiat ab arteriis; sed à cerebri substantia, ut nos docuit Gal. 3. de placitis cap. Ultimo, & lib. de instrumento odoratus in fine." See: Sanctorius 1612a: 267.

[9] "Hinc divinus Gal. lib. de sectis ad eos qui in: docet nullo modo internas aegritudines penetrari sine corporum incisione, sine philosophia, & dialectica: …." See: ibid.: 739.

the framework of large scholarly commentaries that were the result of his teaching of medical theory. As this indicates, the intellectual and the material are entangled with the old and the new, too, adding to the complexities of this constellation that includes, of course, also social dimensions (Valleriani 2017b: vii).

The aim of this study was to investigate this constellation—comprising innovative and traditional, practical and theoretical aspects, as well as their social dimensions—and its dynamic in all its complexities with regard to Sanctorius's undertakings, in order to understand how he developed his innovative ideas and more generally, how innovation occurred within a highly traditional framework in the early modern period. Thus, instead of concentrating solely on the parts of his work that are, or appear to be innovative, and instead of searching for breaks and disruptions, I searched for continuities that position Sanctorius's undertakings between traditional natural philosophical and medical concepts and the transformation, in his day, of views on nature. My interest lay in the epistemic processes which made the use of quantification and measurements in medicine *conceivable* to Sanctorius and which might explain how these methods *made sense* to him in ways that they had not before.

In the early modern period, the clock became not only a mechanical paradigm, but was also used by scholars to express the complexity of their era. The diffusion of the printed book, the journeys of exploration, and the intensifying process of urbanization are only a few of the many developments that characterized this period and contributed to change and incremental transformation encompassing economic, social, political and cultural spheres (Valleriani 2017a: 12–8). Thus, the complexity sensed by early modern scholars in a way mirrors the complexity of the processes by which new knowledge was generated in their era. It was against these complexities and from a broad perspective that Sanctorius's undertakings were considered in this study. Thereby his role, not as a revolutionist, but as an exceptional and creative physician became manifest. This review of Sanctorius in the light of his era hopefully contributes to our understanding of processes of knowledge transformation in the early modern period. It is the major goal of this book.

8.1 Tradition and Innovation: Continuities, Reinterpretation, and Reorganization

The starting point for my investigation of Sanctorius's role in the comprehensive process of the transformation of knowledge that ultimately led to the abandonment of Galenic medicine and to the introduction of a new medical science, based on the use of quantification and measurement in medical research, was to identify, on the one hand, the knowledge that Sanctorius changed and, on the other hand, those parts of Sanctorius's knowledge which triggered the change. In doing so, I was able to show that Sanctorius did not develop his innovative approach to physiology *despite* his adherence to the medical tradition, Galenic medicine, but rather *exactly because* of this tradition, since he developed his methods of quantification and measurement from the core of this "knowledge system" without calling its authority into question.[10] Accordingly, Sanctorius's most famous work, the *De statica medicina*, was based on an ancient concept originating in Galen's works and a principal idea of Galenism—the doctrine of the six non-natural things. Moreover, his concept of *perspiratio insensibilis* and also the fundamental principle on which his weighing procedures were based, namely that health is an ideal balance between ingestion and excretion, were profoundly shaped by the teachings of Hippocrates and Galen, as Sanctorius himself explicitly highlighted. His novel interpretation of the six non-natural things according to their effect on insensible perspiration becomes comprehensible when considering that, in Galenic dietetics, bodily evacuations were closely connected to the processes of digestion and respiration, i.e., to air, food, and drinks, and were influenced by the motion or rest of the body—all of them traditional non-natural factors. The fact that the traditional list of the six non-naturals included "evacuation and repletion" as a non-natural thing itself makes Sanctorius's step appear even more plausible. In light of this, it is no longer surprising that Sanctorius chose this of all concepts to structure the results of his weighing procedures. Indeed, my experiences with the reconstruction of the Sanctorian chair have revealed the problems that Sanctorius must have faced in trying to include all of the six non-natural factors into his measurements. Therefore, it is easy to imagine that it was this traditional dietetic context that gave him the idea of examining *perspiratio insensibilis* in the first place. To go even further, from the perspective of Galenic dietetics, according to which balance and moderation were crucial factors in order to maintain health, the step from the idea of balance to the use of an actual balance seems, at least in retrospect, quite natural. Hence, a closer look at the intellectual context in which Sanctorius developed his novel approach to physiology revealed, firstly, that he drew on existing medical-dietetic traditions and secondly, highlighted the importance of these for his static medicine. It thus refuted the established narrative of the lone genius who developed his novel ideas almost out of the blue.

Through an analysis of the content of the *De statica medicina*, this study has disclosed the way in which Sanctorius developed his new medical idea, the quantification of insensible perspiration, out of a well-established Galenic doctrine. Generally, he followed traditional concepts regarding the influence of the non-natural factors on the body, but reinterpreted them by focusing on their impact on body weight and on the excretion of insensible perspiration. Interestingly, in some instances Sanctorius apparently struggled to integrate his novel quantitative findings into traditional medical theory. So, he discovered, for example, that a high amount of insensible perspiration was expelled during sleep. But according to the Galenic teachings, the third stage of the digestive process, during which bodies perspired insensibly, occurred during waking hours (Sect. 3.3.3). From today's perspective,

---

[10] I follow here the concept of "knowledge system" as developed by the research of Department 1 of the Max Planck Institute for the History of Science (MPIWG) and summarized by Jürgen Renn as "knowledge amalgamated by the connectivity of its elements within their mental, material, and social dimensions" (Renn 2020: 427).

Sanctorius's solution to this problem seems somewhat inconsistent and suggests that the merging of the new and the old, the compromise between innovation and tradition, was sometimes challenging for him. This is further confirmed by the fact that Sanctorius was ready, at times, not only to reinterpret, but even to revise traditional knowledge on the basis of the knowledge which he gained through his weighing procedures, as when it came, for example, to the amount of food that should be ingested for each meal (Sect. 3.3.2).

However, my analysis of Sanctorius's stance on anatomy clearly illustrated that the revision of traditional knowledge was possible for Sanctorius only up to a point: He endorsed recent anatomical findings only as long as they could be accommodated within Galenic theory. The new material, which he integrated into his works, either had to fit within a Galenic framework, or it was refused. In doing so, Sanctorius was in line what historians have found to be the attitude generally prevailing among Medieval and Renaissance learned physicians and anatomists, such as Vesalius or André du Laurens (Wear 1981: 233–7). Galenic medicine was still conceived as providing a reliable framework in which novel elements had to be integrated.

But within this framework, as I was able to show, there was an increasing trend to quantification, ranging from Galen's works to those of other Renaissance scholars. In this connection, it is remarkable that Sanctorius put himself explicitly in the tradition of the ancient authorities of Hippocrates and Galen, while remaining silent on, or refuting altogether, any influence of contemporary scholars on his novel approach—despite the sometimes, striking similarities between their undertakings. Of course, his recourse to the ancients has to be seen in the context of Medical Humanism, a movement that, to put it most simply, was established at the Italian medical universities in the early sixteenth century, and according to which medical theory and practice had reached unparalleled heights among the ancient Greeks, especially through the work of Hippocrates and Galen (Bylebyl 1979: 339).[11] To allude directly to these authorities was thus desirable. But in addition to this, Sanctorius probably worried that reference to more recent works would diminish his originality. When confronted with Obizzi's accusation of plagiarism regarding the work *De staticis experimentis*, published by Cusanus in the fifteenth century, all he had to say was that the Cardinal had not dealt with insensible perspiration. For Sanctorius, this was proof enough to show that he did not take a word from him. Hence, Sanctorius neither denied his knowledge of the work nor explained in detail how his *De statica medicina* differed from the work of Cusanus, except for the focus on insensible perspiration. Yet, as my analysis has revealed, Cusanus had already conceptualized many of the quantitative measurements which Sanctorius later claimed to have realized (Sect. 5.3.2). This and the fact that he published them in a work with a title so similar to Sanctorius's *De statica medicina*—*De staticis experimentis*—provides strong evidence that this is more than just a "genial coincidence" and that by ignoring the similarities Sanctorius hoped to assert his originality.

---

[11] For more information on the medical humanism movement, see e.g., Bylebyl 1979, Grendler 2002: 324–8.

Against this background, it is, in my opinion, most certain that Sanctorius's innovative approach to physiology was inspired by the different forms of quantification
that were developed incrementally within the medical tradition. Sanctorius's quantification efforts, however singular, were, thus, informed by a tradition of thought.
Accordingly, he presented his use of measuring instruments in medicine not as a
break, but as a direct advancement of Galenic medicine. By measuring different
physiological parameters, he declared to have found an answer to a problem which
Galen had been unable to solve: to quantify the latitude of health. In doing so, he
claimed he was able to bring to medicine, which he identified as an art, a new precision which would approximate, if not achieve, certainty. On the epistemic level,
Sanctorius departed here markedly from a tradition according to which certainty
was exclusively reserved for *scientia*. And yet, he did not dismiss the Aristotelian
definitions of *ars* and *scientia*. Similar to the way in which he adopted new findings
of contemporary anatomists, in his own attempt to ascertain Galenic medicine, he
did not abandon the fundamental principles upon which the whole discipline of
medicine rested. What is more, Sanctorius appears to have been in doubt regarding
the degree of certainty that his instruments and measurements could actually provide, since he repeatedly qualified his statements concerning the possibility of gaining a true and certain knowledge of quantity in medicine. Here again tensions occur
as to the importance that Sanctorius ascribed, on the one hand, to his new methods
of quantification and instrumentation and, on the other, to traditional medical methods based on logical reasoning.

Another factor that is important to consider here is Sanctorius's social and intellectual context at the University of Padua. Sanctorius spent several years here—first
as a student, and later as a professor of medicine. Prominent scholars here, in his
time, who took a creative approach to innovation and tradition in the field of medicine, were Andreas Vesalius, Girolamo Fabrici d'Acquapendente, or the famous
English physician and anatomist William Harvey (1578–1657), to name but a few.
Their novel observations and experiences sparked controversy and—just like
Sanctorius—they had to walk the tightrope between innovation and tradition.
Acquapendente, who founded the Anatomical Theater in Padua, was one of
Sanctorius's teachers, and certainly also a role model for him as a student, when it
came to dealing constructively with criticism of the canon. William Harvey likewise
famously developed new approaches rooted in traditional lore, discovering the circulation of blood, for example—and was aware that he would therefore court controversy. Harvey and Sanctorius most likely did not meet in Padua, since Harvey
studied there between 1593 and 1602, at a time when Sanctorius was a practicing
physician probably in Pannonia, Croatia, and Hungary (Sect. 2.2). However, they
both studied under Acquapendete, who passed his experience on to them and influenced Harvey in the development of his natural philosophy.[12] Overall, there was an
intellectual climate in Padua in which the opinion prevailed that it was possible to

---

[12] For more information on William Harvey, his natural philosophy, and his handling of innovation
and tradition, see: French 1994; for Girolamo Fabrici d'Acquapendente's anatomical teaching at
the University of Padua, see: Cunningham 1985.

improve on or *advance* traditional authorities. The establishment in Padua of the Botanical Garden (1545) and the Anatomical Theater (1595), the very first of their kind, demonstrate this openness to new knowledge and the encouragement of innovation. Most likely, this intellectual climate shaped Sanctorius in his approach to traditional and new knowledge.

To sum up, the analysis of the complex constellation in Sanctorius's works between traditional and innovative elements revealed that the Venetian physician drew on traditional concepts when developing his novel ideas. Rather than breaking with the old, he reinterpreted and reorganized it, thereby creating something new. He brought different bodies of knowledge together in a new way: the doctrine of the six non-natural things, the concept of insensible perspiration, and previous ideas on quantification and the concept of latitudes. For him, combining quantification and Galenism was no paradox but made perfect sense. The noticeable ambiguities and inconsistencies in Sanctorius's works indicate that he explored the limits of the existing knowledge system, provoking the traditional explanatory framework without, however, abandoning it. Sanctorius's case therefore illustrates the complex nature of the process whereby the scientific culture and intellectual universe of a medical community began to be transformed. Namely, it sheds light on the mechanisms through which knowledge was reconfigured in the early modern period: through the conceptual reinterpretation and reorganization of existing knowledge (Siraisi 1987: 358; Renn 2020: 427). Therefore, we should not think of Sanctorius as a man divided, with one foot in modernity and one in tradition. Sanctorius understood himself to be a critical and creative physician, engaged in continuing and refining the work of the traditional authorities—Hippocrates, Aristotle, and, above all, Galen. In adding new knowledge to the old, he would not have dreamt of questioning the underlying theoretical bases. He understood his work to be "locally" new, in the specifics of his discoveries, but "globally" continuous with the objectives and methods of Galenic medicine (Distelzweig 2016: 138). In order to further identify those aspects of Sanctorius's knowledge which moved him to reinterpret and reorganize traditional medical knowledge, and thus transform it, I have examined the entanglement of theory and practice in his works, unveiling how Sanctorius's making and doing related to his thinking.

8.2   Theory and Practice—An Uneasy Relation

In the *Commentary on Avicenna* Sanctorius explained that theory had to be confirmed, a posteriori, by practice and that practice could only be understood if it was corroborated, a priori, by theory (Sect. 4.3). As a learned physician, shaped by his medical training at the University of Padua, he considered that experience gained in medical practice had to be paired with authority and reason, with medical theory, in order to be reliable. As we have seen, according to Sanctorius, a physician needed to use his hands and his head, not least to distinguish himself from the ignorant empirics and quacks. Just as innovations had to be reconciled with the basic principles of the medical tradition, so too, practical experiences had to be interpreted according to the theoretical framework of Galenic medicine. Following the Aristotelian theory of knowledge according to which certain knowledge about universal truths could be perceived only by the mind and was hidden from the senses, Sanctorius held that it was reasoning based on medical theory that ultimately led to certain medical knowledge and not experience gained through medical practice. Undoubtedly, medical theory and reasoning played an important part for Sanctorius in the purview of medicine, as was common among contemporary learned physicians (Sect. 6.2.4).

And yet, I have shown in this book that the relation between theory and practice was complex and sometimes even ambiguous in Sanctorius's works. He rejected the division of medicine into theory and practice in the university curricula, on the grounds that medicine, contrary to theory and practice, was a "factive" or operative (*factivus*) art, meaning that its purpose was not truth (*veritas*), as it was for theory, nor action (*actio*), as it was for practice, but instead operation (*opus*), i.e., the preservation and restoration of health. Notwithstanding that medical knowledge contained, according to him, both, contemplation (*contemplatio*) and action, it also differed from both, due to its operative (*factivus*) and restorative (*resarcitivus*) character. Accordingly, he tried to challenge the disciplinary boundaries from within and to reform the teaching of theoretical medicine (*theoria*) by linking his lectures on Avicenna's *Canon* to practical applications, and by confirming theory by evidence drawn from *practica*. The result is a seemingly peculiar mixture of highly traditional theoretical discussions with completely new elements relating to medical practice (Sect. 4.3).

In addition to this, Sanctorius occasionally challenged the Aristotelian theory of knowledge and claimed that certain medical knowledge could be generated on the basis of experience alone—through quantification and measurements. In the preface to the *De statica medicina* Sanctorius wrote that "not only the mind and the intellect perceive sincere and pure truth, but also the eyes and the hands virtually palpate it." Furthermore, he described the work elsewhere as "mathematical medicine" or "static theorems" and explained that his weighing procedures were in the first degree of certainty. Along with the weighing chair, the use of his *pulsilogia*, thermoscopes, and hygrometers could greatly reduce elements of uncertainty in medicine, he held, since they ascertained (*reddimur certi*) the quantity of the deviation of a body from its natural state (Sects. 6.2.2 and 6.2.4).

Thus, by introducing instrumentation, quantification, and measurements into medicine, Sanctorius provided the physician with new "tools" that should help him gain, or at least, approximate certain medical knowledge and, more generally, improve his work. In doing so, he reorganized and reinterpreted the traditional relation between theory and practice in medicine. Especially with the *De statica medicina*, Sanctorius attempted to overcome the division between sensory experience and intellection. The mere idea of rendering visible by means of a mechanical instrument an inner and unseen bodily process which was completely hidden from the senses and thereby claiming to achieve mathematical certainty, shows that Sanctorius was ready to think what was by earlier Aristotelian-Galenic standards unthinkable:

experience and quantification could offer knowledge about universal causes. As a true Galenist, Sanctorius was, however, at pains to distance himself from the "empirical sect" and to emphasize that his novel methods were based on learned medical knowledge. In his aim to improve Galenic medicine, theoretical medical concepts, such as dietetics or the doctrine of the six non-natural things, had necessarily to be the starting point for any inquiry into the uncertainties involved in the medical art. Against this backdrop, the ambiguities regarding Sanctorius's attitude toward the roles of reasoning and experience, of theory and practice in medicine, dissipate and Sanctorius's continuing confidence in the adaptability of the existing knowledge system becomes manifest (Siraisi 1987: 358).

Consequently, the results of my study not only show how the realms of theory and practice (in modern terminology) are related in Sanctorius's works, but also suggest that the interaction of theoretical and practical knowledge played an important part in Sanctorius's generation of new knowledge, in his innovative approach to physiology.[13] Historical accounts of Sanctorius and his work have usually focused on his intellectual activity and thus neglected an important dimension of his endeavors—the material and practical. As has become apparent throughout this book, Sanctorius was not only an erudite university professor with a broad knowledge of the works of ancient as well as contemporary medical writers, but also and especially a diligent practitioner for whom reading books was not enough: references to dissections and surgical operations of his own, his attempt to improve dietetic-therapeutic measures and patient care through instruments such as a movable bath or cupping glasses and, of course, his development of measuring instruments all illustrate that he spent many hours at the bedside of the sick and underscore his practical expertise in a medical fields. Importantly, the aim of the various devices which he developed was to improve daily medical practice. Thus, besides theoretical knowledge related to the doctrine of the non-natural things, the concept of latitudes, or the intellectual conception of static experiments by Cusanus, Sanctorius's practical experiences and the practical knowledge of his time shaped his novel approach to physiology.

I was able to show that, not only medical theory, but also medical practice before Sanctorius involved certain forms of quantification. There was a general awareness of the importance of regulating food intake in quantitative terms in the Renaissance and measuring meals was practiced at the time. The similarity of the *De statica medicina* to contemporary dietetic handbooks like the *Regimina sanitatis* strongly implies that the hygienic practices on which these treatises were based influenced Sanctorius in his quantitative approach to physiology and in the way in which he

---

[13] Practical knowledge is defined here as the knowledge needed to obtain a certain product that follows a defined workflow and results from the experiences of specially trained practitioners—for instance, a mechanical artifact that is produced through a construction procedure, or healing practices based on recipes (Valleriani 2017a: 1). Theoretical knowledge is understood as "knowledge systems with high degrees of systematicity and reflexivity, typically represented by texts in which abstract concepts are represented by controlled vocabularies or symbol systems understandable only with prior knowledge" (Renn 2020: 430).

combined theoretical knowledge with knowledge gained from practice and observation (Sects. 4.1.2 and 5.1).[14] Moreover, contemporary pharmacological practices involved a practical handling of quantities, in which doses were not computed by means of mathematical theories, but determined by hands-on testing paired with text-based knowledge. My analysis revealed that Sanctorius was very familiar with these practices and most probably frequented the *Struzzo* pharmacy in Venice not only to buy medicinal substances but also to exchange knowledge and experiences with the pharmacist Stecchini (Sect. 5.2.3).

Furthermore, the close examination of Sanctorius's measuring instruments has shown that Sanctorius looked beyond the confines of medicine and was attentive to the practical technologies of the time. The investigation of moving water and the engineering problems of river control in the frame of Renaissance practical hydraulics probably inspired him in his development of an early type of a water current meter. For his hygrometers, Sanctorius most certainly drew on common experience, on shared practical knowledge, such as the contraction of hemp cords in moist air. His strong interest in mechanics, illustrated by the different steelyards he devised, was anything but ordinary for a physician. Most certainly, it evolved from Sanctorius's socio-intellectual context, especially the *Ridotto Morosini*, where he met, among others, Galileo. The famous mathematician and engineer-scientist was very engaged in practical and theoretical mechanics at the time, discussing his ideas in the intellectual milieu that he shared with Sanctorius. Moreover, he used two of the same instruments that Sanctorius did—the *pulsilogium* and the thermoscope (Chap. 7).

In view of this, it seems plausible that Sanctorius's interest in numerical aspects of life's phenomena, and his subsequent application of quantification, instrumentation, and experimentation to his medical research and practice was stimulated by the practical knowledge in circulation in the vibrant milieu in which he moved, which already included quantitative aspects relating to dietetics, pharmacology, mechanics, and the use of instruments.

Last but not least, the reconstruction of the Sanctorian chair made more palpable the real world of Sanctorius's medical practice and helped develop a deeper understanding of the practical knowledge this involved. My experiences with the replica further uncovered the way in which the material and technical aspects of Sanctorius's endeavors played an important part in his research process. In order to achieve his goal to measure the *perspiratio insensibilis*, Sanctorius repurposed an old instrument and introduced weighing as a new body technology. An apparently straightforward measuring instrument became thus a complex apparatus. Sanctorius's novel application of the steelyard raised challenges for the mechanical design and use of the instrument. In the process of transforming the steelyard into his weighing chair, and of simultaneously producing new medical knowledge about insensible

---

[14] The fact that the *De statica medicina* centers on prevention rather than cure confirms the findings of Sandra Cavallo and Tessa Storey, that early modern regimens were by no means a static body of knowledge and that ideas about healthy living changed remarkably in the early modern period, unveiling a dynamic culture of preventative medicine (Cavallo and Storey 2013).

perspiration, he meshed theoretical and practical knowledge and determined the course of the research. This may have resulted in his discovery of an additional use of the weighing chair: to determine how to maintain an ideal body weight. My research with the reconstruction provides thus a window onto how the encounter between medicine and mechanics served to generate new knowledge and change how knowledge was actively produced (Sect. 7.5).

To cut a long story short, Sanctorius brought together theoretical and practical knowledge from different contexts—dietetics, pharmacology, mechanics—and fused it in novel combinations. In the process, he reorganized and reinterpreted the medical system of his time, Galenic medicine, exploring to what extent quantification and measurements, as new epistemic tools, could serve to develop and to legitimize the physician's pursuit of medical knowledge. By integrating practical and theoretical knowledge, he decidedly brought medical practice into the realm of medical theory, which is best illustrated by his inclusion of instruments in his teaching of physiology, a highly theoretical discipline, in his day. Still, according to him, the physician was both a philosopher and an artist and thus Sanctorius, the learned Galenist, and Sanctorius, the diligent practitioner, existed side by side. Similarly, his work highlights both the elasticity of Renaissance Galenism and the incremental shifts in views of how the human body functions. His novel approach to physiology was a response not only to changing medical ideas and practices, but also culturally driven—by the contemporary enthusiasm for instruments and measurements and by concerns and habits regarding health advice. After this reflection on the epistemic constellations that led Sanctorius to produce new knowledge, I will now turn more specifically to his use of measuring instruments in medicine and briefly consider what has been learnt from their analysis.

8.3   Quantifying Health

Wind, water currents, pulse, body heat, humidity of the air, insensible perspiration—as this book has shown, the spectrum of parameters that Sanctorius proposed to measure and for which he developed instruments is quite impressive. Over the course of Chap. 7, it was demonstrated that it is often difficult to assess whether Sanctorius's various measuring devices actually worked and could be used in the way he described them; and in the same vein, to what extent Sanctorius used them in his daily medical practice, and on how many different people, frequently remains unclear. With the exception of the weighing procedures with the Sanctorian chair, the Venetian physician hardly ever referred to the numerical outcomes of his measurements. And even in the *De statica medicina*, compared to the overall length of the treatise, only a fraction of aphorisms specify quantitative values.

This lack of numerical data might be explained by the impossibility of producing them. Another aspect to consider in this regard is that Sanctorius often directed his readers to his proposed but never published book *De instrumentis medicis* for more information on his instruments and quantitative observations. However, the scarce remarks on quantitative data might also point to the way in which Sanctorius conceived and used his instruments: as comparators. In this connection, the dissemination of precise quantitative measuring results and the elaboration of uniform scales most probably took a backseat. What was important was that the physician always used the same instrument with the same scale for an individual patient in order to make the measurements comparable and to monitor health trends. Although Sanctorius did not state this explicitly, with regard to his thermoscopes, he pointed to the fact that different instruments would produce different measuring results. He repeatedly indicated that, due to the considerable differences between individual patients, it was very difficult to generalize measurements and make them applicable to many individuals. In view of this, concrete measuring results could indicate general tendencies, but a physician would nonetheless always have to take measurements of his own, of his specific patient and using his specific instruments. Even though Sanctorius explained in the *Commentary on Galen* that he used his *pulsilogium*, the thermoscope, and the weighing chair to assess the condition of a patient whom he had never seen before, he immediately qualified this statement in the next sentence by explaining that he believed, along with Galen, that the exact and specific quantity will not be comprehended by the physician. This implies that Sanctorius used his measuring instruments to generalize, if necessary, but was fully aware of the shortcomings of such a procedure. Against this backdrop, his silence on the quantitative results of his measurements appears in a different light—they were simply of secondary interest (Sanctorius 1612b: 376).

The preceding paragraphs have demonstrated that it is crucial to not isolate Sanctorius's measuring instruments from their original medical context. In fact, it is exactly this context that makes Sanctorius's undertakings special. Notwithstanding that he mostly based his devices on well-known phenomena and techniques such as the oscillation of the pendulum, or weighing, it was he who first *applied* these to medical diagnosis, prognosis, and therapy—be it in thought or in deed. Therefore, the measuring instruments and their use can be understood only when considered in the framework of contemporary Galenic medicine. As was seen, Galen's concept of latitudes provided the starting point for Sanctorius's innovative idea of measuring deviations from the natural, healthy state of a body by means of his four most famous instruments. Although Sanctorius, in doing so, significantly departed from traditional medical theory and practice, his measurements must still always be related to these. Thus, his use of *pulsilogia* was deeply embedded in contemporary pulse theory, according to which the pulse frequency was but one of several parameters that indicated to a physician his patient's state of health. Similarly, the thermoscopes allowed Sanctorius to replace the subjective appreciation of body heat by means of touch, but in the face of traditional fever theory, it was not enough to determine the degree of heat in a patient to diagnose the disease. The impact of measured degrees of air's humidity on the body could only be understood when knowing the individual complexions of people living in a certain region, and even here many differences occurred. In order to define a healthy excretion of insensible perspiration, Sanctorius considered it necessary to examine the effects of the six non-natural things on this bodily evacuation. Hence, on the one hand Sanctorius ascribed a very important role to the quantitative observations which he made with his measuring instruments, but on the other, he considered them as new techniques of Galenic medicine, as complementary methods that mostly added to traditional, qualitative procedures and only occasionally replaced them. Sanctorius did at times aspire to a medical practice more systematically informed by quantification, but as the foregoing analysis has revealed, he did not consistently formulate a quantitative medical program. To be sure, quantitative methods played an important part in his medicine, but they were far from being wholly constitutive of it. Experience, measurements, and instruments served Sanctorius as a means to refine prior theory (Ragland 2017: 505, 527).

In this regard, it should be remembered that Sanctorius designed his instruments to enhance the certainty of medical knowledge with the aim of improving the practical work of physicians. This implies that he addressed his quantitative studies exclusively to colleagues. The inclusion of the illustrations and descriptions of his measuring instruments in the *Commentary on Avicenna*, an extensive medical textbook, further corroborates this assumption. It seems that Sanctorius held that every physician needed his devices in his daily practice. Accordingly, not the devices themselves, but their use was the main focus of attention. Operating these measuring devices called not for details of how they worked, but rather manual skill—this was far more important at the time than technological or mechanical knowledge. This might then explain why Sanctorius did not reveal more technical details of his devices and limited his descriptions of them to purely medical applications. However, in the 1626 edition of the *Commentary on Avicenna*, he stated that he had not published illustrations of his *pulsilogium* in previous works, since to properly convey to the reader how to build the device would require many plates; and this is why he decided to postpone a detailed description and illustration of the instrument until the work *De instrumentis medicis* (Sanctorius 1626: 21). From this it would appear that Sanctorius thought his readers, most of them physicians, capable of understanding the technical and mechanical details of his *pulsilogium*, if they just had enough detailed illustrations. Yet, evidently, Sanctorius and his fellow physicians, too, relied on the support of craftspeople to build the medical measuring instruments. Unfortunately, there are no sources regarding whether Sanctorius had assistants, not to mention their identities, and it is only by analyzing the material, practical dimensions of Sanctorius's work that we can seek to uncover their contribution to his quantitative approach to physiology.

With the weighing chair Sanctorius seems to have gone even a step further. The instrument not only enabled physicians to observe their patients' insensible perspiration of, but also offered laymen the opportunity to find and maintain an ideal weight. To this end, the weight watcher needed neither understand the mechanical properties of the device, nor have expert medical knowledge, as my own experimentation with the reconstruction illustrated. Even though it sometimes seems as if Sanctorius invited his readers in the *De statica medicina* to perform the weighing procedures themselves (Sect. 5.4.1), the treatise reads not so much as a guide to experimentation, but rather as a dietetic handbook for physicians, students, or other well-educated persons fluent in Latin (Sect. 4.1). In all likelihood, it served

Sanctorius to demonstrate his authority and expertise (Steinle et al. 2019: 8). However, eleven years later, with the illustration and description of the weighing chair in the *Commentary on Avicenna*, Sanctorius seems to have been ready to ascribe some authority to his readers: to physicians, the observation of the *perspiratio insensibilis* in their patients, and to patients, the monitoring of their own body weight. In doing so, he ascribed authority also to another actor—the Sanctorian chair.

Turning to the impact of Sanctorius's works, it is interesting to note that a preliminary assessment regarding the reception of his measuring instruments suggests that they did not widely enter medical practice, let alone the household. Testimonies of Sanctorius's pupils and contemporary physicians show that there was a great interest in his devices, and scholars all over Europe made their own copies of them. Yet, these undertakings were often not related to medicine, but to other fields of study. Isaac Beeckman, for example, drew inspiration from Sanctorius's *pulsilogia* for his observations on vibrating chords. Agostino da Mula informed Giovan Francesco Sagredo about Sanctorius's thermoscopes not because the latter was interested in their medical application, but in the instruments *per se*. Even a century later, at the beginning of the eighteenth century, Giovanni Battista Morgagni approved of Sanctorius's quantitative approach to physiology, but I could find no evidence that he himself used a *pulsilogium* or a thermometer in his medical practice. Hygrometers, anemometers and other instruments to measure climatic conditions never gained a strictly medical identity. Only much later did the fever thermometer and bathroom scales find their way into medical practice and the household. All this implies that integrating Sanctorius's measuring instruments into medical practice was a lengthy process; indeed, the revised and novel versions of only some of his instruments eventually succeeded on this path.

Hence, contrary to what the new mechanistic vision of the body that evolved in the seventeenth century might lead one to expect, quantitative measurements with thermometers, balances, or *pulsilogia* did not gain in importance, either in diagnostics or everyday general medical practice. Quite the reverse, in fact: as Volker Hess has shown with regard to the thermometer, the concepts of fever the seventeenth and eighteenth centuries made the measurement of body heat a mere secondary interest (Hess 2000: 19–33). Only in the nineteenth century did the measurements of pulse, body temperature, and weight start to become established as standard diagnostic methods.[15] The conceptual framework in which these practices and their attendant instruments gained importance was, of course, far removed from the one in which Sanctorius first proposed their use. That this may have given rise to an anachronistic reading of the original instruments owing to their roots in the Galenic medicine become obsolete for the later devices, has been often overlooked or neglected by historians. But my study has clarified how important it is to apprehend knowledge related to technical instruments and to the practices surrounding them *in its*

---

[15] Comprehensive studies of why pulse measurement and weighing became widespread or indeed, standard diagnostic methods in medical practice only in the course of the nineteenth century have yet be written. Kümmel (1974) touches upon the issue with regard to pulse measurement, and Frommeld (2019), with regard to bathroom scales.

historical specificity, for only so can one understand and make sense of how knowledge was transformed in earlier times. Thus, to separate Sanctorius's measuring instruments from their Galenic context is to lose their historical meaning. Indeed, the late application of thermometers, scales, and pulse meters to medical practice fits with the more general historiographical observation that practices associated with the body in sickness and in health tend to be particularly resistant to the innovation and change forged by new scientific theories or discoveries (Stolberg 2012: 519).

Contrary to this, with respect to medical research, Sanctorius's quantitative approach to physiology, especially his method of investigating the *perspiratio insensibilis*, immediately fell on fertile ground. Sanctorius's novel ideas paved the way for subsequent scholars to a mathematical and experimental analysis of physiological and pathological phenomena. Dissatisfied with the concepts of traditional Galenic medicine, these scholars were inspired by Sanctorius's novel approach in their creation of new medical theories. In the process, Sanctorius's name became inextricably linked to the quantitative investigation of the *perspiratio insensibilis* and his research method, weighing, was used until as late as the twentieth century.[16] It seems ironic, in hindsight, that Sanctorius set the stage for a new medical science in which medical practice remained largely unchanged—the very opposite of what he intended.

References

Ainsworth, Thomas. 2020. Form vs. Matter. In *The Stanford Encyclopedia of Philosophy*, ed. Edward N. Zalta. https://plato.stanford.edu/archives/sum2020/entries/form-matter/. Accessed 21 April 2020.

Benedict, Francis G., and Howard F. Root. 1926. Insensible Perspiration Its Relation to Human Physiology and Pathology. *Archives of Internal Medicine* 38: 1–35.

Bertoloni Meli, Domenico. 2016. Machines of the Body in the Seventeenth Century. In *Early Modern Medicine and Natural Philosophy*, ed. Peter Distelzweig, Benjamin Goldberg, and Evan R. Ragland, 91–116. Dordrecht: Springer.

Bodnar, Istvan. 2018. Aristotle's Natural Philosophy. In *The Stanford Encyclopedia of Philosophy*, ed. Edward N. Zalta. https://plato.stanford.edu/archives/spr2018/entries/aristotle-natphil/. Accessed 21 Apr 2020.

Bylebyl, Jerome J. 1979. The School of Padua: Humanistic Medicine in the Sixteenth Century. In *Health, Medicine and Mortality in the Sixteenth Century*, ed. Charles Webster, 335–370. Cambridge\London: Cambridge University Press.

Cavallo, Sandra, and Tessa Storey. 2013. *Healthy Living in Late Renaissance Italy*. Oxford: Oxford University Press.

Cunningham, Andrew. 1985. Fabricius and the 'Aristotle project' in Anatomical Teaching and Research at Padua. In *The Medical Renaissance of the Sixteenth Century*, ed. Andrew Wear, R.K. French, and Iain M. Lonie, 195–222. Cambridge\New York: Cambridge University Press.

---

[16] See: Benedict and Root 1926, Jores 1930, Newburgh et al. 1931, Wiley and Newburgh 1931, Müller 1991: 264, 275.

Distelzweig, Peter. 2016. "Mechanics" and Mechanism in William Harvey's Anatomy: Varieties and Limits. In *Early Modern Medicine and Natural Philosophy*, ed. Peter Distelzweig, Benjamin Goldberg, and Evan R. Ragland, 117–140. Dordrecht: Springer.

Farina, Paolo. 1975. Sulla formazione scientifica di Henricus Regius: Santorio Santorio e il "De statica medicina". *Rivista critica di storia della filosofia* 30: 363–399.

French, Roger K. 1994. *William Harvey's Natural Philosophy.* Cambridge: Cambridge University Press.

Frommeld, Debora. 2019. *Die Personenwaage: Ein Beitrag zur Geschichte und Soziologie der Selbstvermessung.* Bielefeld: transcript Verlag.

Grendler, Paul F. 2002. *The Universities of the Italian Renaissance.* Baltimore: The John Hopkins University Press.

Grmek, Mirko D. 1990. *La première révolution biologique: Réflexions sur la physiologie et la médecine du XVIIe siècle.* Paris: Payot.

Hatfield, Gary. 2003. *Routledge Philosophy Guidebook to Descartes and the Meditations.* London: Routledge.

Hess, Volker. 2000. *Der wohltemperierte Mensch: Wissenschaft und Alltag des Fiebermessens (1850–1900).* Frankfurt\New York: Campus Verlag.

Jores, A. 1930. Perspiratio insensibilis: I. Mitteilung. *Zeitschrift für die Gesamte Experimentelle Medizin* 71: 170–185.

Koslicki, Kathrin. 1997. Four-Eighths Hephaistos: Artifacts and Living Things in Aristotle. *History of Philosophy Quarterly* 14: 77–98.

Kümmel, Werner Friedrich. 1974. Der Puls und das Problem der Zeitmessung in der Geschichte der Medizin. *Medizinhistorisches Journal* 9: 1–22.

Lokhorst, Gert-Jan. 2018. Descartes and the Pineal Gland. In *The Stanford Encyclopedia of Philosophy*, ed. Edward N. Zalta. https://plato.stanford.edu/archives/win2018/entries/pineal-gland/. Accessed 17 Apr 2020.

Mazzolini, Renato G. 1994. Mechanische Körpermodelle im 16. und 17. Jahrhundert. In *Technomorphe Organismuskonzepte: Modellübertragungen zwischen Biologie und Technik*, ed. Wolfgang Maier and Thomas Zoglauer, 113–133. Stuttgart: frommann-holzboog.

Miller, Valentine Rodger, and Reese P. Miller. 1982. *René Descartes: Principles of Philosophy.* Dordrecht: Springer.

Müller, Ingo Wilhelm. 1991. *Iatromechanische Theorie und ärztliche Praxis im Vergleich zur galenistischen Medizin.* Stuttgart: Franz Steiner Verlag.

Newburgh, Louis Harry, F.H. Wiley, and F.H. Lashmet. 1931. A Method for the Determination of Heat Production over Long Periods of Time. *The Journal of Clinical Investigation* 10: 703–721.

Ragland, Evan R. 2017. "Making Trials" in Sixteenth- and Early Seventeenth-Century European Academic Medicine. *Isis* 108: 503–528.

Renn, Jürgen. 2020. *The Evolution of Knowledge: Rethinking Science for the Anthropocene.* Princeton\Oxford: Princeton University Press.

Deer Richardson, Linda. 1985. The Generation of Disease: Occult Causes and Diseases of the Total Substance. In *The Medical Renaissance of the Sixteenth Century*, ed. Andrew Wear, R.K. French, and I.M. Lonie, 175–194. Cambridge\New York: Cambridge University Press.

Robinson, H.M. 1974. Prime Matter in Aristotle. *Phronesis* 19: 168–188.

Sanctorius, Sanctorius. 1603. *Methodi vitandorum errorum omnium, qui in arte medica contingunt, libri quindecim.* Venice: Apud Franciscum Barilettum.

———. 1612a. *Commentaria in Artem medicinalem Galeni*, Vol. I. Venice: Apud Franciscum Somascum.

———. 1612b. *Commentaria in Artem medicinalem Galeni*, Vol. II. Venice: Apud Franciscum Somascum.

———. 1625. *Commentaria in primam Fen primi libri Canonis Avicennae.* Venice: Apud Iacobum Sarcinam.

———. 1626. *Commentaria in primam Fen primi libri Canonis Avicennae.* Venice: Apud Iacobum Sarcinam.

———. 1634. *Ars Sanctorii Sanctorii de statica medicina et de responsione ad Staticomasticem.* Venice: Apud Marcum Antonium Brogiollum.

Siraisi, Nancy. 1987. *Avicenna in Renaissance Italy: The Canon and Medical Teaching in Italian Universities After 1500.* Princeton: Princeton University Press.

Steinle, Friedrich, Cesare Pastorino, and Evan R. Ragland. 2019. Experiment in Renaissance Science. In *Encyclopedia of Renaissance Philosophy*, ed. Marco Sgarbi, 1–15. Cham: Springer.

Stolberg, Michael. 2012. Sweat. Learned Concepts and Popular Perceptions, 1500–1800. In *Blood, Sweat and Tears: The Changing Concepts of Physiology from Antiquity into Early Modern Europe*, ed. H.F.J. Horstmanshoff, Helen King, and Claus Zittel, 503–522. Leiden: Brill.

Valleriani, Matteo. 2017a. The Epistemology of Practical Knowledge. In *The Structures of Practical Knowledge*, ed. Matteo Valleriani, 1–19. Cham: Springer International Publishing.

———. 2017b. Foreword. In *The Structures of Practical Knowledge*, ed. Matteo Valleriani, v–viii. Cham: Springer International Publishing.

Van Lunteren, Frans. 2016. Clocks to Computers: A Machine-Based "Big Picture" of the History of Modern Science. *Isis* 107: 762–776.

Wear, Andrew. 1981. Galen in the Renaissance. In *Galen: Problems and Prospects*, ed. Vivian Nutton, 229–262. London: The Wellcome Institute for the History of Medicine.

Wiley, F.H., and Louis Harry Newburgh. 1931. The Relationship Between the Environment and the Basal Insensible Loss of Weight. *The Journal of Clinical Investigation* 10: 689–701.

Appendices

© The Author(s) 2023

323

T. Hollerbach, *Sanctorius Sanctorius and the Origins of Health Measurement*,
https://doi.org/10.1007/978-3-031-30118-6

Appendix I: Annotator Used to Study Sanctorius's Books

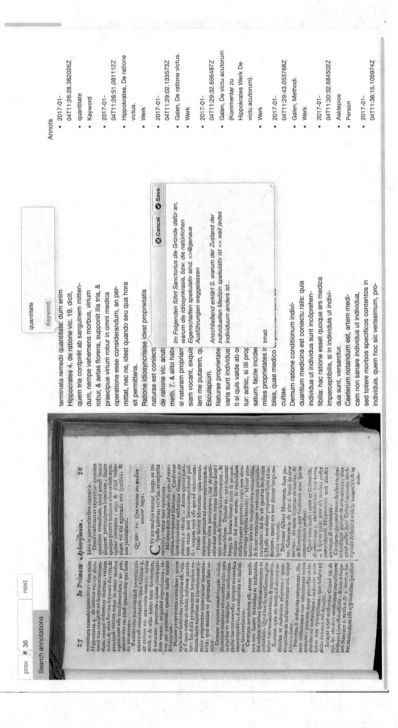

Brief Description

Here, a screenshot of the digital annotator that I used to study Sanctorius's published works shows a sample page, namely from the first edition of his *Commentary on Hippocrates*: the digitized image of the original page can be seen on the left, while the transcribed and searchable version of this same original page can be seen on the right. The words highlighted in yellow are my annotations, which become visible at the touch of a mouse and can be opened for editing with a click. This is illustrated by two examples: the word "quantitate," which I marked and categorized as a keyword; and the term "ratione idiosyncrisiae," for which I entered (in German) a brief summary of the (German) content of the next passages and accordingly categorized the annotation by subject matter, as in: "Inhalt" (content). In the far-right column, all the annotations that I made on the sample page from the *Commentary on Hippocrates* are listed in chronological order and their content is noted in full. The button in the top left-hand corner can be used to search through all the annotations that I made in a book and the tool directly above it, to scroll through the book page by page.

Appendix II: Editions and Translations of the *De statica medicina*

It is difficult to gain an overview of the numerous editions of the *De statica medicina*. Among the previous scholars who endeavored to compile a complete list of them, Arturo Castiglioni, and L. S. Ettari and M. Procopio succeeded at least in providing considerable detail (Castiglioni 1931: 783 f.; Ettari & Procopio 1968: 70–4). The following compilation of editions and translations of the *De statica medicina* is based on information gathered in online library catalogues using the *Karlsruher Virtueller Katalog* (KVK) and with the support of the Library of the Max Planck Institute for the History of Science, with slight modifications based on my archival research in Padua and Venice. The editions and translations marked with an asterisk (*) are those mentioned by Castiglioni and/or Ettari and Procopio, but which I was unable to find. I cannot claim the list is exhaustive, but I hope it illustrates the book's broad diffusion, particularly in its first 150 years.

Editions

- 1614, Venice: *Apud Nicolaum Polum*
- *1614, Venice: *Apud Marcum Antonium Brogiollum*
- *1615, Venice
- *1616, Leipzig; Venice
- *1617, Venice

- 1624, Leipzig: *Sumpt. Zachariae Schüreri & Matthiae Götzen. Excud. Gregor Ritzsch*
- *1626, Leipzig
- 1634, Venice: *Apud Marcum Antonium Brogiollum*
- 1642, Leiden: *Apud Davidem Lopes de Haro*
- *1650, Leiden: *Apud David Lopez*; The Hague: *Apud Adr. Vlacq*
- *1652, Leiden
- 1657, The Hague: *Vlacq*
- 1660, Venice: *Apud Franciscum Brogiollum*
- 1664: The Hague: *Ex typographia Adriani Vlacq*
- *1664, Venice
- *1666, Venice
- 1670, Leipzig: *Apud Haered. Schüreri, Götzianorum, & Joh. Fritzschium*
- *1679, Leipzig
- 1690, Lyons: *Antonius Cellier*
- *1694, Bologna
- *1700, London, with a commentary by Martin Lister
- 1701, London: *Impensis Sam Smith & Ben Walford,* with a commentary by Martin Lister
- *1701, Padua
- 1703, Leiden: *Cornelius Boutesteyn*
- *1703, Padua
- 1704, Rome: *Typis Bernabò. Sumpt. Haered. L'Hulliè*, with the *Canon* by Giorgio Baglivi
- *1705, Liège; Leiden; London
- 1710, Padua: *Typis Jo. Baptistae Conzatti*
- *1710, Strasbourg
- 1711, Leiden: *Cornelius Boutesteyn*
- 1713, Leiden: *Cornelius Boutesteyn*; Strasbourg: *Lerse* (in Henninger, Johann Sigismund: *Quadriga Scriptorum Diaeteticorum Celebriorum*).
- *1713, Padua
- 1716, London: *Typis Gul. Bowyer, impensis Gul. Innys*; Padua
- *1723, Padua
- 1725, Paris: *Apud Natalem Pissot*
- 1726, Leipzig: *Apud Joh. Sigism. Straussium* (in *Collectio Scriptorum Medico-Diaeteticorum*)
- 1728, Leiden: *Apud Didericum Haak, Samuelem Luchtmans*; Ferrara: *Apud Didericum Haak, Samuelem Luchtmans*; Leipzig: *Johann Samuel Heinsius* (in *Commentationes de diaeta eruditorum*); Padua: *Typis Jo. Baptistae Conzatti*
- *1730, Padua
- 1737, Leipzig: *Teubner* (in *De diaeta Humanae Naturae ad conservandam et prorogandam vitam demonstrata ex optimis physiologicis principiis*)
- 1742, Padua: *Typis Seminarii apud Jo: Manfrè*
- 1749, Venice: *Domenico Occhi*
- 1753, Duisburg: *Ovenius*

- *1758, Leiden
- 1759, Venice: *Typis Jo. Baptistae Novelli*
- *1760, Leiden
- *1761, Venice
- 1762, Leipzig: *Teubner*
- *1763, Venice
- *1768, Venice
- 1770, Paris: *Cavelier,* with a commentary by A. C. Lorry
- *1778, Paris
- 1784, Naples: *Apud Vincentium Ursinum: expensis Josephi de Lieto*
- *1842, Glasgow
- 1950, Florence: *Santoriana, A. Vallecchi*, with an introduction by Evaristo Lebàn
- 2001, Florence: *Giunti*, with an introduction by Giuseppe Ongaro.

Many of the eighteenth-century editions of the *De statica medicina* included a commentary by Giorgio Baglivi (*Canones de medicina solidorum*) and/or by Martin Lister. Here, only the first editions published with one or both of the two commentaries are indicated.

Translations

- Italian—*1704: Rome; *1727: Padua, by C.F. Cogrossi; 1743, 1749, *1761, 1784: Venice, by the Abbott Chiari; 2001: Florence, by Giuseppe Ongaro.
- English—1676: London by J. D.; *1678, 1712, 1720, 1723, 1728, 1737: London, by John Quincy.
- French—1695: Lyons, by M. Alemand; 1722, *1723, 1725, 1726: Paris, by M. Le Breton.
- German—1736: Bremen, by J. Timm.
- Dutch—1684: Amsterdam, by Steven Blankaart.

Printed in the United States
by Baker & Taylor Publisher Services